SECOND EDITION

The **GEOGRAPHY** COLORING BOOK

Wynn Kapit

LONGMAN

An Imprint of Addison Wesley Longman, Inc.

New York • Reading, Massachusetts • Menlo Park, California • Harlow, England
Don Mills, Ontario • Sydney • Mexico City • Madrid • Amsterdam

Printer and Binder: Courier Companies, Inc.
Cover Printer: Phoenix

Please visit our website at http://longman.awl.com

ISBN 0-321-03281-0

2345678910—CRS—0100

DEDICATION

To my wife, Lauren, and to our sons, Eliot and Neil

ACKNOWLEDGEMENTS

Thanks to the staff of Gerry Ichikawa's TypeStudio in Santa Barbara for their usual fine typography. They include Gerry, Terry Sturz, and Jill Breedon.

Special thanks to Jill, an expert proofreader and copy editor who *really* knows her geography.

And thanks to Rod Rolle for the cover photograph.

Preface
How to Use and Color This Book
A Glossary of Geographical Terminology

CONTENTS

EUROPE

ASIA

OCEANIA

AFRICA

POLAR REGIONS

HISTORIC LAND EMPIRES

FLAGS OF THE WORLD

WORLD THEMATIC MAPS

INDEX / DICTIONARY / QUIZ

PREFACE TO THE SECOND EDITION

When the first edition was published in 1991, communism had collapsed in Eastern Europe, and the Soviet Union, Yugoslavia, and Czechoslovakia were breaking apart. There was considerable uncertainty as to which direction the nations emerging from behind the the Iron Curtain would take. Could they make the transition from controlled, socialist societies to democratic, free-market economies? Except for the strife in the former Yugoslavia, the move toward freedom, although difficult, has been free of violence.

In addition to the new maps and text describing nations that once were the republics of the former Soviet Union and Yugoslavia, this second edition updates the descriptive text accompanying the countries of the rest of the world. Included in each description are facts concerning size, population, capital cities, types of government, official languages, religions, exports, and climates. There are comments on the unique qualities of a particular country, which might concern size, population, economics, land formations, cultural facts, famous citizens, or matters of historic, military, and political significance. I have tried to relate countries to the events of the day to provide the reader with greater insight into the news of the world.

Three fascinating features have been added to this second edition:

1. A section on the flags of all the nations, placed on maps, near their country of origin. These five plates, when colored in, are striking in appearance and tell a good deal about the history and significance of many of the flags. A visual quiz is made possible, with the reader given the opportunity to name the unidentified nations.

2. A section of maps showing the boundaries of eight historic land empires, from the Persian to the Ottoman. The empires are colored over modern maps to make it apparent which of today's nations have been influenced by past empires. Each map is accompanied by an informative history of the empire.

3. An expanded index that now includes a dictionary and the possibility of two kinds of quizzes. Each index entry is followed by a brief definition. By concealing either the entry names or the column of definitions, you can either try to think of the entry name by reading its definition or you can create your own definitions as you go down the list of index entries.

The contents of the book, combined with the proven effectiveness of learning by coloring, provide an unbeatable way to enjoyably learn a great deal of useful information. Every effort has been made to present the material in ways that promote understanding and retention, such as repeated reminders of comparative sizes, and placing subjects in a larger context. Even without coloring, the book is as an easily accessible world reference book—but coloring is what makes it memorable and enjoyable.

You may wish to color the World Thematic Maps in the back of the book. (Plates 55-60) before starting the book itself. Even if you don't color them, I recommend you read the text, which provides an introduction and background for many of the subjects and terms that appear frequently in the book. Such matters include weather and climate, wind patterns and ocean currents, varieties of vegetation and land use, population and race, and languages and religions.

But before you plunge in, I urge you to carefully read "How to Use and Color This Book." Spending a few minutes reading through this material will enable you to get the most out of this book.

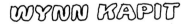

WYNN KAPIT

HOW TO USE AND COLOR THIS BOOK

Please take the few minutes needed to read through these instructions and recommendations. They will enable you to derive much more from this book than you might get if you were to plunge in without guidance. In fact, it would be wise for you to refer back to this page a number of times, as it is difficult to digest all these instructions in a single reading. This information will be more significant once you have had some practice with the coloring process. Most of it is just a matter of common sense and will become quite obvious once you get started, but there are certain things to look out for, and symbols to take note of.

WHAT MATERIALS TO USE

I recommend that you use fine-pointed felt-tip pens or colored pencils. Do not use crayons. Twelve colors, including a medium gray, should give you enough variety. Some plates will require more than 12 colors. In those cases you will have to use the same color more than once. The more colors you have at your disposal, the better your results will be and the more fun you will have in doing the book. If you have access to an art supply store that sells pens or pencils separately, you will be able to buy mostly lighter colors. Light colors allow the surface detail on the maps to show through. If you are limited to a conventional color selection, use the lightest colors on the largest countries. Dark colors used on large areas tend to dominate the page.

HOW THIS BOOK IS ARRANGED

This book is made up of 61 individual plates. Each plate is usually a two-page spread, with a large map on the left page and the text printed on the right-hand page. The names of the countries appear in outlined letters on the text page; color each name with the same color you use to color the respective country.

The book is divided into sections, each dealing with a separate continent. The first plate in each section is a political map, introducing you to the countries of that continent. The next plate displays the physical features of the continent: major rivers, mountains, mountain peaks, or land regions. The remaining plates of the section cover the individual countries, grouped according to region (northwestern, southcentral, etc.). On these plates you will generally color a country three times: (1) in outline form on the large map, (2) in a small drawing presented for size comparison, and (3) in the context of a continental view. A glance at the cover will show you how the system works.

WHAT THE SYMBOLS MEAN

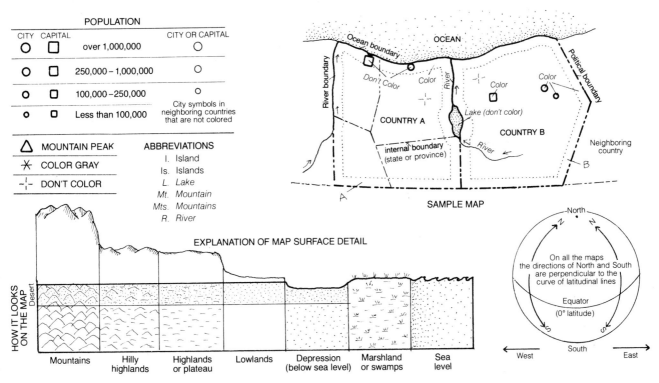

POPULATION

	CITY	CAPITAL		CITY OR CAPITAL
	○	☐	over 1,000,000	○
	○	☐	250,000 – 1,000,000	○
	○	☐	100,000 – 250,000	○
	○	☐	Less than 100,000	City symbols in neighboring countries that are not colored

△ MOUNTAIN PEAK

✳ COLOR GRAY

-¦- DON'T COLOR

ABBREVIATIONS
I. Island
Is. Islands
L. Lake
Mt. Mountain
Mts. Mountains
R. River

SAMPLE MAP

Ocean boundary — OCEAN — Political boundary
River boundary
Don't Color — Color
Color — Color
Lake (don't color)
COUNTRY A
internal boundary (state or province)
COUNTRY B
Neighboring country
River
A — B

EXPLANATION OF MAP SURFACE DETAIL

HOW IT LOOKS ON THE MAP / Desert

Mountains | Hilly highlands | Highlands or plateau | Lowlands | Depression (below sea level) | Marshland or swamps | Sea level

North
On all the maps the directions of North and South are perpendicular to the curve of latitudinal lines
Equator (0° latitude)
West — South — East

HOW TO COLOR

When a city (represented by a circle or square) falls within the colored outline, leave it uncolored. When it falls within the uncolored interior, color it with the same color you used on the outline (see the cover of this book). Do the same for mountain peaks (triangles). Be aware that the circles are not meant to suggest the actual physical size of the cities, but rather their relative population size (see above, "What the Symbols Mean"). The population sizes given for some of the major cities may seem unusually small, because the numbers refer to the city itself and not to the metropolitan area in which it is located.

Occasionally you will come across a large lake that separates the borders of two countries or states. In most cases, the borders usually meet in the center of the lake, but for purposes of showing the lake clearly, the borders are drawn around the edge of the lake and you are asked to leave the lake uncolored.

When coloring these plates, take time to look at the surface detail of the country you are coloring. Look to see what the neighboring countries are. Are the major cities confined to certain regions? Can you figure out why? Follow the directional flow of the major rivers (marked by tiny directional arrows). Do they play a role in population distribution? Can you predict the climate of the country by its terrain, its distance from the Equator or the Poles, or its proximity to a major body of water? Always take note of the scale of distances on each map—they vary with the plates.

If you don't begin at the front of the book, you should at least start on the first plate of whatever section you choose. Color only areas on the map that are within dark outlines. The area to be colored might be the outline of a country (the space between the dark border and the dotted parallel line. Where a country should be entirely colored, don't color the interior circles or squares (representing major cities and capitals). A group of similar areas, such as a group of islands, should be the same color, even if not all of them are labeled.

Use a different color for each subscript letter. If you run out of colors then it is all right to repeat them, but try to avoid using the same color on adjacent countries (you can accomplish this by coloring a country before you color its name on the opposite page). Take special note of two symbols: the asterisk (✳) tells you to color gray anything that it is attached to, whether a country or just some outlined heading. When something is labeled with the "do not color" sign (-¦-), it means just that.

Pay attention to color notes (CN) wherever they appear. They are meant to clear up any ambiguities, or to call your attention to something about the coloring procedure. You may also wish to read about the subject before starting to color it. The coloring process will tend to be more meaningful if you know something about whatever it is that you are coloring.

A GLOSSARY OF GEOGRAPHICAL TERMINOLOGY

Before you begin coloring the actual plates you may wish to warm up with this introduction to geographical terminology. If you don't have the 21 colors needed to color A–U, feel free to repeat as many of them as needed.

Begin by coloring the word "archipelago," labeled "A," and use the same color on the part of the illustration below that has the same label.

Note that each caption ends with a well-known example of the word under discussion. These examples are set in italics. Also set in italics are other geographical terms that are related, in some way, to the word which is being defined.

ARCHIPELAGO A
ATOLL B
BAY C
CANYON D
CAPE E
CONTINENTAL DIVIDE F
DELTA G
ESTUARY H
FJORD I
GLACIER J
GULF K
HEADWATERS L
ISLAND M
ISTHMUS N
LAGOON O
MESA P
OCEAN CURRENTS Q
PENINSULA R
PLATEAU S
REEF T
STRAIT U

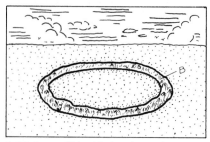

A circular coral island that encloses a *lagoon*. Atolls are usually formed on top of submerged *volcanoes*. *Bikini Atoll in the Marshall Islands of the Pacific Ocean, a US atomic test site.*

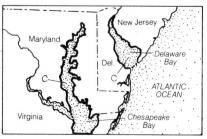

A body of water that penetrates a coastline. It is generally wider in the middle. It is usually smaller than a *gulf*, but larger than a *cove*. *Delaware and Chesapeake Bays.*

A deep, narrow depression in the earth's surface, often having a river running through it. Canyons are also known as *gorges*. *Ravines* are not quite as deep. *The Grand Canyon in northwest Arizona.*

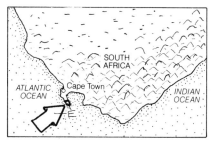

A point of land extending into the sea. It is usually smaller than a *peninsula*. A mountainous cape is called a *promontory* or a *headland*. *The Cape of Good Hope off the South African coast.*

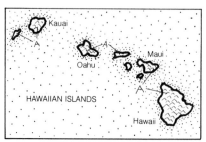

Either a group of islands or a body of water that has many islands in it. *The Hawaiian Islands; the Aegean Sea off the coast of Greece.*

The highest point of a continent, from which the direction of river flow is determined. *The Great Divide is the name given to the crest of the Rocky Mountains, which sends rivers east and west.*

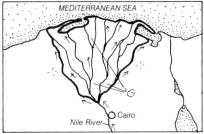

The triangular-shaped land found at the mouth of some large rivers. So much soil is transported by the river that the coastal waters cannot wash it all away. *The Nile Delta on the Mediterranean Sea.*

Upper river springs, streams, and tributaries. Head-waters can refer to *continental divides* or *watersheds*. *Watershed* also describes a region drained by a river. *The Alps have been called the headwaters of Europe.*

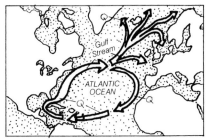

Ocean streams that are propelled by prevailing winds and earth rotation. They flow clockwise in the northern hemisphere and counterclockwise below the equator. *The Atlantic's Gulf Stream.*

An ocean inlet that merges with the mouth of a river. The estuary's salinity varies according to river flow and ocean tides. *The Río de la Plata, separating Argentina from Uruguay.*

A body of land completely surrounded by water. It is smaller than a *continent* but larger than a *cay*, a *key*, or certainly a *large rock*. *Greenland is the world's largest island.*

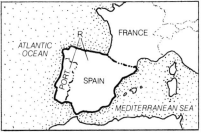

A mass of land almost entirely surrounded by water. It is usually connected to the mainland by a narrow neck. *The Iberian Peninsula in Europe, home to Spain and Portugal.*

A narrow, winding ocean inlet that penetrates a coastal mountain range. The steep cliffs that line its route make a fjord (fiord) one of nature's grandest sights. *Norway's Sogne Fjord is the world's longest.*

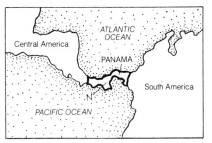

A narrow strip of land, with water on both sides, that connects two larger land masses. *The Isthmus of Panama connects Central America and South America.*

A broad expanse of generally high and flat land, also called a *tableland*. Plateaus can rise up from a lower area, or can be level regions within a mountain range. *Most of Spain is the Meseta Plateau.*

A river of ice, moving slowly down a mountain slope or outward from its central mass. It stops where the leading edge melts faster than the forward rate of movement. *Vatnajökull in Iceland is Europe's largest.*

A small body of water separated from the larger sea by a barrier of *sand or coral reefs*. It can either be adjacent to a coastline or surrounded by an *atoll*. *Mirim Lagoon off the coast of Brazil.*

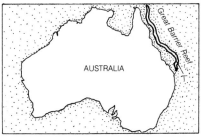

A narrow, low ridge of rock, or more commonly of coral, that is connected to a coast *(fringing reef)* or lies off a coast *(barrier reef)*. The Great Barrier Reef, off the northeast coast of Australia.

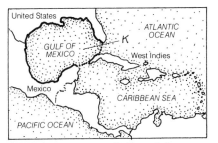

A part of an ocean or sea that is partially enclosed by a curving coastline. A more fully enclosed body of salt water could be called a *sea*. *The Gulf of Mexico.*

A tall, flat-topped mountain with steep vertical sides. Erosion-resistant mesas are left standing after all else has gone. *Buttes are small mesas. Monument Valley in Utah has 1,000 ft. (305 m) mesas.*

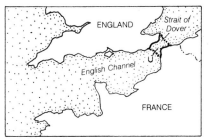

A narrow passage of water connecting two larger bodies of water. A *channel* is wider than a *strait*. If it is shallow, it is called a *sound*. The English Channel becomes narrower at the Strait of Dover.

CONTINENTS OF THE WESTERN HEMISPHERE *

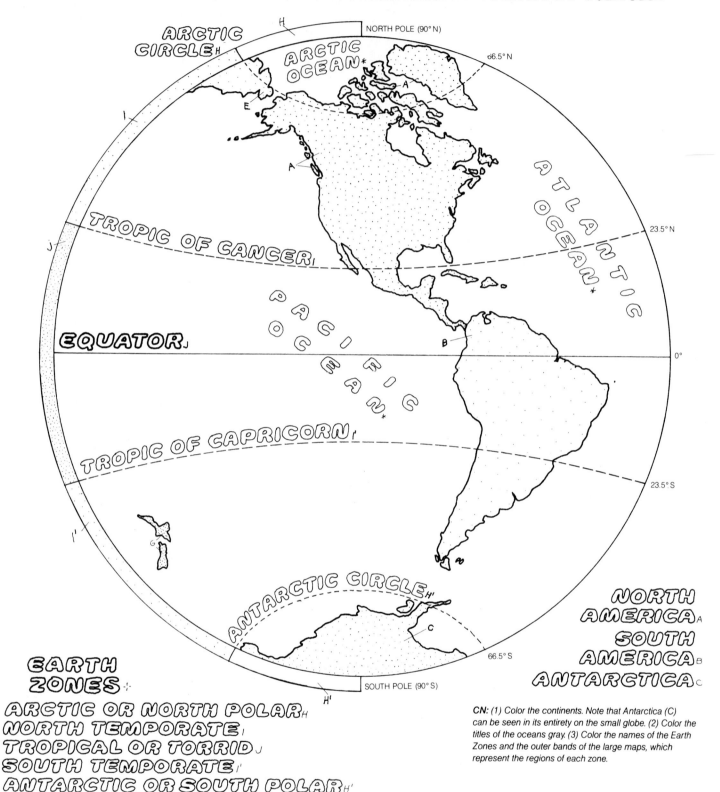

ARCTIC CIRCLE H

ARCTIC OCEAN *

NORTH POLE (90° N)

66.5° N

ATLANTIC OCEAN *

TROPIC OF CANCER I

23.5° N

PACIFIC OCEAN *

EQUATOR J

0°

TROPIC OF CAPRICORN I'

23.5° S

ANTARCTIC CIRCLE H'

66.5° S

SOUTH POLE (90° S)

EARTH ZONES ∹

ARCTIC OR NORTH POLAR H
NORTH TEMPORATE I
TROPICAL OR TORRID J
SOUTH TEMPORATE I'
ANTARCTIC OR SOUTH POLAR H'

NORTH AMERICA A
SOUTH AMERICA B
ANTARCTICA C

CN: *(1) Color the continents. Note that Antarctica (C) can be seen in its entirety on the small globe. (2) Color the titles of the oceans gray. (3) Color the names of the Earth Zones and the outer bands of the large maps, which represent the regions of each zone.*

The four seasons occur only in the temperate zones. The *Arctic Circle,* 23.5° from the *North Pole,* is the southern boundary of the *Arctic or North Polar Zone.* In this zone, the sun fails to rise during the winter months. The sun stays below the horizon for one day at the Arctic Circle and for six months at the North Pole. During the summer, the sun fails to set for a comparable period of time. The *Tropic of Capricorn* is the southern boundary of the Tropical Zone and the southernmost parallel (23.5° S) where the sun appears overhead (noon of the winter solstice; see p. 44). This line also marks the northern border of the *South Temperate Zone,* which is limited to the south by the *Antarctic Circle* (23.5° from the *South Pole*); this is also the northern boundary of the *Antarctic Zone or South Polar Zone or Region.*

Earth zones are defined by imaginary lines of latitude circling the globe, parallel to the *Equator.* The latitude lines shown above are not parallel because of the type of map projection used (they are parallel in the small global view to the right). The *Tropical or Torrid Zone* is the largest and hottest. The sun is always shining directly over some part of this zone. It is bounded by the *Tropics of Cancer and Capricorn.* The *Equator* passes through the center of the Tropical Zone, halfway between the two poles, dividing the earth into *Northern and Southern Hemispheres.* The northern boundary of the Tropical Zone is the *Tropic of Cancer,* the northernmost parallel (23.5° N latitude) where the sun shines directly overhead (noon of the summer solstice; see p. 44). About 75% of the earth's population lives in the *North Temperate Zone.*

CONTINENTS OF THE EASTERN HEMISPHERE *

NORTH POLE (90° N)

66.5° N

ARCTIC OCEAN *

OCEAN *

PACIFIC OCEAN *

23.5° N

ATLANTIC

0°

INDIAN

OCEAN *

23.5° S

66.5° S

SOUTH POLE (90° S)

EUROPE D
ASIA E
AFRICA F
OCEANIA G

Continents are large land masses with adjacent islands, surrounded or nearly surrounded by water. The seven continents cover slightly less than 30% of the earth's surface. The rest of our "water planet" is covered by four *oceans* and many *seas* (shallower extensions of oceans, partially surrounded by land).

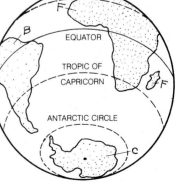

EQUATOR

TROPIC OF CAPRICORN

ANTARCTIC CIRCLE

CONTINENT	% OF TOTAL	LAND AREA	% OF TOTAL	POPULATION
Asia	29.5	17,230,000 sq. mi. (44,625,700 km²)	60.4	3,500,000,000
Africa	20.0	11,700,000 sq. mi. (30,279,600 km²)	12.9	750,000,000
North America	16.3	9,400,000 sq. mi. (24,346,680 km²)	8.2	465,000,000
South America	11.8	6,000,000 sq. mi. (17,871,400 km²)	5.7	330,000,000
Antarctica	9.6	5,400,000 sq. mi. (13,986,000 km²)	—	—
Europe	6.5	3,810,000 sq. mi. (9,867,900 km²)	12.4	720,000,000
Oceania	5.2	3,300,000 sq. mi. (8,547,000 km²)	0.4	28,000,000

OCEAN	% OF TOTAL	OCEAN AREA	MAXIMUM DEPTH
Pacific	49.2	64,100,000 sq. mi. (165,890,800 km²)	36,170 ft. (11,027 m)
Atlantic	24.6	32,220,000 sq. mi. (83,385,360 km²)	30,200 ft. (9,207 m)
Indian	22.0	28,900,000 sq. mi. (74,793,200 km²)	24,440 ft. (7,451 m)
Arctic	4.2	5,300,000 sq. mi. (13,716,400 km²)	—

MOVEMENT OF CONTINENTS⨂

NORTH AMERICA A
SOUTH AMERICA B
ANTARCTICA c
EUROPE D
ASIA E
AFRICA F
OCEANIA G

CN: Use the same colors you used for the continents on Plate 1. Use a very light color for the earthquake zones (I). (1) Color the four maps on this page; complete each one before doing the next. (2) At the top of the opposite page, color the tiny triangles representing volcano sites. (3) Color the earthquake zones (areas covered with light parallel lines)

200 MILLION YEARS AGO⨂

Around 200 million years ago there was a single continent called Pangaea (Greek for "all earth"). The Tethys Sea, ancestor of the Mediterranean, partially divided this land mass. It is assumed that the continents making up Pangaea previously migrated from other unknown locations.

100 MILLION YEARS AGO⨂

About 100 million years ago, Pangaea divided into two masses: Laurasia (North America, Europe, and Asia) and Gondwanaland (South America, Africa, Australia, and Antarctica). The Atlantic and Indian Oceans were expanding. India was heading for the Asian continent . The continents were all moving northward, and Greenland was starting to break away from North America.

TODAY⨂

Today, we find Australia far from Antarctica. India is now an integral part of Asia, although it is still grinding under the Asian continent, causing the Himalayan Mountains to rise. The Atlantic Ocean is still expanding, and Africa and South America are well within the Equatorial region. Note how far north of the Equator the continents have drifted. This explains why evidence of past tropical vegetation can be found in northern regions.

50 MILLION YEARS FROM TODAY⨂

If current trends continue, in 50 million years Australia will be part of Asia, the Mediterranean Sea will be greatly reduced by the encroachment of Africa, the Atlantic Ocean will be wider, and part of California will be headed for Alaska.

VOLCANO SITES

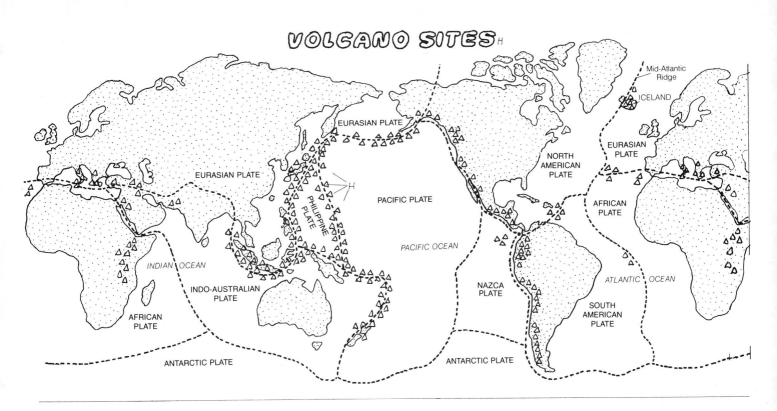

For many years, scientists wondered why coal deposits, the decay products of tropical vegetation, were located in northern regions. Why was there evidence of glacial formation in tropical Africa? Why do the opposing coastlines of South America and Africa look as if they could fit together?

Early in this century, Alfred Wegener, a German scientist, attempted to explain these phenomena. According to his theory, the continents have changed locations over millions of years, and they are still in motion because they rest on moving sections of the earth's crust called tectonic plates. These plates act as huge rafts floating on the earth's mantle, which in turn is moved by convection currents deep within it. The plates are 19–25 miles (30–40 km) thick; the mantle is about 100 times thicker. The 9 plates shown on this page are the largest of 18 known plates. Plate movement generally follows major earthquakes; it averages 3/4 in. (2 cm) to 2 in. (5 cm) per year. Cracks in the mantle allow magma (molten rock) to rise to the surface in volcanic eruptions, and new plate material is created. This process is occurring is the Mid-Atlantic Ridge, near Iceland. Newly formed North American and Eurasian plates are expanding the Atlantic Ocean.

Moving plates have to go somewhere; they often collide with other plates. When such collisions occur, a crumpled plate will be the source of new mountain material. The tallest mountains in the world, the Himalayas in north ern India, are still growing as the Indo-Australian plate pushes under the Eurasian plate. Plates can also slide past each other; the junction between the moving plates is called a fault. The San Andreas fault, which knifes along the coast of California, separates the Pacific plate (which carries San Francisco and Los Angeles) from the North American plate. In 30 million years, the Pacific plate will move 400 miles (640 km) northward and Los Angeles will replace San Francisco opposite Oakland in the fault-divided San Francisco Bay Area. In 50 million years, Los Angeles will be on its way to Alaska (see lower map on opposite page).

Events along the borders of moving plates release enormous energy in the form of earthquakes and volcanoes. The coastlines and islands on the rim of the Pacific Ocean are in a violently active geological region called the "Ring of Fire." Another major earthquake and volcano zone runs westward from Southeast Asia through China, southwest Asia, and southern Europe.

EARTHQUAKE ZONES

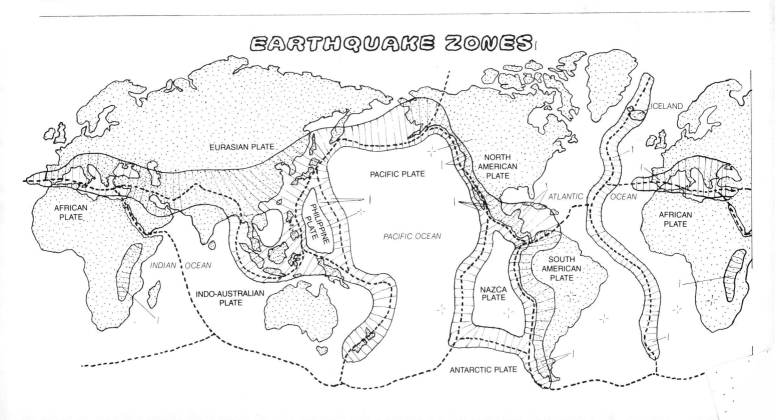

NORTH AMERICA: THE COUNTRIES

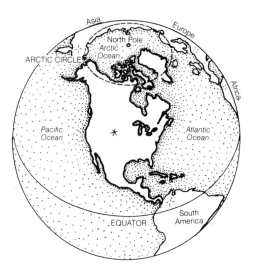

GREENLAND A / GODTHÅB (DENMARK)
CANADA B / OTTAWA
UNITED STATES C / WASHINGTON, D.C.
MEXICO D / MEXICO CITY

CENTRAL AMERICA
BELIZE E / BELMOPAN
GUATEMALA F / GUATEMALA CITY
EL SALVADOR G / SAN SALVADOR
HONDURAS H / TEGUCIGALPA
NICARAGUA I / MANAGUA
COSTA RICA J / SAN JOSÉ
PANAMA K / PANAMA CITY

WEST INDIES
CUBA L / HAVANA
HAITI M / PORT-AU-PRINCE
DOMINICAN REPUBLIC N / SANTO DOMINGO
PUERTO RICO O / SAN JUAN (US)
JAMAICA P / KINGSTON
TRINIDAD & TOBAGO Q / PORT OF SPAIN
OTHER ISLANDS R

North America stretches from deep within the Arctic Circle to close to the Equator. Note that the continent lies almost entirely west of South America.

The map below shows the extent of European domination of North America prior to France's loss of its mainland possessions. France was to cede to England its lands east of the Mississippi and sell its midwestern holdings through the Louisiana Purchase of 1803, which doubled the size of the U.S. The U.S. would acquire additional land from Spain, Mexico, and Russia by means of war, treaty, and purchase.

North America, which includes Greenland, Central America, and the West Indies, covers 9,400,000 square miles (24,346,000 km²). It is the third-largest continent (after Asia and Africa). Its population of 465 million is the fourth largest (after Asia, Europe, and Africa). Population density in North America increases the farther south one goes; Canada, the continent's largest and most northern nation, has only a tenth the population of the United States.

Most North Americans are descended from people who immigrated within the past 400 years from all other continents except Antarctica. The majority came from Europe. The true natives, named "Indians" by Columbus, who believed he had landed in the East Indies, numbered about 5 million. Their ancestors came from Asia 20,000–40,000 years ago, crossing a land bridge uncovered during the last Ice Age, when the formation of ice lowered the sea. The bridge is now under the Bering Strait. Eskimos arrived more recently, having made the same journey around 6,000 years ago. Today, most people living south of the U.S. are of mixed European and Indian ancestry and are called "Mestizos." Blacks living in the U.S. and the West Indies are descended from slaves brought from Africa.

The Vikings were the first Europeans to set foot on North America. After Columbus's arrival some 500 years later, exploration, colonization, and exploitation began in earnest. In the early 1500s, the Spanish, with superior weaponry and military cunning, easily conquered the natives of the south-western United States, Mexico, Central America, and the West Indies. A century later, the French explored Canada and the central U.S. as far south as New Orleans. Both nations introduced the Roman Catholic religion. English Protestants colonized northern Canada and the eastern seaboard of the U.S., the site of the 13 original colonies. Though British rule ended some 200 years ago, the English left a permanent impression on the language, religion, and culture of the U.S. Along with the French and the Dutch, the British continue to have many possessions among the islands of the West Indies, the last vestiges of colonization of North America.

18TH CENTURY OWNERSHIP OR INFLUENCE

ENGLISH S
FRENCH T
SPANISH U
RUSSIAN V
NORWEGIAN & DANISH W

4

RUSSIA   ASIA

NORTH POLE

NORWAY   EUROPE   BRITISH ISLES

ARCTIC OCEAN *

CHUKCHI SEA

BERING SEA

Bering Strait

St. Lawrence (U.S.)

BEAUFORT SEA *

GREENLAND

ICELAND

ARCTIC CIRCLE

BAFFIN BAY *

DAVIS STRAIT

LABRADOR SEA *

ALASKA

GULF OF ALASKA

YUKON TERRITORY

NORTHWEST TERRITORY

NUNAVUT (As of April, 1999)

Hudson Strait

BRITISH COLUMBIA

ALBERTA

SASKATCH-EWAN

HUDSON BAY *

NEWFOUNDLAND

GULF OF ST. LAWRENCE

PACIFIC OCEAN

MANITOBA

CANADA

ONTARIO

QUEBEC

NEW BRUNS.

NOVA SCOTIA

WASH.

MONT.

N. DAK.

ME.

ORE.

ID.

WYO.   S. DAK.

MINN.

WIS.

MICH.

N.Y.

VT.   N.H.

MASS.   CONN.   R.I.

NEV.

UTAH

COLO.

NEB.

IOWA

ILL.   IND.

OHIO

PA.

MD.   N.J.   DEL.

Bermuda (Gt. Brit.)

CALIF.

ARIZ.

N. MEX.

KAN.

MO.

KY.

VA.

N. CAR.

UNITED STATES

OKL.

ARK.

TENN.

S. CAR.

GA.

TROPIC OF CANCER

BAJA CALIFORNIA

Gulf of California

TEXAS

MISS.   ALA.

LA.

FLA.

BAHAMA ISLANDS

ATLANTIC OCEAN *

MEXICO

GULF OF MEXICO *

CUBA

GREATER ANTILLES

HAITI   D.R.

P.R.

LESSER ANTILLES

JA.

CARIBBEAN SEA *

BE.

GUA.   HON.

EL S.   NIC.

C.R.   PAN.

Panama Canal

VENEZUELA

COLOMBIA

0   500   1,000   *   1,500 miles

2,400 km

EQUATOR

Galapagos Is. (Ecuador)

ECUADOR   PERU

BRAZIL

NORTH AMERICA: THE PHYSICAL LAND ⟨·⟩

CANADIAN SHIELD 2.
EASTERN UPLANDS 3.
CENTRAL PLAINS 4.
WESTERN MOUNTAINS 5.
COASTAL LOWLANDS 6.

PRINCIPAL RIVERS ⟨·⟩

ARKANSAS A
CHURCHILL B
COLORADO C
COLUMBIA D
FRAZER E
MACKENZIE F
SLAVE G
PEACE H
MISSISSIPPI I
MISSOURI J
NELSON K
OHIO L
RED M
RIO GRANDE N
ST. LAWRENCE O
N. & S. SASKATCHEWAN P
SNAKE Q
YUKON R

PRINCIPAL MOUNTAIN RANGES ⟨·⟩

APPALACHIAN MTS. S
CASCADE RANGE T
COAST MTS. U
ROCKY MTS. V
SIERRA MADRE (E) W (W) W'
SIERRA NEVADA X

PRINCIPAL LAKES ⟨·⟩

THE GREAT LAKES ⟨·⟩
  L. SUPERIOR O¹
  L. HURON O²
  L. MICHIGAN O³
  L. ERIE O⁴
  L. ONTARIO O⁵
GREAT BEAR L. Y
GREAT SLAVE L. G¹
GREAT SALT L. Z
L. WINNIPEG P¹
L. NICARAGUA 1.

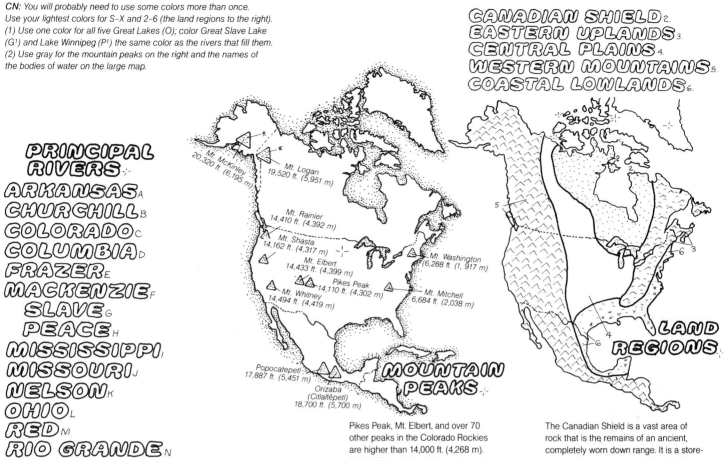

MOUNTAIN PEAKS ⟨·⟩

Mt. McKinley 20,320 ft. (6,195 m)
Mt. Logan 19,520 ft. (5,951 m)
Mt. Rainier 14,410 ft. (4,392 m)
Mt. Shasta 14,162 ft. (4,317 m)
Mt. Elbert 14,433 ft. (4,399 m)
Pikes Peak 14,110 ft. (4,302 m)
Mt. Whitney 14,494 ft. (4,419 m)
Mt. Washington 6,288 ft. (1,917 m)
Mt. Mitchell 6,684 ft. (2,038 m)
Popocatepetl 17,887 ft. (5,451 m)
Orizaba (Citlaltépetl) 18,700 ft. (5,700 m)

LAND REGIONS ⟨·⟩

Pikes Peak, Mt. Elbert, and over 70 other peaks in the Colorado Rockies are higher than 14,000 ft. (4,268 m). Mt. McKinley is the tallest peak on the continent. Popocatepetl is the tallest of the many active volcanoes in Mexico and Central America.

The Canadian Shield is a vast area of rock that is the remains of an ancient, completely worn down range. It is a storehouse of largely untapped minerals. The thin layer of soil over it can support forests, but not farming. The water of Hudson Bay fills a large depression in the Shield.

North America, the continent with the longest coastline, has a widely varied landscape and climate: frozen wastes in Greenland and northern Canada, evergreen forests in central Canada and the western U.S., towering peaks in the Rocky Mountains, barren deserts in the southwestern U.S., palm-covered islands in the Caribbean Sea, and steamy jungles in Central America. Even the shape of the continent is unusual: a broad expanse of 4,900 miles (7,840 km) from Alaska to Newfoundland narrows to a mere 30 miles (48 km) at the Isthmus of Panama. Variations in temperature between the polar north and the tropical south can exceed 200° F (93° C). Parts of Greenland (85% of which is covered with ice) and the Yukon have recorded temperatures as low as –105° F (–76° C). Death Valley, California, the lowest point on the continent, at 282 feet (86 m) below sea level, has baked in 134° F (57° C) heat.

The continent has three basic physical regions: an Eastern Uplands of very old, low, worn-down mountains; a much younger, steeper, and more rugged series of Western Mountains that cover a third of the continent, from Alaska to Central America; and a broad, flat area in between, which is further broken down into the Central Plains, Canadian Shield, and Coastal Lowlands areas (see the small map, marked Land Regions).

There are three large river systems: (1) The Mississippi-Missouri complex forms the continent's longest river (3,872 miles; 6,195 km), flowing south to the Gulf of Mexico. (2) The Great Lakes (including Lake Superior, the world's largest freshwater lake) and the St. Lawrence River and Seaway flow to the Atlantic Ocean. (3) The Mackenzie and Nelson systems flow to northern Canadian waters. Most river flow in North America is directed by the crest of the Rocky Mountains, the Great Divide. (Such a watershed region is called a continental divide.) North American rivers flow west to the Pacific, east to the Atlantic, north to Hudson Bay or the Arctic Ocean, or south to the Gulf of Mexico, depending on which side of the Great Divide the headwaters of these rivers.

GREENLAND (DEN.)

ARCTIC CIRCLE

60°N

LABRADOR SEA*

DAVIS STRAIT*

BAFFIN BAY*

Thule
Kuvdlorssuaq
Jacobshavn
Godthab
Julianehab

Clyde
Baffin
Iqaluit

HUDSON STRAIT
Ivujivik
Ungava Bay

LABRADOR
Ft. George

St. John's
St. Pierre
St. Miquelon (France)
GRAND BANKS
Halifax

GULF OF ST. LAWRENCE
Anticosti I.
Charlottetown
Fredericton
MAINE
N.H. Boston
VT. MASS.
CONN.
N.Y.

ATLANTIC OCEAN

50°N

Quebec
Montreal
Ottawa
Toronto
L. Ontario
Buffalo
Hamilton
L. Erie
MICH.

Queen Elizabeth Is.
Banks I.
Victoria
Resolute
Gioa Haven
Hall Beach
FOXE BASIN
Southampton
Belcher

BEAUFORT SEA*

Amundsen Gulf
Great Bear L.
Yellowknife
Great Slave L.
Slave R.
Back R.

Churchill
Churchill R.
Ft. Severn
Severn R.
James Bay
Ft. Albany
Albany R.

HUDSON BAY*

Sault Ste. Marie
Sudbury
L. Huron
MICH.
L. Michigan

ALASKA (US)
Yukon R.

Mackenzie R.
MACKENZIE MTS.

WOOD BUFFALO NAT'L PK.
L. Athabasca
Peace R.
Athabasca R.

Reindeer L.

Nelson R.
L. Winnipeg
L. Manitoba
Winnipegosis
Winnipeg

Thunder Bay
L. Superior
L. of the Woods
MINN.
WIS.

Edmonton
N. Saskatchewan R.
Saskatchewan R.
S. Saskatchewan R.
Prince Albert
Saskatoon
Regina

N. DAK.
S. DAK.

UNITED STATES

Mt. Logan 19,850 ft (6,050 m)
Whitehorse
Yukon R.
Dawson Creek
Prince George
ROCK MTS.

JASPER NAT'L PK.
BANFF NAT'L PK.
Calgary
Lethbridge

MONT.
WYO.

600 miles
960 km
400
200
0

COAST MTS.
Prince Rupert
Mt. Waddington 13,260 ft (4,043 m)
Frazer R.
Vancouver
Victoria
Seattle
WASH.
Queen Charlotte Is.
Vancouver

IDAHO
NEV.
ORE.
CALIF.

PACIFIC OCEAN*

NORTH AMERICA: CANADA & GREENLAND *

CN: (1) Color a province or territory, its name, and its outline on the large map. (2) Color the outline of Greenland on the large map and then on the globe to the right (on which Canada and its northern islands are to be colored gray).

PROVINCES *

ALBERTA A EDMONTON *

BRITISH COLUMBIA B VICTORIA *

MANITOBA C WINNIPEG *

NEW BRUNSWICK D FREDERICTON *

NEWFOUNDLAND E ST. JOHN'S *

NOVA SCOTIA F HALIFAX *

PRINCE EDWARD I o G CHARLOTTETOWN *

ONTARIO H OTTAWA *

QUEBEC I QUEBEC *

SASKATCHEWAN J REGINA *

TERRITORIES *

NORTHWEST TERRITORIES K YELLOWKNIFE *

YUKON TERRITORY L WHITEHORSE *

NUNAVUT M IQUALUIT *

(As of April, 1999)

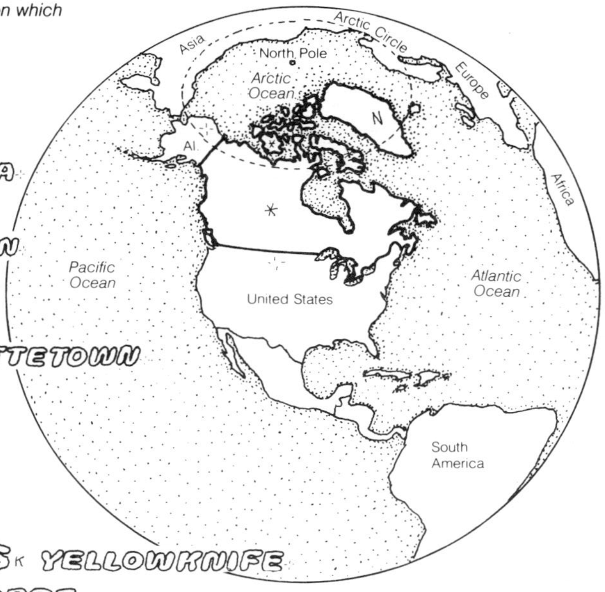

GREENLAND N

Area: 840,000 sq. mi. (2,175,600 km²). **Population:** 59,400. **Capital:** Godthåb, 11,000. **Government:** Self-governing province of Denmark. **Language:** Danish, Eskimo dialect. **Exports:** Fish products. **Climate:** Frigid, but summers are above freezing. ▢ The world's largest island is also the coldest inhabited region. The population is confined to the southwest coast, which is not as cold as the ice-covered interior. Mountains fringe a giant sheet of ice, 1–2 miles (1.6–3.2 km) thick, which covers 85% of the island. The glacial ice, created under the weight of many layers of accumulated snow, is pushed into the sea, where it breaks up into icebergs that menace North Atlantic shipping. Greenlanders are descendants of Canadian Eskimos and Danish settlers (Vikings) who arrived in 982. Eric the Red coined the name "Greenland" to entice immigrants from home. He may have been misled by the coastline, which turns green during the cool summer. The generally frigid weather is at least partially responsible for the world's highest suicide rate. When the fish aren't running, unemployment is rampant. There is hope that the recent discovery of gold will bring prosperity.

ATLANTIC PROVINCES

QUEBEC

ONTARIO

TERRITORIES

PRAIRIE PROVINCES

BRITISH COLUMBIA

REGIONS *

Texas

CANADA *

Area: 3,850,000 sq.mi.(9,971,500 km²). **Population:** 30,700,000. **Capital:** Ottawa, 720,000. **Government:** Constitutional monarchy. **Language:** English 65%, French 20%; 15% speak both. **Religion:** Roman Catholic 47%, Protestant 40%. **Exports:** Timber, newsprint, autos, machinery, fish products, grains, asbestos, nickel, zinc. **Climate:** West coast is mild; southern and eastern regions have warm summers and very cold winters; north is frigid. ▢ Canada, which is about the size of Europe, is the world's second-largest country (after Russia). It spans North America from the Atlantic to Alaska, and shares with the United States the world's longest unprotected border. Though 7% larger than the U.S., Canada has only a tenth of the population. Most Canadians live within a 200-mile (320 km) strip along the southern border with the U.S.

Despite the significant growth of minority groups, Canada remains a nation of two cultures: English and French (both are official languages.) Though France gave up its Canadian holdings to England in 1763, the French minority has resisted assimilation. today; separatist groups are moving French-dominated Quebec toward independence.

Canada has vast undeveloped resources. A wide band of forests spanning the nation enables Canada to be the world's leading producer of pulp and paper. It has the most lakes and rivers of any country (about a third of the world's freshwater supply), which provide transportation routes, irrigation, and hydroelectric power. Canada shares four of the five Great Lakes with the U.S. It has 9 lakes over 100 miles (160 km) long and 35 that are more than 50 miles (80 km) long. The Mackenzie-Peace is the longer of the two major river systems, but far more important is the St. Lawrence River Seaway network. which gives Atlantic access to most of Canada's population and industrial centers.

Canada is an independent federation of 10 self-governing provinces and 2 territories, but Canadians regard the Queen of England as their queen. Canada belongs to the English Commonwealth of Nations—a worldwide association of former colonies and current dependencies. Canada consists of six regions: (1) *The Atlantic or Maritime Provinces* are *New Brunswick, Newfoundland, Nova Scotia, and Prince Edward Island.* These smallest provinces were settled first. They are the heart of a huge east coast fishing industry. The Bay of Fundy, separating New Brunswick and Nova Scotia, is famous for its 70 foot (21 m) tides. The scenic fjords on the coast of Newfoundland must have looked like home to the Viking explorers, who in 1000 A.D. established a short-lived colony called Vinland. (2) *Quebec* is the largest province and the only one in which the French language and the Roman Catholic religion represent the majority. More French-speaking people live in Montreal than in any other city except Paris. Quebec is almost entirely covered by the Canadian Shield (see diagram on p. 4). Mining and timber are the major industries. (3) *Ontario,* the site of Ottawa, the Canadian capital, is the most populous and industrialized province. Ontario's capital, Toronto, has nearly 3 million people and is the nation's center of commerce and manufacturing. The many mines of Ontario contribute to Canada's world leadership in nickel and zinc production. (4) *The Prairie Provinces—Alberta, Saskatchewan, and Manitoba—*are home to enormous wheat farms and cattle ranches. Alberta has the world's largest national park (Wood Buffalo), and two of the most scenic (Banff and Jasper). Winnipeg, the "Chicago" of Canada, is the capital of Manitoba and a major transportation hub. (5) *British Columbia,* Canada's Pacific coast, is the most beautiful province, with its thick forests, snowcapped mountains, and fiord-lined coast. Timber and fishing are its major industries. Vancouver is the nation's busiest seaport. Timber and fishing are its major industries. Vancouver is the nation's busiest seaport. A third of the land area but have only 1% of the people. In April, 1999, the new territory of Nunavut will be created out of the Northwest Territory. Larger than Alaska, it will be the first major political region in North America governed by aborigines. Inuits (Canadian Eskomos) make up 85% of the area's population of 26,000. About 370,000 native Indians live to the south, across the entire nation. The North's brutal climate has not prevented extensive mining and oil exploration. Gold made the Klondike region of the Yukon famous, but is no longer the region's most valued mineral.

Canada has vast natural resources and unlimited land for population increase. The standard of living is equal to that of the U.S. Just as the French-speaking minority strives to preserve its identity, Canadians as a whole are fearful of being swallowed up culturally and economically by their colossal neighbor to the south.

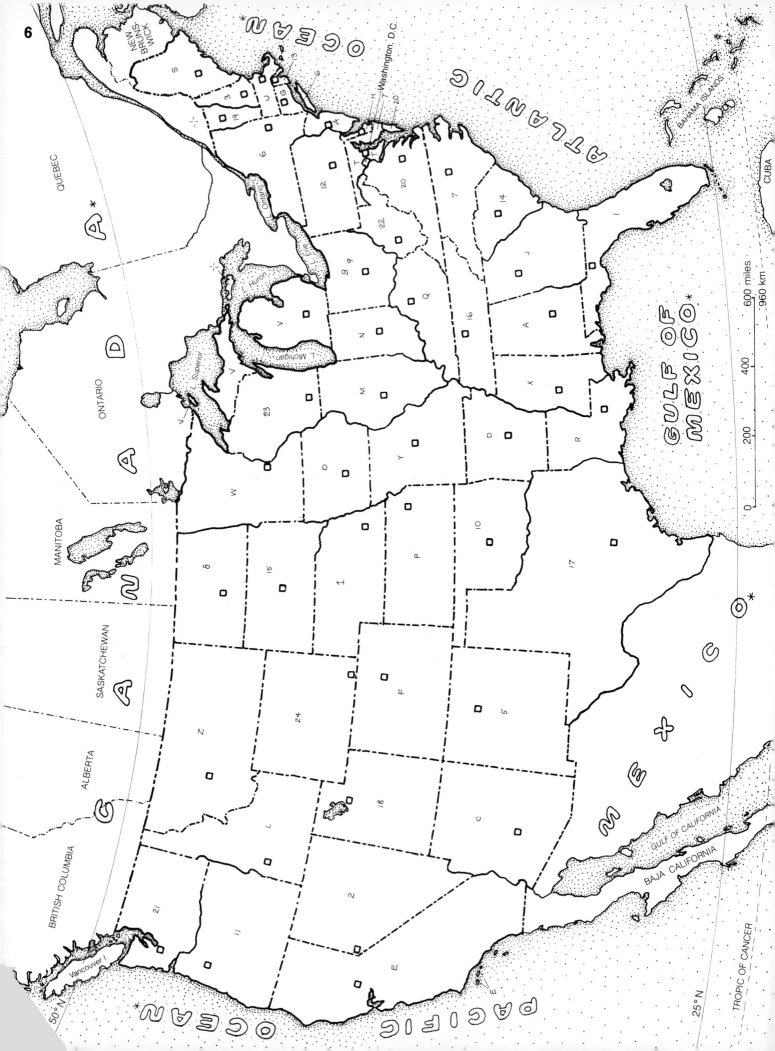

6

QUEBEC

NEW
BRUNS-
WICK

ONTARIO

CANADA*

MANITOBA

SASKATCHEWAN

ALBERTA

BRITISH COLUMBIA

50°N

Vancouver I.

L. Ontario

L. Erie

L. Huron

L. Michigan

L. Superior

ATLANTIC OCEAN

Washington, D.C.

BAHAMA ISLANDS

CUBA

GULF OF
MEXICO*

GULF OF CALIFORNIA

BAJA CALIFORNIA

MEXICO*

25°N

TROPIC OF CANCER

PACIFIC OCEAN

0   200   400   600 miles
960 km

S

3
9

13
9

G
U

H
T
20

T

12

22

9 9

N

V

V

23

M

O

O

W

8

15

1

24

Z

L

21

11

E

18

2

F

5

C

P

10

17

Y

D

X

R

16

Q

A

J

7

14

I

NORTH AMERICA: UNITED STATES

Area: 3,620,000 sq.mi (9,412,000 km²). **Population:** 272,000,000. **Capital:** Washington, D.C., 625,000. **Government:** Republic. **Language:** English; many ethnic dialects. **Religion:** Protestant 33%; Roman Catholic 22%. **Exports:** Machinery, aircraft, electronics, coal, iron and steel, chemicals, textiles, cotton, grains, soybeans, corn, produce. **Climate:** Temperate, except for northern and midwestern extremes; Pacific coast is mild; southwest is dry with very hot summers; Gulf coast is subtropical. ☐ During the 20th century the United States became the world's most affluent and powerful nation. It is fourth in size (after Russia, Canada, and China), and fourth in

population (after China, India, and Russia). The U.S. has benefited from having enormous natural resources, an ideal location for trade, vast regions of fertile land and favorable weather, friendly neighboring countries, protective ocean barriers, and economic system that rewards individual effort, a stable and democratic government, and, finally, the existence of personal, religious, and economic freedom for its people, including the millions of highly motivated, recent immigrants. At the peak of U.S. prosperity, the value of its manufactured, mined, and grown products far exceeded that of any nation—6% of the world's population was producing 50% of its wealth. In the 1970s, the U.S. began to face growing competition from Europe and Asia. In addition, the cost of waging war, both hot and cold, against international communism had drained its resources. But by the late 1980s, communism collapsed in Europe, and with it the Soviet Union; the U.S. no longer had a major enemy. In the late 1990s, the booming Asian economies suffered an unexpected economic collapse. The U.S., with a restructured, highly efficient manufacturing sector, once again found itself in the role of the world's most powerful nation.

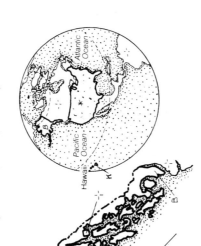

ARCTIC OCEAN

CANADA

RUSSIA

Bering Strait

BERING SEA

Hawaii

Atlantic Ocean

Pacific Ocean

ALABAMA_A / MONTGOMERY
ALASKA_B / JUNEAU
ARIZONA_C / PHOENIX
ARKANSAS_D / LITTLE ROCK
CALIFORNIA_E / SACRAMENTO
COLORADO_F / DENVER
CONNECTICUT_G / HARTFORD
DELAWARE_H / DOVER
FLORIDA_I / TALLAHASSEE
GEORGIA_J / ATLANTA
HAWAII_K / HONOLULU
IDAHO_L / BOISE
ILLINOIS_M / SPRINGFIELD
INDIANA_N / INDIANAPOLIS
IOWA_O / DES MOINES
KANSAS_P / TOPEKA
KENTUCKY_Q / FRANKFORT
LOUISIANA_R / BATON ROUGE
MAINE_S / AUGUSTA
MARYLAND_T / ANNAPOLIS
MASSACHUSETTS_U / BOSTON
MICHIGAN_V / LANSING
MINNESOTA_W / ST. PAUL
MISSISSIPPI_X / JACKSON
MISSOURI_Y / JEFFERSON CITY
MONTANA_Z / HELENA
NEBRASKA_1 / LINCOLN
NEVADA_2 / CARSON CITY
NEW HAMPSHIRE_3 / CONCORD
NEW JERSEY_4 / TRENTON
NEW MEXICO_5 / SANTE FE
NEW YORK_6 / ALBANY
NORTH CAROLINA_7 / RALEIGH
NORTH DAKOTA_8 / BISMARCK

OHIO_9 / COLUMBUS
OKLAHOMA_10 / OKLAHOMA CITY
OREGON_11 / SALEM
PENNSYLVANIA_12 / HARRISBURG
RHODE ISLAND_13 / PROVIDENCE
SOUTH CAROLINA_14 / COLUMBIA
SOUTH DAKOTA_15 / PIERRE
TENNESSEE_16 / NASHVILLE
TEXAS_17 / AUSTIN
UTAH_18 / SALT LAKE CITY
VERMONT_19 / MONTPELIER
VIRGINIA_20 / RICHMOND
WASHINGTON_21 / OLYMPIA
WEST VIRGINIA_22 / CHARLESTON
WISCONSIN_23 / MADISON
WYOMING_24 / CHEYENNE

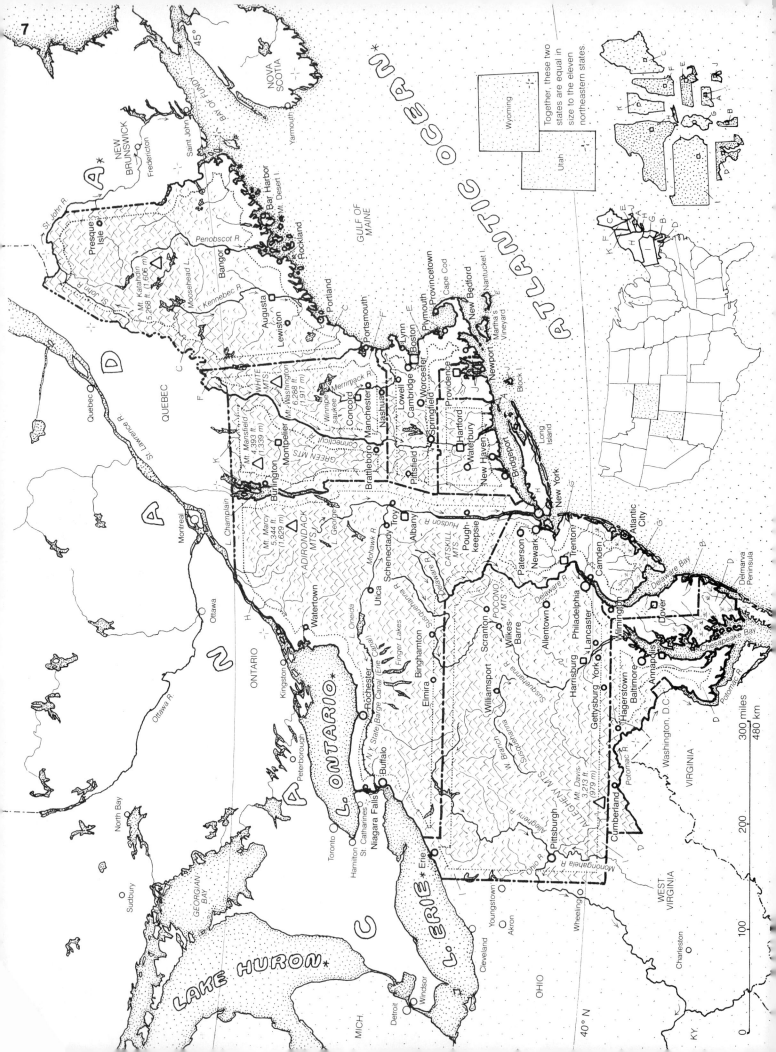

NORTH AMERICA: NORTHEASTERN U.S. ✣

All eleven northeastern states were among the original 13 English colonies. The six New England states are Connecticut, Maine, Massachusetts, New Hampshire, Rhode Island, and Vermont. Their historic past is evident in the names, monuments, schools, museums, and architecture of the area. The region is famous for the brilliant color of its autumn foliage. Northern New York, Vermont, New Hampshire, and Maine have long, cold winters.

Adjacent to New England are the mid-Atlantic states—New York, New Jersey, and Pennsylvania. This major industrial region has the greatest population concentration in the nation. A megalopolis of major cities runs from Boston to Washington, D.C. Farther south are the culturally Southern states of Delaware and Maryland. These former slave-owning states remained in the Union during the Civil War, but loyalty among their populations was sharply divided.

Mountains that cover much of the Northeast are part of the ancient Appalachian chain, whose peaks have been worn smooth over the passage of time. The only flatland is the Atlantic Coastal Plain of eastern Maryland, Delaware, and southeastern New Jersey. Farther north, the plain narrows down to beaches and dunes, which give way to rugged cliffs and fiords when the coastline reaches Maine.

CONNECTICUT_A

Area: 5,009 sq.mi.(12,963 km²). **Population:** 3,280,000. **Capital:** Hartford, 140,000. **Economy:** Aircraft engines, nuclear submarines, tobacco, dairying, insurance. □ The Connecticut (don't pronounce the middle "c") River is New England's longest. It separates Vermont from New Hampshire and flows south, through Massachusetts and Connecticut, on its way to Long Island Sound. In wooded and hilly Connecticut, the river valley is a low, fertile region where the most expensive tobacco (for wrapping cigars) is shade grown. Connecticut was known as the birthplace of many inventions and innovations that contributed to both the American and industrial revolutions. An industrial area around Waterbury was called the "arsenal of the nation." Hartford has been the insurance industry capital for 200 years. Yale is located in New Haven, and the Coast Guard Academy is in the active seaport of New London. The affluent southwest corner is home to many New York City commuters.

DELAWARE_B

Area: 2,045 sq.mi.(5,292 km²). **Population:** 747,000. **Capital:** Dover, 29,000. **Economy:** Chemicals, poultry, food products, fishing. □ Sharing the Delmarva Peninsula with Maryland and Virginia, this second-smallest state spans less than 35 mi.(56 km) at its widest point. Delaware is the state with the lowest average elevation. The only hills are in the northern industrial region of Wilmington (73,000), the largest city. It is the home of Du Pont, the chemical giant that has dominated Delaware economically and politically for two centuries. Wilmington has been called the "Chemical Capital of the World." Delaware is known for liberal laws regulating business. Many U.S. companies (half of the top 500) are Delaware corporations, although located elsewhere. In 1638, Delaware was the Swedish colony of "New Sweden"; it was taken over by the Dutch and then the British, who made it part of Pennsylvania.

MAINE_C

Area: 33,260 sq.mi.(86,076 km²). **Population:** 1,252,000. **Capital:** Augusta, 22,000. **Economy:** Paper products, seafood, potatoes, blueberries, shipbuilding. □ The northern half of Maine, the most easterly state, is completely surrounded by Canada. Maine is a major producer of paper products; 90% of the state is covered by forests, mostly owned by paper companies. Hordes of visitors come to the scenic, rocky coast each summer. Mt. Desert Island is the site of Acadia, New England's only national park. Maine has so many islands that its name refers to the "mainland." The busy port of Portland (65,000) is the largest city. Potatoes are the principal crop, and the famed "Maine lobster" is the prize catch of the fishing industry.

MARYLAND_D

Area: 10,470 sq.mi.(27,096km²). **Population:** 5,062,000. **Capital:** Annapolis, 37,600. **Economy:** Trade, electronics, chemicals, food products, tobacco, poultry. □ The Chesapeake Bay, a bountiful provider of seafood (particularly crabs, clams, and oysters), nearly divides Maryland in two. The eastern shore is a flat, fertile vegetable-producing area. The larger western shore is densely populated, especially near Baltimore and Washington D.C. The land becomes progressively hillier toward the Appalachians. Baltimore (740,000) is one of the nation's busiest seaports and a center of trade and manufacturing. The city is famous for endless rows of attached brick houses with white marble steps. Nearby is Annapolis, the state capital, the site of the U.S. Naval Academy. The northern boundary of Maryland was once called the Mason-Dixon line, which divided the slave states of the South from the abolitionist North.

MASSACHUSETTS_E

Area: 8,257 sq.mi.(21,369 km²). **Population:** 6,150,000. **Capital:** Boston, 590,000. **Economy:** Electronics, precision products, fishing, textiles, cranberries. □ Small only in size, Massachusetts is a leader in finance, trade, culture, industry, medicine, and education—almost all of which are centered in the greater Boston area, along with half the population. Cambridge is the site of Harvard, the nation's oldest university, and many other leading colleges, including the Massachusetts Institute of Technology. Boston was founded in 1630, only 10 years after the Pilgrims landed at Plymouth Rock on Cape Cod Bay. Colonial architecture, museums, monuments, and restored villages are visible reminders of the state's historic past. The Cape Cod peninsula and the islands of Martha's Vineyard and Nantucket are popular summer resorts on the Atlantic coast. Massachusetts is no longer a major producer of clothing, textiles, and shoes; those mills and factories, which first went south and then overseas, have been replaced by electronics and engineering companies.

NEW HAMPSHIRE_F

Area: 9,380 sq.mi.(24,275 km²). **Population:** 1,200,000. **Capital:** Concord, 36,500. **Economy:** Wood products, electronics, granite, shoes, dairying. □ The "Granite State" is known for its large deposits of many varieties of granite. New Hampshire is almost completely mountainous except in the southeastern Merrimack River valley, in which Concord, the capital, and Manchester (100,000), the largest city, are located. The state barely has a coastline—only 13 mi.(21 km) separate Maine from Massachusetts. In the dramatic White Mountains of the north is the tallest peak in the northeast, Mt. Washington (6,288 ft.,1,917 m). Around its summit swirled the highest winds ever clocked: 231 mph (370 km/h).

NEW JERSEY_G

Area: 7,790 sq.mi.(20,160 km²). **Population:** 8,085,000. **Capital:** Trenton, 90,000. **Economy:** Chemicals, pharmaceuticals, food products, fishing. □ The "Garden State" is named for its flower-filled hothouses, fruit orchards, and productive truck farms. But New Jersey is also a major industrial state. It has the nation's highest population density, and the greatest percentage (95%) of town and city dwellers. Many residents commute to Philadelphia and New York. Even so, the state does not have a single major city. Newark (285,000) is the largest. New Jersey is the shipping hub of the Northeast; its ports in New York Harbor handle almost all of the container traffic. Atlantic City, with its gambling casinos, heads the list of the state's popular seaside resorts.

NEW YORK_H

Area: 52,730 sq.mi.(136,465 km²). **Population:** 18,142,000. **Capital:** Albany, 107,000. **Economy:** Manufacturing, finance, trade, clothing, produce, dairying. □ Though overtaken by California in population and manufacturing output, New York remains the national leader in business and trade. New York City, the "Big Apple," with over 7 million people, is the business capital of the U.S., if not the world. It is also the center of finance, trade, fashion, advertising, publishing, music, theater, and the arts. New York's dominance in commerce began with the opening of the Erie Canal in 1825. The canal, linked to the Mohawk and Hudson Rivers, gave midwestern cities on the Great Lakes access to the Atlantic Ocean. The Dutch settlers originally called the area "New Netherlands" and the city "New Amsterdam." They built a wall across lower Manhattan for protection against the Indians and the British; "Wall Street" is now America's financial center. The famous Empire State Building has been topped twice, by the towers of the World Trade Center (each one 1,350 ft., 412 m). Rural New York is a fertile farming region. At the Canadian border are the Niagara Falls, which are grinding their way up the Niagara River toward Lake Erie at the rate of 1-3 in.(2.5-8 cm) per year.

PENNSYLVANIA_I

Area: 46,040 sq.mi.(119,151 km²). **Population:** 12,150,000. **Capital:** Harrisburg, 52,900. **Economy:** Manufacturing, iron and steel, coal. food products, dairying and mushrooms. □ William Penn founded Pennsylvania as a religious sanctuary. It is called the "Keystone State" because of its central location within the original 13 colonies. Philadelphia (1,600,000), the largest city, was the nation's first capital and the economic, political, and cultural center of the colonies. It is still a major river port and a center for trade, business, culture, and education. The Pittsburgh area is the heart of a huge steel and coal industry. Some of the nation's richest farmland is found in the southeastern area—the Pennsylvania Dutch (a mispronunciation of "Deutch," meaning German) country, home of the Amish and Mennonite religious sects.

RHODE ISLAND_J

Area: 1,214 sq.mi.(3,142 km²). **Population:** 995,000. **Capital:** Providence, 161,000. **Economy:** Jewelry, silverware, textiles, and poultry (Rhode Island Red hens). □ Roger Williams was banished from Massachusetts in 1636 for being too liberal. He founded Rhode Island as a sanctuary for political and religious freedom—engendering an enduring distrust of big government and an adherence to the separation of church and state. Rhode Island is an important manufacturing center, known for its jewelry and silverware. It has the highest proportion of industrial workers; only about 1% of the population is involved in agriculture. Island-filled Narragansett Bay is a recreational area that occupies a large part of the state. Newport, at its mouth, is best known for yacht races, music festivals, and the most palatial summer estates in the nation. Providence, the capital, has over 65% of the state's population.

VERMONT_K

Area: 9,610 sq.mi.(24,870 km²). **Population:** 600,000. **Capital:** Montpelier, 8,250. **Economy:** Timber products, precision manufacturing, maple syrup, granite, marble, asbestos, dairying. □ Vermont, the least populous state east of the Mississippi, has the smallest percentage of city dwellers (30%). Burlington (39,600) is the largest city. Lakes and rivers define most of Vermont's boundaries. It is the only northeastern state without an Atlantic coastline. The scenic Green Mountains run through the center of the state. The name "Vermont" comes from the French "vert mont," meaning "green mountain." Stone quarries in these mountains provided most of the granite and marble used in building the nation's cities. Extensive forests supply many products, including the pure maple syrup for which Vermont is famous.

NORTH AMERICA: SOUTHEASTERN U.S.

Long hot summers, mild winters, heavy rainfall, and fertile land make the Southeast a major agricultural producer. Cotton, tobacco, and peanuts are the leading crops. The South, which also includes Texas, Louisiana, Arkansas, and Oklahoma, was once known as the "land of cotton." After World War II, the emphasis shifted to manufacturing; northern companies arrived in search of cheap, non-union labor. New England's textile mills moved closer to their source of cotton. The end of racial segregation removed a stigma long associated with the South. The availability of air conditioning further encouraged migration from the North. Only the West has had more growth. Industrialization and the arrival of "Yankees" have not diminished the South's traditional charm.

The southeastern states have thick, fast-growing softwood forests that produce 40% of the nation's timber. Except for the Florida peninsula, the region slopes eastward from Appalachian crests, down across hilly plateaus (the Piedmont) to the flat Atlantic coastal plain. The major cities are congregated in the Piedmont because of the rushing rivers that provide hydroelectric power. The rivers of the South, particularly the Mississippi and its tributaries, play a vital role in commerce, farming, and recreation. Almost all of the large lakes in this region were created by the damming of rivers.

Washington, D.C. (District of Columbia), the nation's capital, is not part of any state. It has a local government, but it is under the control of Congress. Located on the Maryland side of the Potomac River (it was once a part of Maryland), Washington is 69 sq.mi.(179 km²) in area and has a population of 625,000. The metropolitan Washington area is much larger and includes parts of Maryland and Virginia.

ALABAMA A

Area: 51,609 sq.mi.(133,667 km²). **Population:** 4,372,000. **Capital:** Montgomery, 190,000. **Economy:** Timber and paper, iron and steel, cotton, peanuts, pecans, soybeans, textiles. □ Because the ingredients of steel production (iron, coal, and limestone) are mined locally, Birmingham (270,000) has been the iron, steel, and heavy industry center of the South. The agricultural industry had revolved around cotton until a pest, the boll weevil, nearly wiped out an entire cotton harvest. Farmers were forced to diversify and now grow a wide variety of crops. This has led to prosperity and security, prompting the state to erect a public monument in honor of the boll weevil. Montgomery was the first capital of the Confederacy. One hundred years later, it was the site of a bus boycott, led by Martin Luther King, which accelerated the civil rights movement. The Marshall Space Flight Center in Huntsville has developed many of the rockets and missiles used in America's space program.

FLORIDA B

Area: 58,664 sq.mi.(151,939 km²). **Population:** 14,800,000. **Capital:** Tallahassee, 110,000. **Economy:** Tourism, citrus, cattle, phosphates, vegetables, sugarcane, electronics. □ Florida, a long (450 mi., 720 km), low, flat peninsula jutting into subtropical seas, is the one of the fastest-growing states. The balmy climate makes year-round agriculture possible. While retirees have given the state the nation's oldest population, tourists bring their kids to see Disney World in Orlando; the Kennedy Space Center at Cape Canaveral; St. Augustine (the first European settlement in the U.S., in 1565); the Florida Keys (a chain of coral islands); and Everglades National Park. The latter is a large, swampy region whose land and water are being devoured by Florida's building boom. Lake Okeechobee is the second-largest freshwater lake (after Lake Michigan) completely within the U.S. Jacksonville (700,000) is physically the largest city in the country. South Florida is dominated by Miami (365,000)—Latin-flavored as a result of Cuban immigration. Florida has a huge citrus industry; nearly all frozen orange juice is processed there. Most of the country's phosphate, a chief ingredient of fertilizer, is mined in Florida. Northern Florida is one of the country's leading cattle regions.

GEORGIA C

Area: 58,876 sq.mi.(152,489 km²). **Population:** 7,600,000. **Capital:** Atlanta, 400,000. **Economy:** Timber, aircraft, marble, cotton, tobacco, soybeans, peanuts, poultry, peaches. □ Georgia was named after King George II. It is the largest state east of the Mississippi and one of the fastest growing. Atlanta, the capital, is the center of business, trade, and finance in the entire Southeast. Many varieties of trees, particularly pines, cover 70% of the state. These pines produce half the world's supply of naval stores (tar, resin, and turpentine). Southeastern Georgia is a flatland covered by marshes and swamps (the Okeefenokee is a lush wildlife refuge); bayous (narrow river outlets); and grassy plains called savannas. The city of Savannah is Georgia's third largest (140,000) and a seaport of historic importance. Undamaged by the Civil War, Savannah is a fine example of antebellum architecture.

KENTUCKY D

Area: 40,410 sq.mi.(104,662 km²). **Population:** 3,940,000. **Capital:** Frankfort, 26,000. **Economy:** Coal, tobacco, timber, bourbon whiskey, racehorses. □ The name "Bluegrass State" comes from the blue tint of the grassy hills of the populated north central region. Most of Kentucky is a hilly plateau that gradually slopes westward to the Mississippi. Along with the breeding of the best racehorses and the making of fine bourbon whiskey, Kentucky is first in the U.S. in coal production and second in tobacco products. The main tourist attraction is Mammoth Cave, the largest system of caves (300 mi.,480 km) in the world. Louisville (275,000), the largest city, is an important inland port on the

Ohio River, and the home of the Kentucky Derby. Fort Knox is reputed to be the storehouse of all the gold owned by the U.S. government.

MISSISSIPPI E

Area: 47,718 sq.mi.(123,494 km²). **Population:** 2,710,000. **Capital:** Jackson, 200,000. **Economy:** Cotton, timber and paper, naval stores, soybeans, oil. □ Mississippi was governed by the French, British, and Spanish rule before becoming a state. Cotton production once made it one of the most prosperous states, but it took nearly 100 years to recover from the devastation it suffered in the Civil War. There was virtually no industry before World War II. Today, the state is one of the nation's most ambitious reforestation programs. Especially fertile farmland is found in alluvial regions (soil deposited by previous floods) along the Mississippi and in the "Black Belt," a strip of dark, rich soil that runs across Mississippi and Alabama. Pond-raised catfish is a major industry. The Gulf Coast is a popular resort area and home to the shrimp industry. Scenes of Mississippi's rural life have been fictionalized by William Faulkner.

NORTH CAROLINA F

Area: 52,586 sq.mi.(136,1988 km²). **Population:** 7,550,000. **Capital:** Raleigh, 220,000. **Economy:** Tobacco, textiles furniture, timber, cotton. □ North Carolina is the nation's largest producer of tobacco, textiles, and wooden furniture. In the western Blue Ridge Mountains is the tallest peak east of the Mississippi, Mt. Mitchell (6,684 ft., 2,038 m). The hilly, fertile Piedmont region holds the major population and industrial centers as well as the tobacco farms. To the east lie the savannas and swamps of the Atlantic Coastal Plain. Offshore is a chain of narrow islands, sandbars, and reefs, similar to those on Florida's east coast but far more treacherous. Shifting sands and unseen reefs have wrecked many a passing ship. Cape Hatteras has been called the "Graveyard of the Atlantic." Kitty Hawk, a little to the north, is the stretch of sand made famous by the first flight of the Wright brothers in 1903.

SOUTH CAROLINA G

Area: 31,112 sq.mi.(80,580 km²). **Population:** 3,785,000. **Capital:** Columbia, 100,000. **Economy:** Tobacco, textiles, paper products, chemicals. □ The busy seaport of Charleston, the second-largest city (85,000), was founded in 1670. Here one can still experience the charm and appearance of the "Old South," where aristocratic plantation owners lived a lifestyle patterned after that of English nobility. Their determination to retain a slave-based economy prompted South Carolina to become the first state to secede from the Union. The opening shots of the Civil War were fired on Ft. Sumter in Charleston harbor. Today, the state's rapidly expanding industrial economy is assisted by hydroelectric power, generated by 12 rivers rushing down from the "up country" in the west to the "low country" in the east. The central location of Columbia, the capital, was a compromise between rival settlers of these two "countries."

TENNESSEE H

Area: 42,110 sq.mi.(108,981 km²). **Population:** 5,500,000. **Capital:** Nashville, 525,000. **Economy:** Chemicals, food products, vehicles, cotton, tobacco, textiles □ Tennessee is divided into three parts by the Tennessee River, unusual in the way it returns to Tennessee after swinging south through Alabama. The Tennessee Valley Authority (TVA) has erected 32 dams along the river to create reservoirs and recreational lakes, and to provide hydroelectric power for Tennessee and seven neighboring states. Cotton and soybeans are the chief crops in the flat and fertile western region. Memphis, on bluffs above the Mississippi River, is the largest city (620,000) and a busy inland port. Nashville, the capital and second-largest city, is the country music capital of America (home of the Grand Ole Opry). Oak Ridge was the site of the nation's first nuclear reactor, which provided material for the first atomic bomb.

VIRGINIA I

Area: 40,817 sq.mi.(105,716 km²). **Population:** 6,845,000. **Capital:** Richmond, 210,000. **Economy:** Chemicals, tobacco, tourism, shipbuilding, textiles, foods, timber. □ Sir Walter Raleigh named Virginia for Queen Elizabeth I, the "Virgin Queen." Virginia's history dates back to the founding of Jamestown in 1607, the first English settlement in North America. Other tourist attractions are Williamsburg, a restored colonial village; Mt. Vernon and Monticello, the homes of George Washington and Thomas Jefferson (Virginia was the birthplace of six other presidents); Arlington National Cemetery; the Blue Ridge Mountains; the Shenandoah Valley; and the many Civil War shrines (Virginia was the principal battleground). Proximity to the sea plays an important role in the state's strong economy. Norfolk, the second-largest city (270,000) after Virginia Beach (400,000) as well as the major seaport, and Newport News (175,000) are centers for shipbuilding and U.S. naval installations.

WEST VIRGINIA J

Area: 24,181 sq.mi. (62,629 km²). **Population:** 1,838,000 **Capital:** Charleston, 58,000. **Economy:** Coal, iron and steel, timber, chemicals, glassware, marbles. □ Most of West Virginia's population and manufacturing centers are either on or close to the Ohio River. A greater percentage of West Virginia is covered by mountains than any other state. It has the highest average elevation of any state east of the Mississippi. The rugged terrain has bred a highly independent-minded citizenry. John Brown led his famous anti-slavery rebellion in 1859 at Harpers Ferry (now a national park). When the Civil War began, western Virginia residents, who had strong anti-slavery feelings, broke away from secessionist Virginia to form their own state, and then rejoined the Union. West Virginia's economy depends largely upon coal mining and industrial production. Indiscriminate strip mining and lumbering have ravaged many parts of this scenic region. Steps have been taken to restore the landscape.

NORTH AMERICA: SOUTHCENTRAL U.S.

ALASKA & HAWAII

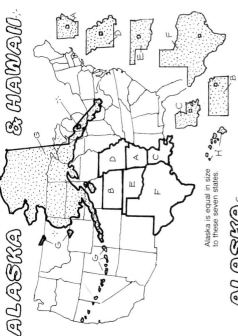

Of the six states shown above, Kansas and Missouri are usually considered Midwestern. The other four are classified as Southern; they all fought for the Confederacy. Oklahoma was still an Indian Territory, but quite a few tribes fought alongside the rebels.

Oil has been the driving force behind the economies of Texas, Oklahoma, and Louisiana, with the discovery of large oil deposits gaining in importance. Alaska became a major oil producer with the discovery of large oil reserves on its north coast. The economies of these states tend to be dependent upon the international price of oil.

Pacific Ocean

Tropic of Cancer

ARKANSAS A

Area: 53,103 sq.mi.(137,537 km²). **Population:** 2,570,000. **Capital:** Little Rock, 175,000. **Economy:** Timber, rice, broiler chickens, oil, natural gas, bauxite. □ Arkansas (ark' in saw) is best known for its scenery and outdoor attractions, especially the natural springs. The therapeutic waters of over 40 springs in the city of Hot Springs draw millions of visitors each year. The springs are actually in Hot Springs National Park, most of which is within the city limits. Arkansas leads the nation in rice and broiler chicken production. The state has America's only diamond mine—visitors may keep whatever diamonds they find. The eastern part of the state, along the Mississippi, has fertile, alluvial soil. Rich farmland can also be found in the northern Ozark region. Arkansas has one of the highest rural populations in the nation.

KANSAS B

Area: 82,264 sq.mi.(212,899 km.). **Population:** 2,626,000. **Capital:** Topeka, 120,000. **Economy:** Wheat, cattle, oil, natural gas, aircraft. □ The nation's leading wheat producer has a colorful western past: cattle drives, gunfighters, saloons, and wagon trails. Railroad terminals in Abilene and Dodge City were the destination for many cattle drives coming from as far away as Texas. The geographical center of the 48 states is in central Kansas, near the Nebraska border. The major population centers are in the eastern half of the state, which has rich alluvial soil and adequate rainfall. In the drier west, there are very few trees, and the land is a flat plain gradually ascending to 4,000 ft.(1,220 m) at the foothills of the Rockies. Land conservation, irrigation, dams, and reservoirs have revived Kansas from the "dust bowl" condition of the 1930s. Wichita, the largest city (310,000), is the nation's number-one producer of private aircraft.

LOUISIANA C

Area: 48,522 sq.mi.(125,575 km²). **Population:** 4,392,000. **Capital:** Baton Rouge, 210,000. **Economy:** Oil and natural gas, chemicals, salt, sulphur, soybeans, sugar, shrimp. □ The huge Delta region was formed by silt (particles of soil suspended in water) donated by states located up the Mississippi River. A massive network of dikes and levees (banks) has been built along the river to protect against flooding. The state's extensive Gulf coastline includes hundreds of inlets, islands, sandbars, bayous, and marshes. This area supports huge pelican and egret populations, and is the winter home for half the wild ducks and geese in North America. New Orleans (487,000), the largest city, lies 100 miles upriver from the Gulf of Mexico. It is a major shipping center and one of America's favorite tourist attractions with its fine restaurants, "Old World" architecture, street art, Dixieland music, and the Mardi Gras. Another "foreign" side of Louisiana is that the state's civil law is based upon France's Napoleonic Code. Many people in the southern half of the state are either Cajuns or Creoles, and most of them are Roman Catholic. Cajuns, who speak a French dialect as well as English, are descendants of early Canadian settlers from Nova Scotia. Creoles are people of French or Spanish ancestry. Northern Louisiana is ethnically different from the South, as it was settled primarily by white Anglo-Saxon Protestants from the other states.

MISSOURI D

Area: 69,685 sq.mi.(180,345 km²). **Population:** 5,454,000. **Capital:** Jefferson City, 33,680. **Economy:** Transportation equipment, lead, soybeans, corn, meatpacking. □ Because of its central location and access to the nation's largest rivers, Missouri (muh zoor' ee) was a major 19th century trade and transportation center; the Santa Fe and Oregon Trails led west from the town of Independence. The Pony Express serviced the western half of the country, from St. Joseph, on Missouri's western border, to California. Close to the Mississippi riverfront in St. Louis (400,000) is the nation's tallest monument, the 630 ft. (192 m) Gateway Arch. It is dedicated to the city's historic role as the gateway to the West. St. Louis and Kansas City (440,000) are also major air, rail, and trucking centers, as well as important inland ports. Missouri is a big producer of transportation equipment: autos, planes, railroad cars, buses, and aerospace products. The hilly Ozark Plateau region is a popular recreation area and the center for the nation's largest lead-mining industry.

OKLAHOMA E

Area: 69,920 sq.mi.(180,953 km²). **Population:** 3,350,000. **Capital:** Oklahoma City, 450,000. **Economy:** Oil and natural gas, cattle, wheat. □ "Oklahoma" is an Indian term meaning "land of red people." But with the firing of a single pistol shot in 1889, which signaled the start of a massive land grab by white settlers, the Indian population lost most of their treaty rights. Those settlers who raced out to claim their land before the gun was fired were called "Sooners"; the nickname now applies to all Oklahomans. In Oklahoma, oil derricks can be seen everywhere, including on the front lawn of the state capitol in Oklahoma City. Tulsa (370,000), the second-largest city, has been called the "oil capital of the world." The city became an important inland port when an enormous navigation project on the Arkansas River opened up commercial traffic to the Mississippi. Grasslands support a huge beef industry, but the soil has never recovered from its pre-dust bowl fertility. The effects of the droughts of the 1930s and the emigration of thousands of "Okies" from the state were described by John Steinbeck in "The Grapes of Wrath."

TEXAS F

Area: 267,338 sq.mi.(691,872 km²). **Population:** 19,560,000. **Capital:** Austin, 475,000. **Economy:** Oil and natural gas, cattle, cotton, sulphur, machinery, electronics, machinery. □ The Alamo was a San Antonio chapel in which Texas revolutionaries (including Jim Bowie and Davy Crockett) were wiped out by a Mexican army. In 1836, the rallying cry "remember the Alamo" led a Texas army to victory and independence from Mexico. Statehood followed nine years later. Although Alaska displaced Texas as the largest state, Texans can still boast of being number one in oil, natural gas, sulphur, asphalt, and gypsum. Texas also has the most ranches and farms, which raise the most cattle, horses, sheep, wool, and cotton; the King Ranch alone is larger than Rhode Island. Texas ranks third, after California and New York, in the value of the goods it produces. Houston (1,650,000), a major seaport, is the trade and manufacturing center for this region of the country. The Johnson Space Center is located close by. Dallas (1,100,000) and Ft. Worth (500,000) are the heart of the oil, banking, and insurance industries, and together with other population centers they form one of the nation's largest metropolitan areas.

ALASKA G

Area: 586,400 sq.mi.(1,517,603 km²). **Population:** 635,000. **Capital:** Juneau, 38,000. **Economy:** Oil, natural gas, timber, fishing, gold. □ Alaska means "great land" in Aleut—and it surely lives up to its name. It is more than twice the size of Texas; it has a longer coastline than all other states combined; and it has the tallest mountain peak in North America (Mt. McKinley, 20,320 ft., 6,194 m). In 1867, Secretary of State Seward pressured Congress to buy "Russian America" for $7.2 million. Even at 2 cents an acre (5 cents a hectare), it was referred to as "Seward's Folly." The U.S. and Russia are separated by the 50 mi.(80 km)-wide Bering Strait. Alaska is America's last frontier—a mostly uninhabited region with a broad range of untapped mineral and natural resources. Its fishing industry (principally salmon) is the nation's largest. The winters are generally frigid, except in the southern and southeastern regions, which have temperatures comparable to many northern U.S. cities. Juneau, the capital, alongside British Columbia, can be reached only by air or sea. Barrow, on the barren Arctic coast, is the largest Eskimo village in the world. East of it, on Prudhoe Bay, a major oil discovery was made in 1968. A pipeline was built to transport oil 800 mi.(1,280 km) to Valdez, a town on Prince William Sound in the Gulf of Alaska. A devastating oil spill occurred there in 1989. Alaska's worst weather and most hostile landscape is in the Aleutian Islands, an archipelago extending westward for 1,700 mi.(2,720 km). Japanese troops occupied the two outermost islands during World War II—their only invasion of North America.

Alaska is equal in size to these seven states.

HAWAII H

Area: 6,450 sq.mi.(16,706 km²). **Population:** 1,230,000. **Capital:** Honolulu, 380,000. **Economy:** Military installations, tourism, pineapples, sugar. □ Entirely created by volcanoes, Hawaii (huh wy' ee) is a 1,500 mi.(2,400 km) archipelago of 130 islands in the mid-Pacific, southeast of the mainland and about the same distance from San Francisco as is New York City. The islands were settled by Polynesians from Southeast Asia around 750 A.D. They were discovered in 1778 by the English explorer James Cook, who named them the Sandwich Islands. Hawaii has the broadest ethnic mix of any state, including the highest percentage of Asians (58%). Most Hawaiians live on the five largest islands. Hawaii followed Alaska in entering the Union in 1959. Beautiful weather, a lush tropical environment, native hospitality, and jet travel have created a thriving tourist trade. Honolulu, on the island of Oahu, is the business center and has almost a third of the state's population. Also on Oahu is Pearl Harbor, the seaport and naval base that was attacked by the Japanese in 1941, bringing the U.S. into World War II. Hawaii, the "Big Island," was formed by five volcanoes; two are still active. Maui has the world's widest inactive crater (7 mi.,11.2 km). A mountaintop on Kauai, the greenest island, is the wettest spot on the planet, with 460 in.(1,168 cm) of rain annually.

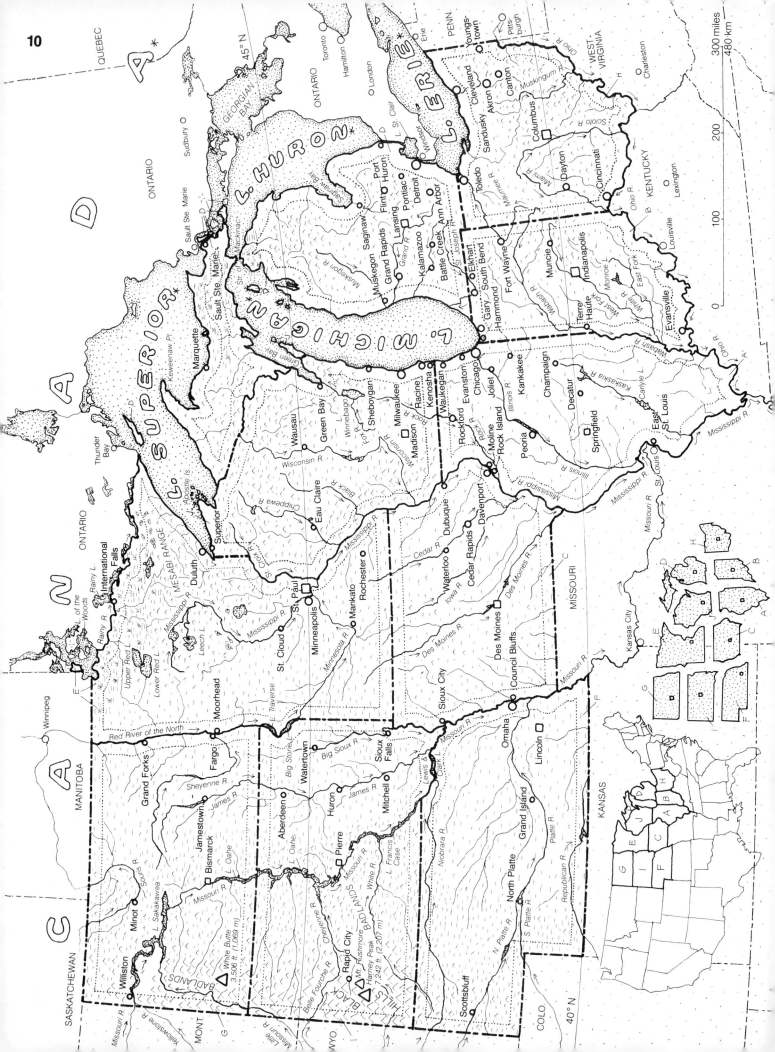

NORTH AMERICA: NORTHCENTRAL U.S.

These states, along with Kansas and Missouri (p. 9), make up the American Midwest, one of the world's most productive agricultural areas. The central location gives it access to transportation on the major rivers (Mississippi, Missouri, and Ohio), the Great Lakes, and St. Lawrence Seaway, and to extensive rail and road networks. The lower tier of states makes up the "corn belt." Much of America's most valuable crop is grown here to feed cattle that are sent to feedlots in the Midwest for fattening before slaughter. Wheat, the most important human food crop, dominates the western fringe of states. The northern states specialize in "spring wheat" while those to the south grow "winter wheat." In addition to the two wheat belts, there is a "dairy belt" across the northcentral states. Wisconsin, the leading dairy producer, and the four states east of the Mississippi are industrial power-houses as well as major agricultural producers. These states are ideally situated between the iron deposits of northern Minnesota, Wisconsin, and Michigan and the coal deposits of the Appalachian Mountains in the east. Interstate commerce via the Great Lakes was expanded to include international trade when the completion of Canada's St. Lawrence Seaway opened a route to the Atlantic (see map on p. 6). Except for northern Minnesota, Wisconsin, and Michigan, this region has very few trees. Much of it is rolling hills or flat prairie that gradually ascends to 5,000 ft (1,524 m) at its western edge.

ILLINOIS A

Area: 56,400 sq.mi.(145,963 km²). **Population:** 11,960,000. **Capital:** Springfield, 106,000. **Economy:** Corn, soybeans, heavy machinery, food products, hogs, coal, electronics. □ As a leading agricultural and industrial state, Illinois (ill uh noy) produces the most corn, soybeans, meat products, farm and road-building equipment, and diesel engines. Chicago (2,900,000), the nation's transportation hub, handles more passengers and freight through its air, ship, rail, truck, and bus terminals than any other city in North America. With the opening of the St. Lawrence Seaway, Chicago became the world's busiest inland port. It is the trade, financial, and cultural center of the Midwest. Its 110-story Sears Tower is the world's tallest building (1,454 ft., 443 m), and O'Hare is the busiest airport. Over 60% of the state's population lives in the greater Chicago area. The Fermi National Accelerator Lab, a leader in atomic research, is in the suburb of Batavia. Illinois is nearly surrounded by water: Lake Michigan and the Mississippi, Illinois, Ohio, and Wabash Rivers.

INDIANA B

Area: 36,280 sq.mi.(93,893 km²). **Population:** 5,850,000. **Capital:** Indianapolis, 735,000. **Economy:** Coal, iron and steel, electrical and food products, corn, soybeans, hogs, limestone. □ Though only 38th in size, Indiana ranks among the top agricultural and industrial states. The 50 mi (80 km) Lake Michigan riverfront of Gary and Hammond is one of the world's great industrial regions—a center for iron and steel, chemicals, oil refineries, and general manufacturing. Indianapolis, the capital and business center, is a major livestock distribution hub with many interstate truck routes passing through it. The city is the home of the Memorial Day automobile racing classic, the Indianapolis 500. Notre Dame University is located in South Bend. Nearby Elkhart is the leading producer of band and orchestral instruments. Most of the limestone used in the nation's buildings is cut in quarries located in southern Indiana.

IOWA C

Area: 56,274 sq.mi.(145,693 km²). **Population:** 2,875,000 **Capital:** Des Moines, 198,000. **Economy:** Corn, hogs, soybeans, food products, farm machinery. □ The golden, gently rolling hills made famous in the paintings of Grant Wood reflect the agricultural richness of this state. Its deep, rich, topsoil—from past glacial activity—amounts to 25% of the nation's best farmland. Though it is medium-sized in area and population, Iowa ranks second only to California in corn food production. The state alternates with Illinois as the leading corn producer. It is first in hogs, second in soybeans, and among the leaders in cattle (most of which are brought into the state to be "corn-fattened" before slaughter). A quarter of the food production is exported abroad.

MICHIGAN D

Area: 58,220 sq mi (150,731 km²). **Population:** 9,625,000. **Capital:** Lansing, 130,000. **Economy:** Autos, machinery, iron ore, oil, food processing, salt, and tourism. □ Surrounded by four of the five Great Lakes, Michigan is almost all coastline. The state consists of two peninsulas: the larger, the shape of a mitten, contains most of the population, factories, and farms; only 3% of the population lives in the iron-rich upper peninsula. At its tip is one of the busiest canals in the world: the Sault Ste. Marie, which connects Lake Superior with Lake Huron. The northern forests (with over 10,000 lakes) and an extensive coastline make Michigan a leading recreational area. More cars and trucks are produced in Detroit (1,050,000), Flint, Lansing, and Pontiac than in any comparable region. In the processing of grains, Battle Creek is known for breakfast cereals. Much of the nation's freshwater fish comes from Michigan's many inland fisheries.

MINNESOTA E

Area: 84,401 sq.mi.(218,514 km²). **Population:** 4,650,000. **Capital:** St. Paul, 275,000. **Economy:** Dairying, grains, iron ore, timber, electronics. □ The Indian name "Minnesota" means "sky-blue water," which befits a land of many rivers, streams, swamps, waterfalls, and over 15,000 lakes. Glacial action that flattened large sections of the state was not quite as effective in the hilly, iron-rich, northeastern "Superior Uplands" (the lower tip of the rocky Canadian Shield). In that region, the Mesabi Range, the largest open-pit mine in the world, has been extracting iron ore for over 80 years. Although the premium ore is gone, newer technology has made it feasible to continue mining inferior grades. The ore is shipped, along with timber and grains, out of Duluth and Superior (Wisconsin), two of the busiest ports on the Great Lakes. Half of the population (which includes the most Scandinavians in the U.S.) lives in the twin cities of Minneapolis (375,000) and St. Paul, modern centers of business and culture. The famed Mayo Clinic is located in Rochester. Grain and dairy production has earned Minnesota the title "Bread and Butter State."

NEBRASKA F

Area: 77,354 sq.mi.(200,270 km²). **Population:** 1,680,000. **Capital:** Lincoln, 195,000. **Economy:** Wheat, corn, cattle, hogs, food products, alfalfa, oil. □ The "Cornhusker State" actually ranks third in corn production, but first in alfalfa and second (after Texas) in cattle. As much as 95% of the land is devoted to agriculture. The corn-growing farms are located in the wetter eastern half of the state. Wheat and cattle are raised in the west. Numerous varieties of grass for cattle forage are grown throughout the state. Omaha (340,000), the largest city, is a major livestock trading center and a hub for finance and insurance. Nebraska is the only state with a single-house legislature; its members are elected without any party affiliation. In the 19th century, many Nebraskans provided services for pioneers passing through the state on their way west. The famous Oregon Trail ran along the Platte and North Platte Rivers.

NORTH DAKOTA G

Area: 70,700 sq.mi.(183,042 km²). **Population:** 653,000. **Capital:** Bismarck, 52,000. **Economy:** Oil, lignite coal, wheat, flax, sunflower seeds, barley, oats. □ In this sparsely populated state, only four cities have more than 25,000 residents, with Fargo (80,500) being the largest. Lack of growth is not surprising, given a climate that ranges from blazing hot and dusty summers to frigid winters with paralyzing blizzards. North Dakota is the nation's leading producer of lignite (a soft coal), spring and durum wheat (the latter is used in pasta), flax seed (pressed into linseed oil), sunflower seeds, and barley. Coal and oil deposits are mined in the rugged western region. The narrow Red River Valley, on the eastern border, is a remarkably fertile ancient lakebed, settled in the 19th century by Norwegians and Germans. The Garrison Dam on the Missouri created Lake Sakakawea, 178 mi (285 km) long.

OHIO H

Area: 41,228 sq.mi.(106,739 km²). **Population:** 11,262,000. **Capital:** Columbus, 660,000. **Economy:** Machinery, aircraft parts, iron and steel, coal, rubber products, corn, soybeans. □ Despite its modest size, Ohio's strategic location, coal reserves, and fertile soil have made it a major industrial and agricultural state. It is among the leaders in the production of rubber products, machine tools, iron and steel, aircraft parts, general manufacturing, corn, soybeans, and hogs. The "Buckeye State" was covered with buckeye trees before the land was cleared for farming. Cleveland (510,000) is no longer larger than Columbus (625,000), the capital. Cincinnati (370,000), on the Ohio River, draws commuters and shoppers from Indiana and Kentucky.

SOUTH DAKOTA I

Area: 77,114 sq.mi.(199,648 km²). **Population:** 758,000. **Capital:** Pierre, 13,000. **Economy:** Food processing, tourism, wheat, cattle, sheep, gold. □ Next to agriculture, tourism is the largest industry. Visitors come to see the Black Hills (the highest mountains east of the Rockies), the Badlands (grotesque formations of rock and clay in a desert environment), and the four presidential heads—Washington, Lincoln, Jefferson, and Theodore Roosevelt—carved in the granite face of Mt. Rushmore. Also of interest is Deadwood, once the most lawless town on the frontier, and a magnet for gold prospectors, gamblers, dancehall girls, and gunslingers. Gold is still mined in the Black Hills, the spiritual home of the Sioux Indians, who were driven out by invading miners in the 1870s. Descendants of the Sioux have received reparations from Congress for the land taken from them. The state is so sparsely settled that Sioux Falls (101,000) is the largest—only three cities to exceed 25,000. The Missouri divides the state into a rugged western part and a flat, fertile eastern half. The construction of four large dams on the river in the 1930s created a source of irrigation and hydroelectric power.

WISCONSIN J

Area: 56,155 sq.mi.(145,386 km²). **Population:** 5,250,000. **Capital:** Madison, 195,000. **Economy:** Dairying, engines, turbines, food, paper, beer. □ Wisconsin is called America's dairy—it produces 40% of the nation's cheese—but the state's economic wealth is in industry: paper products, engines for outboards and lawnmowers, beer brewing, and food processing. The northern region—the base of Michigan's northern peninsula—is a popular recreational area. Despite indiscriminate lumbering of the past, a highly successful forestry program has restored much of Wisconsin's forests, which cover half the state. Milwaukee (650,000), the largest city and industrial center, is the nation's beer-brewing capital and, not surprisingly, home to the nation's largest German-American community. A German creation, the kindergarten, was introduced to America over 100 years ago in a small Wisconsin village. The state has been the most socially progressive, passing legislation regarding jobs, health, and welfare.

Except for Florida, and the few states with oil-driven economies, the West is the fastest-growing part of America. Ever since World War II, there has been a steady population shift from the Northeast and Midwest to the West and the South. The western states, except Washington and Oregon, are deserts that depend on irrigation projects. The West has the nation's most dramatic landscape: the tallest mountains, highest plateaus, deepest canyons, wettest rainforests, and driest deserts. The region is bracketed by the Coast Ranges, Cascades, and Sierra Nevada Mountains on its western edge and the massive Rocky Mountains to the east. Between them lie the high basins and plateaus sometimes referred to as the Intermountain Area, whose flat lands are partially covered by smaller mountain ranges.

ARIZONA A

Area: 113,950 sq.mi.(295,130 km²). **Population:** 4,570,000. **Capital:** Phoenix, 1,000,000. **Economy:** Electronics, manufacturing, copper, metals, cotton, cattle, tourism. □ Reservoirs, irrigation, and air conditioning have transformed a barren desert into a booming industrial and agricultural economy. Most of the population lives and works in the hotter and drier southern half of the state. Tucson (410,000) and Phoenix are agricultural and industrial centers. Arizona's Hopi, Navajo, and Apache tribes make up the nation's largest Indian population (120,000). Near Phoenix is America's first apartment house, Casa Grande, an 800-year-old, four-story adobe (sun-dried brick) structure. The majestic Grand Canyon is the product of 6 million years of erosion by the Colorado River (with help from rain and snow runoff). The rocks at the bottom of the nearly 1 mile (1.6 km) deep canyon are 2 billion years old.

CALIFORNIA B

Area: 158,700 sq.mi.(411,033 km²). **Population:** 32,200,000. **Capital:** Sacramento, 375,000. **Economy:** Aircraft, space equipment, electronics, oil, produce, and cotton. □ The manufacturing and agricultural output of the most populous state rank it as the world's sixth largest economy. The mild, Mediterranean climate permits year-round agriculture in the wide, fertile, and irrigated 500 mi.(800 km) Central Valley. California is the leading industrial state as well as producer of fruits, nuts, vegetables, cotton, and flowers. Los Angeles (3,500,000) is sunny, colorful, flat, sprawling, and plagued by smog. It is the nation's second-largest city, the most multi-racial, and the top manufacturing center. The part called Hollywood is the entertainment capital of the world. San Francisco (775,000) is L.A.'s opposite, confined to a small peninsula with breathtaking hills, cable cars, Victorian houses, chilly summer fog, and the Golden Gate Bridge. The fastest-growing cities are San Diego (1,275,000) and San Jose (800,000). California's varied landscape includes a dramatic coastline, snowcapped mountains, fertile valleys, dense forests, and hot deserts. The tallest peak in the 48 states is Mt. Whitney (14,494 ft., 4,418 m). Sixty miles away is Death Valley, the lowest (-282 ft., -86 m) and hottest place in the Western Hemisphere. California's redwoods are the world's tallest trees, the sequoias are the largest, and the 4,000-year-old bristlecone pines are the oldest.

COLORADO C

Area: 104,200 sq.mi.(269,878 km²). **Population:** 4,000,000. **Capital:** Denver, 475,000. **Economy:** Oil, coal, precision manufacturing, minerals, cattle, tourism. □ With an average elevation of 6,800 ft.(2,073 m), Colorado is the highest state. It has over half of the 50 Rocky Mountain peaks taller than 14,000 ft.(4,268 m). The headwaters of the Colorado and Rio Grande Rivers and tributaries of the Missouri originate in these mountains. Manufacturing, agriculture, and tourism have replaced mineral wealth as the basis for the state's expanding economy. Colorado's mountains have enormous deposits of shale oil. Because population centers are on the dry, eastern slopes, water is brought through mountain tunnels in order to tap the greater runoff on the western side of the Great Divide. Denver is the commercial hub, with over 3/4 of the state's population living in its greater metropolitan area. The Air Force Academy is in Colorado Springs; in a nearby mountain is the headquarters of the North American Air Defense Command.

IDAHO D

Area: 83,560 sq.mi.(216,420 km²). **Population:** 1,275,000. **Capital:** Boise, 150,000. **Economy:** Potatoes, timber, silver, food processing, minerals. □ Scenic beauty abounds in this sparsely populated, mountainous state, whose northern panhandle has been designated a wilderness preserve. Tourists are attracted to the thousands of lakes and streams, caverns, high waterfalls, and steep canyons—Hells Canyon, on the Snake River, is deeper than the Grand Canyon. Most cities and farms are located on the highly irrigated Snake River Plain in the south. Here the nation's largest potato crop is grown and the largest Mormon community, outside of Utah, resides. A huge navigation project along the Snake River has opened a sea route from Idaho to the Pacific Ocean via the Columbia River. Idaho is the leading producer of silver, phosphate rock, and molybdenum (used in hardening steel).

MONTANA E

Area: 147,250 sq.mi.(381,377 km²). **Population:** 920,000. **Capital:** Helena, 25,500. **Economy:** Oil, coal, minerals, wheat, cattle. □ Montana means "mountain" in Spanish, but 3/5 of the state lies in the eastern high plains where wheat, cattle, and sheep are the dominant industries. The state's most dramatic scenery is at the Waterton-Glacier International Peace Park on the Canadian border, where some 60 glaciers are on the move. Most of the earliest white settlers of the state came in search of gold and silver; the smaller cities in this thinly populated state began as Rocky Mountain mining towns.

The largest cities, Billings (85,000) and Great Falls (57,000), are located in the high plains region. Rich in gold, silver, and precious stones, Montana has been called the "Treasure State," but oil has become the major source of revenue. Images of the "cowboy and Indian" past are perpetuated by rodeos and traditional Indian ceremonies.

NEVADA F

Area: 110,550 sq.mi.(286,325 km²). **Population:** 1,735,000. **Capital:** Carson City, 45,000. **Economy:** Gambling, tourism, gold, minerals, manufacturing. □ The fastest-growing state is also the driest, because the Sierra Nevada Mountains screen out Pacific-bred storms. Nevada is totally dependent on irrigation projects. But there is no shortage of alcohol consumed by millions of visitors lured by gambling, entertainment, liberal divorce and marriage laws, and legalized prostitution. Ghost towns are a reminder of the 19th century rush for gold and silver. Mountain ranges, running north and south, line Nevada's high basin, which averages 5,000 ft.(1,524 m). The U.S. owns 87% of the land and operates test centers for nuclear energy and weaponry.

NEW MEXICO G

Area: 121,620 sq.mi.(314,996 km²). **Population:** 1,790,000. **Capital:** Santa Fe, 60,000. **Economy:** Oil, natural gas, coal, uranium, electronics, cattle, sheep, tourism. □ This ruggedly beautiful state has a long human history: a stone age civilization, centuries of pueblo (village) dwellers, Spanish occupation, Mexican rule, Confederate occupation, and pre-statehood, U.S. territorial status. Santa Fe, the capital, uses the nation's oldest government building, built by the Spanish in 1609. The city's adobe architecture is unique among American cities. The first road in what is now the U.S. was built in 1581; it ran from Sante Fe to Mexico City. The U.S. government conducts space and nuclear energy operations around Albuquerque (420,000), the state's largest city and manufacturing center. New Mexico has huge uranium deposits. The first atomic bomb was created in Los Alamos and exploded near Alamogordo.

OREGON H

Area: 97,040 sq.mi.(251,337 km²). **Population:** 3,290,000. **Capital:** Salem, 115,000. **Economy:** Timber products, wheat, food products, electronics. □ Two distinct climates characterize Oregon. West of the towering Cascades, the weather is mild and moist. In this region lie Oregon's vast forests and the fertile Willamette River Valley, with its major cities, industries, and productive farmland. Oregon is the nation's leading timber state. The eastern 2/3 of Oregon consists of a dry plateau subject to wide variations in temperature. Irrigation has made this a productive agricultural area. Portland (500,000), the largest city, is an important port on the Columbia River, which provides much of the Northwest's hydroelectric power. Some of Oregon's scenic attractions are Crater Lake (the country's deepest lake, occupying the crater of an extinct volcano), dramatically steep gorges on the Columbia and Snake Rivers, Pacific beaches with rugged cliffs, and the snow-capped Mt. Hood.

UTAH I

Area: 84,912 sq.mi.(219,922 km²). **Population:** 2,100,000. **Capital:** Salt Lake City, 163,000. **Economy:** Oil, coal, heavy equipment, electronics, minerals. □ Utah is either desert or mountain. The state's wealth is in mineral deposits. Less than 10% of the land is arable. Most of the population resides on the only fertile land—a narrow strip lying between the Great Salt Lake and the Wasatch Range. The western part of the state, in the Great Basin, is the site of ancient Lake Bonneville. This area holds rock-hard salt flats and the Great Salt Lake, six times saltier than the ocean. Without outlets, fresh water flowing into the lake evaporates, leaving salt residues. The capital, Salt Lake City, is the headquarters of the Mormon Church, whose members make up 70% of the state's population. The city was founded in 1847 by the Mormon leader, Brigham Young. Utah's statehood was delayed by Congress for nearly 50 years until the church banned polygamy; the Mormon birthrate is still twice the national average. Near the Arizona border are two natural wonders: Monument Valley, with its giant red sandstone formations 1,000 ft.(305 m) above the desert floor, and Bryce Canyon, noted for the brilliant colors and bizarre shapes of its huge eroded rock structures.

WASHINGTON J

Area: 68,190 sq.mi.(176,612 km²). **Population:** 5,687,000. **Capital:** Olympia, 27,500. **Economy:** Timber, aircraft, shipbuilding, fruit, fishing. □ The Puget Sound region contains hundreds of islands, natural harbors, forests, Mt. Rainier, two mountain ranges, and Mt. St. Helens, a volcanic peak that has had several eruptions since 1980. West of Puget Sound are the wet and primitive Olympic Mountains. Seattle (550,000), the largest city, is just one of many Washington locations for the aerospace giant, the Boeing Company. East of the Cascades is a large plateau formed by ancient lava deposits. This arid region has been made into a productive farming area with water and electricity supplied by dams on the Columbia River. Grand Coulee is the largest cement dam in the U.S. Washington grows more apples and hops (used in brewing beer) than any other state.

WYOMING K

Area: 97,904 sq.mi.(253,571 km²). **Population:** 500,000. **Capital:** Cheyenne, 51,000. **Economy:** Oil, natural gas, coal, uranium, cattle, sheep. □ The ninth-largest state has the fewest people. Sheep and cattle raising are the major industries; herds are still driven by cowboys, many of whom are from Latin America. Despite its "Wild West" reputation, Wyoming is called the "Equality State"; it was the first to grant women the right to vote, to hold public office, and to serve on juries. It even had the first woman governor, who, in 1925, was allowed to finish out her deceased husband's term. Two of America's most beautiful national parks are in Wyoming's Rockies: Yellowstone, the oldest and largest, and Grand Teton. Casper (51,000) is the state's largest city.

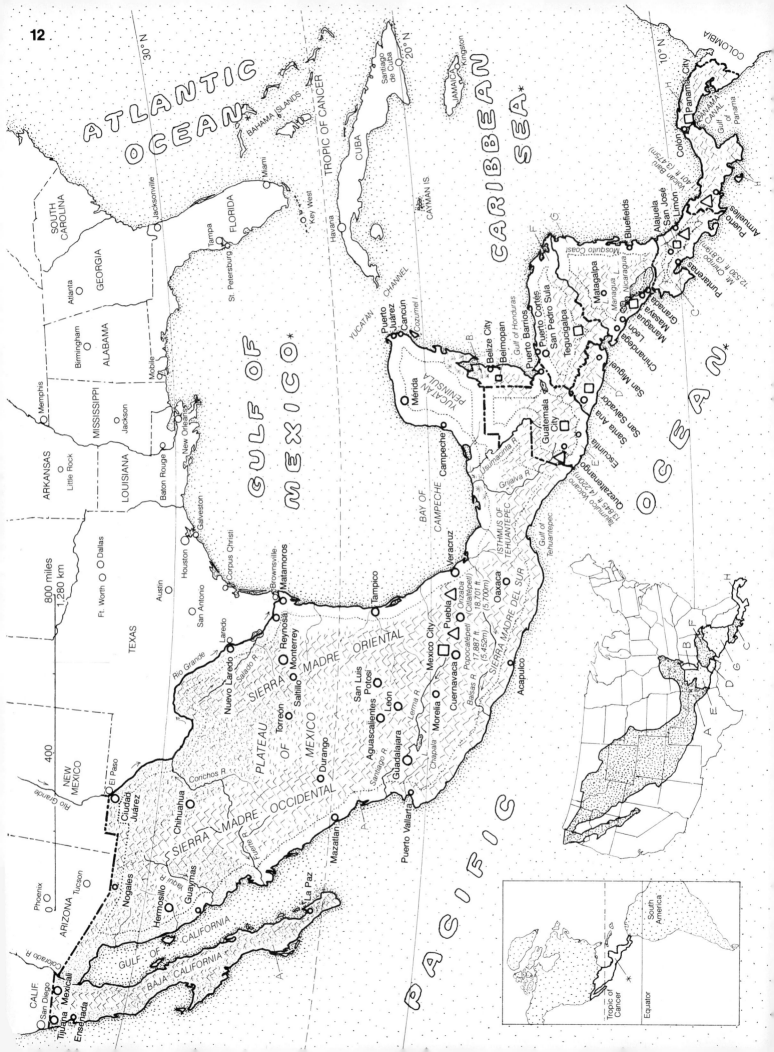

12

ATLANTIC OCEAN*

SOUTH CAROLINA

GEORGIA

Jacksonville

FLORIDA

Tampa

St. Petersburg

Miami

Key West

Havana

CUBA

Santiago de Cuba

TROPIC OF CANCER

30° N

20° N

BAHAMA ISLANDS

JAMAICA

Kingston

CAYMAN IS.

Atlanta

Birmingham

ALABAMA

Memphis

MISSISSIPPI

Jackson

Mobile

New Orleans

Baton Rouge

LOUISIANA

ARKANSAS

Little Rock

GULF OF MEXICO*

YUCATAN CHANNEL

YUCATAN PENINSULA

Mérida

BAY OF CAMPECHE

Campeche

Cozumel I.

Cancún

Puerto Juárez

CARIBBEAN SEA*

Belize City

Belmopan

Gulf of Honduras

Puerto Barrios

Puerto Cortés

San Pedro Sula

Tegucigalpa

Matagalpa

Managua

Managua L.

Nicaragua L.

Bluefields

Mosquito Coast

León

Chinandega

Granada

Masaya

Nicaragua

San Miguel

Santa Ana

San Salvador

Escuintla

Guatemala City

Quezaltenango

Tajumulco Volcano 13,845 ft (4,220m)

Usumacinta R.

Grijalva R.

Gulf of Tehuantepec

ISTHMUS OF TEHUANTEPEC

Panama City

PANAMA CANAL

Gulf of Panama

Colón

Volcán Barú 11,401 ft (3,475m)

Mt. Chirripó 12,530 ft (3,819m)

Puerto Armuelles

Limón

San José

Alajuela

Puntarenas

COLOMBIA

10° N

PACIFIC OCEAN*

ARIZONA

Phoenix

Tucson

Nogales

CALIF.

San Diego

Tijuana

Mexicali

Ensenada

NEW MEXICO

El Paso

Ciudad Juárez

Colorado R.

Rio Grande

GULF OF CALIFORNIA

BAJA CALIFORNIA

La Paz

Mazatlán

Puerto Vallarta

Guadalajara

Chapala

Morelia

Acapulco

Cuernavaca

SIERRA MADRE DEL SUR

Oaxaca

Popocatépetl 17,887 ft (5,452m)

Orizaba (Citlaltépetl) 18,701 ft (5,700m)

Puebla

Mexico City

León

Aguascalientes

Guanajuato

San Luis Potosí

Durango

Lerma R.

Santiago R.

Balsas R.

Veracruz

Tampico

SIERRA MADRE ORIENTAL

PLATEAU OF MEXICO

SIERRA MADRE OCCIDENTAL

Chihuahua

Fuerte R.

Yaqui R.

Hermosillo

Guaymas

Conchos R.

Durango

Torreón

Saltillo

Monterrey

Reynosa

Nuevo Laredo

Matamoros

Brownsville

Salado R.

Rio Grande

Laredo

San Antonio

Austin

TEXAS

Corpus Christi

Houston

Galveston

Ft. Worth

Dallas

800 miles

1,280 km

400

TROPIC OF CANCER

South America

Equator

Tropic of Cancer

NORTH AMERICA: MEXICO & CENTRAL AMERICA

The seven Central American countries, south of Mexico, form a land bridge between North and South America. All but Belize and El Salvador have both Pacific and Atlantic coastlines. From mid-Mexico south, the climate is tropical: summery year-round except at higher elevations (where most of the population lives). Mountain ranges run the length of northern Mexico; the Sierra Madre Oriental in the east and the Sierra Madre Occidental in the west converge to form a high volcanic chain that continues through Central America. Violent earthquakes, hurricanes, and active volcanoes threaten the region. Past volcanic activity has enriched the soil, which with ample rainfall and consistent warmth produces fine mountain-grown coffee; down on the steaming coast grow bananas, the region's second-largest crop.

Except for the English influence in Belize, the language, culture, and religion of these nations reflect 300 years of Spanish domination, beginning in the early 1500s. Franciscan friars converted the Indians to Roman Catholicism and paved the way for their absorption into Spanish life. The indigenous populations were nearly wiped out by killings, enslavement, and, particularly, the European diseases for which they lacked immunity. "Mestizos," people of mixed Indian and Spanish heritage, make up most of the population. The few remaining pure-blooded Indian tribes live in isolated areas.

Most of these nations became independent from Spain in the early 19th century. Nearly 200 years later, the land, wealth, and political power are almost all still in the hands of the same wealthy families (the "landed aristocracy"), while the vast majority are desperately poor. The U.S. has historically regarded this region as its sphere of influence, intervening in the affairs of these countries whenever it has felt a threat to its economic interests or political security.

MEXICO A

Area: 761,602 sq.mi.(1,972,549 km²). **Population:** 98,600,000. **Capital:** Mexico City, 10,000,000. **Government:** Republic. **Religion:** Roman Catholic. **Exports:** Oil, vehicles, steel, chemicals, silver, coffee, cotton, sisal, chicle. **Climate:** Temperate to tropical. □ Most of the world's largest Spanish-speaking population live between the two Sierra Madre ranges, on a high triangular plateau. In the northwest, separated from mainland Mexico by the Gulf of California, is Baja (Lower) California, a long (800 mi.,1,280 km), narrow peninsula of mountains, deserts, and beaches. Mexico's large mestizo population (70%) are very proud of its Indian heritage. The 20th-century Mexican muralists Rivera, Orozco, and Siqueiros have drawn heavily upon the ancient Indian muralist tradition. Long before the conquistadores destroyed the Aztecs, Mexico was home to the advanced Mayans and Toltecs, who built pyramids rivaling those in Egypt. Mexico City's metropolitan area (21,000,000) is the world's largest, fastest growing, and smoggiest. It was built on the site of the Aztec capital. Mexico is growing so rapidly that over half the population is under the age of 20. In the early 20th century, the nation had a revolution that set in place institutions for meaningful social, economic, and political reforms. Unfortunately, these measures have yet to be fully implemented, as one political party has dominated Mexican politics since 1929. The 1990s have seen economic collapse and recovery, armed insurrection, and signs of political reform. Mineral-rich Mexico is the leading producer of silver; it may have the world's largest oil reserves. It is also the leading producer of sisal, a hemp fiber used in rope, and chicle, the basic ingredient of chewing gum—substances that grow in the rain-forests of the Yucatan peninsula. Only 12% of Mexico is cultivated, but a wide variety of crops are grown; much produce is sold to U.S. winter markets.

BELIZE B

Area: 8,860 sq.mi.(22,947 km²). **Population:** 230,000. **Capital:** Belmopan, 4,700. **Government:** Constitutional monarchy. **Language:** English; some Spanish. **Religion:** Roman Catholic 60%; Protestant 40%. **Exports:** Sugar, timber, citrus, bananas, seafood. **Climate:** Tropical. □ In the southeast corner of the Yucatan, Belize (beh leez'), formerly British Honduras, is the only English-speaking country in this region; Spain did not value its dense jungles and swampy coastlines. Belmopan, the tiny capital, was placed inland for protection against hurricanes. Belize City (50,000) is an important port for tropical hardwoods (mahogany and rosewood). Half the people are black or mulatto (mixed black and white); a fifth are direct descendants of the Mayans. Tensions have eased since Guatemala ceased laying claim to Belize.

COSTA RICA C

Area: 19,620 sq.mi.(50,816 km²). **Population:** 3,600,000. **Capital:** San José, 325,000. **Government:** Republic. **Religion:** Roman Catholic. **Exports:** Coffee, bananas, timber, food products. **Climate:** tropical. □ Costa Rica has the region's highest standard of living, the highest percentage of mestizos (97%), the highest literacy rate, the greatest percentage of small landholders, and the longest orderly succession of democratic governments. It is the only Latin American country without a standing army, a factor that may be responsible for its stability, since the chief threat to democracy in Latin America is usually the military. The nation occupies the western third of the Isthmus of Panama and lies on a plateau that is ideal for growing coffee.

EL SALVADOR D

Area: 8,204 sq.mi.(21,249 km²). **Population:** 5,800,000. **Capital:** San Salvador, 465,000. **Government:** Republic. **Language:** Spanish. **Religion:** Roman Catholic. **Exports:** Coffee, cotton, sugar, timber, textiles, foodstuffs. **Climate:** Tropical. □ This mountainous country is the smallest and most densely populated in the region, and the only one without an Atlantic coastline. El Salvador is more industrialized than its neighbors, but the creation of new jobs cannot keep pace with the expanding population. Most of the cities and farms are located in the central highlands region, where coffee is the principal cash crop. Over 90% of the people are mestizos, 3% are Indian, and 5% are the white, ruling land-owners. The great disparity between rich and poor created a revolutionary movement that controlled many parts of the interior. A peace treaty was signed in 1992, formally ending the 12-year civil war.

GUATEMALA E

Area: 42,048 sq.mi.(108,904 km²). **Population:** 12,000,000. **Capital:** Guatemala City, 1,250,000. **Government:** Republic. **Language:** Spanish; many Indian dialects. **Religion:** Roman Catholic; Protestant. **Exports:** Coffee, bananas, timber, cotton, chicle. **Climate:** Tropical. □ Guatemala has the largest percentage (55%) of pure-blooded Indians in the region. Most are direct descendants of the great Mayan culture, which lasted nearly 2,000 years and ended mysteriously around 900 AD. Deep in the northern lowland jungles are the ruins of Tikal, a Mayan city of stone buildings and pyramids. Today, Indians live in their ancestral villages while the country is run by the westernized, mestizo Guatemalans ("ladinos"), who are concentrated in the southern highlands. Sitting on a high plateau is the capital, Guatemala City, the largest city in Central America. It has been wrecked three times in this century by devastating earthquakes; the highlands are also subject to eruptions from some of the 27 volcanoes. For the past 30 years, a series of military governments has waged a low-level war against guerrilla forces protesting the inequitable ownership of land. Despite the election of a civilian President, the violence continues. A third of the Guate-

malans have been converted to Protestantism by American evangelists. The Catholic Church in Latin America has been accused of being too sympathetic to the poor, and in countries with left-wing movements, the church has been a target of the military. Evangelicals in Latin America have been most successful in Guatemala, where a former military leader became a "born-again" Christian.

HONDURAS F

Area: 43,270 sq.mi.(109,479 km²). **Population:** 5,900,000. **Capital:** Tegucigalpa, 1,000,000. **Government:** Republic. **Language:** Spanish. **Religion:** Roman Catholic. **Exports:** Bananas, coffee, timber, minerals, cattle. **Climate:** Tropical □ If any Latin American country deserves the title the "Banana Republic," it is Honduras, the poorest country in the region. Very large, mostly American-owned plantations are located on the fertile and humid Caribbean coast. Here, the nation's only railroads are used for hauling bananas to coastal ports. Honduras, with large unplanted areas of cultivable land and large reserves of untapped mineral deposits, has significant economic potential. Tegucigalpa, the capital, began as a rich silver-mining town in the mountains.

NICARAGUA G

Area: 57,440 sq. mi.(148,770 km²). **Population:** 4,600,000. **Capital:** Managua, 700,000. **Government:** Republic. **Language:** Spanish. **Religion:** Roman Catholic. **Exports:** Coffee, cotton, coffee, sugar, bananas, meat. **Climate:** Tropical. □ The largest country in Central America is triangular and has three distinct regions: the Mosquito Coast of swamps and rainforests; the mountains of the central highlands; and the fertile, hilly Pacific region, which holds Central America's largest lake, Lake Nicaragua. The lake is home to the world's only freshwater sharks, which evolved from sharks that were trapped when a volcanic eruption sealed off their bay from the ocean. The capital, Managua, the major cities, and the most productive farms are located in the Pacific region. Managua was destroyed by earthquakes twice in this century. The population of Nicaragua is 85% mestizo, 10% mulatto, and 5% Indian. The Miskito Indians are mixed-blooded descendants of black slaves brought to Nicaragua during the early British rule of the Caribbean coast. In 1979, after 30 years of repression, Anastasio Somoza was overthrown by the left-wing Sandinista party. Fearing a communist foothold on the continent, the U.S. organized a band of "Contras" to wage war against the Sandinistas. Though the rebels could not gain popular support, 10 years of war forced the Sandinistas to yield to U.S. demands for free elections. In 1990, a coalition party defeated the Sandinistas.

PANAMA H

Area: 29,205 sq.mi.(75,641 km²). **Population:** 2,800,000. **Capital:** Panama City, 475,000. **Government:** Republic. **Language:** Spanish. **Religion:** Roman Catholic. **Exports:** Bananas, coffee, mahogany, shrimp. **Climate:** Tropical. □ On average, 33 ships a day pass through the Panama Canal, earning Panama the title "Crossroads of the World." Through this natural gap in the mountains, the Spanish used mules to pack Inca gold brought up from the west coast of South America. The U.S. helped create the nation of Panama as well as the canal. In 1903, Panama, with American support, asserted its independence from Colombia, which was opposed to the canal. The U.S.-built canal, 50 mi.(80 km) long, was opened in 1914. In 1979, the U.S. agreed to give Panama control of the Canal Zone, a strip 10 mi.(16 km) wide that crosses the isthmus. The U.S. is to relinquish control of the canal in 1999. Most Panamanians live and work in the Canal Zone. Panama City is on the Pacific end, Colón (78,200) is on the Caribbean. In the late 1980s, the U.S. pressured Panama to overthrow General Manuel Noriega, who had seized power and threatened U.S. interests. In 1989, the U.S. invaded Panama and brought Noriega back to Miami, where he was convicted of drug-running and imprisoned in the U.S.

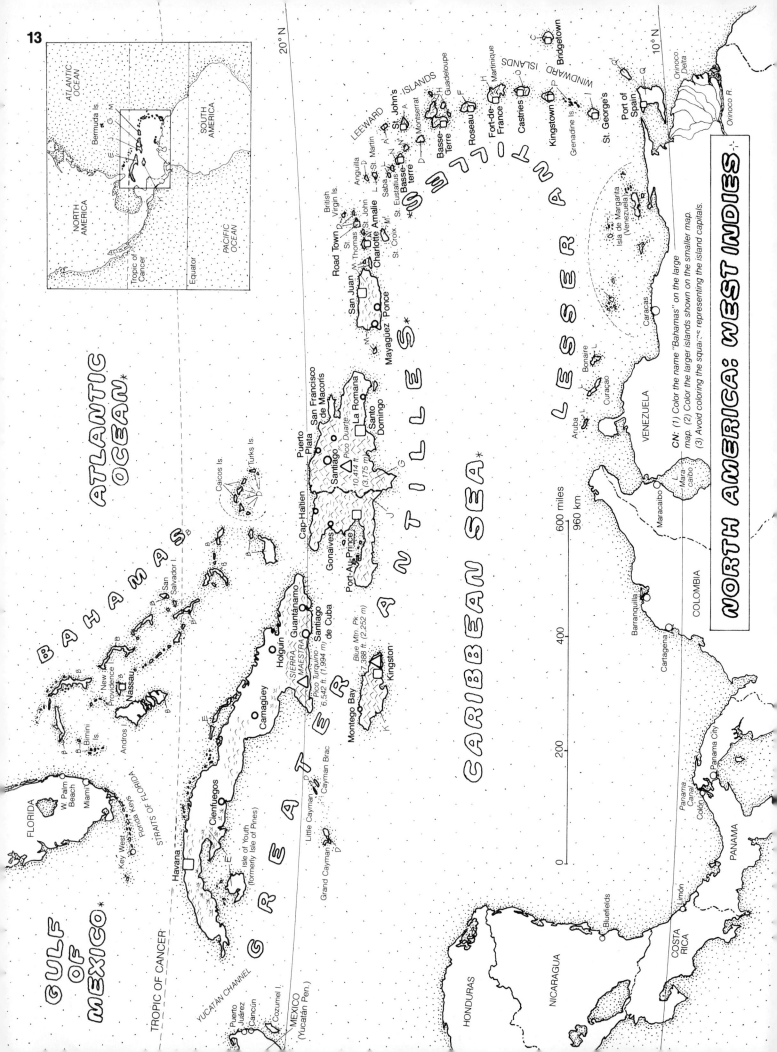

Believing he had reached the East Indies, Christopher Columbus named these islands the "Indies." Columbus made four voyages to the West Indies from 1492 to 1500, naming many of the islands and claiming them for Spain. Spain used these islands as a base for the exploration and plunder of the New World. A century later, other nations arrived to pirate Spain's treasure-laden galleons and contest the ownership of its colonies. Natives of the region were virtually wiped out by the Europeans. Those that did not succumb to Old World diseases died from overwork, beatings, and executions. Africans were forced to replace them on farms and plantations. Except in Cuba and Puerto Rico, the West Indies are populated mainly by pure or mixed-blooded (mulatto) descendants of black slaves. Catholicism dominates the former Spanish and French colonies; Protestantism is the religion on islands owned by the British, Dutch, Danes, and the U.S. Residents of Haiti practice voodoo. Many of these islands gained their independence after World War II.

The West Indies archipelago contains over 7,000 mostly uninhabited islands, cays (small islands), coral reefs, and rocks, which form the northern and eastern boundaries of the Caribbean Sea. The islands fall into three groups: (1) The Bahamas are coral islands off the coast of Florida. (2) The Greater Antilles, which include the larger islands of Cuba, Hispaniola (shared by Haiti and the Dominican Republic), Jamaica, and Puerto Rico, are worn-down peaks of sunken mountains. (3) The Lesser Antilles are an arc of smaller volcanically formed islands. Trade winds, blowing from the east, keep the temperature of this tropical region comfortable. Agriculture flourishes in volcanic or alluvial soil. Sugar is king, followed by bananas. On most of these islands, tourism has become the major industry.

ANTIGUA A & BARBUDA A'

Area: 172 sq.mi.(445 km²). **Population:** 64,000. **Capital:** St. John's, 27,000. **Government:** Constitutional monarchy. **Language:** English. **Religion:** Protestant. **Exports:** Sugar, cotton, rum, light industry. □ Antigua, the larger of the two islands, includes the capital of St. John's and 98% of the population.

BAHAMAS B

Area: 5,382 sq.mi.(13,939 km²). Population: 280,000. **Capital:** Nassau, 140,000. **Government:** Constitutional monarchy. **Language:** English. **Religion:** Protestant. **Exports:** Tropical fruit, fish products, petrochemicals. □ Fifty miles (80 km) from Florida are the Bahamas, over 2,000 islands and coral reefs, only 29 of which are inhabited. Columbus may have landed in the New World on San Salvador Island. After the American Revolution, southern planters loyal to England fled to the Bahamas with their slaves. The country is still functioning as a haven for foreigners, but now it attracts bank deposits.

BARBADOS C

Area: 166 sq.mi.(430 km²). Population: 260,000. **Capital:** Bridgetown, 7,500.**Government:** Constitutional monarchy. **Language:** English. **Religion:** Mostly Protestant. **Exports:** Sugar, molasses, rum, fish. □ Barbados, one of the world's most densely populated countries, is the most easterly of the islands. Barbados has been called "England in the Tropics." Residents speak with a British accent, play cricket, and drive on the left side of the road.

BRITISH TERRITORIES D

Anguilla: 35 sq.mi.(91 km²). Population, 11,150; Capital, The Valley. *Bermuda:* 21 sq.mi.(53 km²). A group of islands north of the West Indies, about 850 mi.(1,360 km) east of Charleston, South Carolina. Population, 63,000; Capital, Hamilton. *British Virgin Islands:* 59 sq.mi.(153 km²). Population, 12,500; Capital, Road Town. *Cayman Islands:* 100 sq.mi.(259 km²). Population, 37,700; Capital, Georgetown. *Montserrat:* 38 sq.mi.(98 km²). Population, 13,500; Capital, Plymouth. *Turks & Caicos Islands:* 166 sq.mi.(430 km²) Population, 7,200; Capital, Cockburn Town.

CUBA E

Area: 44,219 sq.mi.(114,527 km²). Population: 11,100,000. **Capital:** Havana, 2,000,000. **Government:** One-party socialist republic. **Language:** Spanish. **Religion:** Roman Catholic. **Exports:** Sugar, tobacco, nickel, citrus, fish. □ Only 90 miles (144 km) from Florida is the largest island in the West Indies. Cuba grows the world's largest export crop of sugar, and a famous tobacco. After seeing natives smoking cigars, Columbus brought tobacco to Europe. The majority of Cubans are of Spanish descent. Fidel Castro's communist regime (the only one in the Western Hemisphere) has eradicated the poverty, hunger, disease, and illiteracy that characterize so much of Latin America. Although democracy and political freedom remain curtailed, greater economic freedom has been allowed to deal with the crisis caused by the end of Soviet aid and by the crippling 30-year-old American embargo on trade with Cuba. The Pope's visit in 1998 brought hope for fewer restrictions on personal freedoms.

DOMINICA F

Area: 295 sq.mi.(764 km²). **Population:** 66,000. **Capital:** Roseau, 16,000. **Government:** Republic. **Language:** English. **Religion:** Mostly Roman Catholic. **Exports:** Bananas, coconuts. □ Dominica was named for Sunday (Domingo), the day that Columbus set foot on its soil. The extremely rugged terrain includes dense rain forests filled with exotic wildlife. Several hundred Caribs, the only remaining Indians native to the Caribbean, live here on a small reservation. The Caribs were a fierce tribe who migrated to the West Indies from South America; the word "cannibal" is derived from their name.

DOMINICAN REPUBLIC G

Area: 18,750 sq.mi.(48,563 km²). **Population:** 8,000,000. **Capital:** Santo Domingo, 2,200,000. **Government:** Republic. **Language:** Spanish. **Religion:** Roman Catholic. **Exports:** Sugar, coffee, cocoa, food products, gold, nickel. □ Christopher Columbus is believed to be buried in the capital, Santo Domingo, the oldest European city in the Western Hemisphere. Its university was established in 1535. The city was the headquarters tor the early Spanish exploration of Latin America. The nation occupies two-thirds of the rugged island of Hispaniola, which it shares with Haiti (it was part of Haiti until 1844).

FRENCH TERRITORIES H

The French Antilles include: *Guadeloupe:* A group ot 8 islands, 685 sq.mi.(1,774 km²). Population, 416,000; Capital, Basse-Terre. *Martinique:* 426 sq.mi.(1,103 km²). Population, 410,000; Capital, Fort-de France.

GRENADA I

Area: 133 sq.mi.(344 km²). **Population:** 96,000. **Capital:** St. George's, 4,500. **Government:** Constitutional monarchy. **Language:** English; French patois. **Religion:** Roman Catholic; Protestant. **Exports:** Bananas, nutmeg, mace, sugar. □ Grenada, the "Isle of Spice," made headlines in 1983 when the U.S. invaded the island and overthrew the Marxist government. The U.S. was concerned that Grenada could spread communism throughout Latin America.

HAITI J

Area: 10,715 sq.mi.(27,752 km²). **Population:** 6,800,000. **Capital:** Port-Au-Prince, 650,000. **Government:** Republic. **Language:** Haitian Creole; French. **Religion:** Voodoo; Roman Catholic. **Exports:** Coffee, sugar, cocoa. □ Haiti, formerly the early Spanish colony of Hispaniola, was seized by French pirates and later became a French possession. In 1804, plantation slaves revolted and Haiti became the Caribbean's first independent state and the world's first black republic. Corrupt and oppressive dictators (two ot the more recent were father and son: "Papa Doc" and "Baby Doc" Duvalier) have made Haiti the poorest and most illiterate nation in the Western Hemisphere.

JAMAICA K

Area: 4,244 sq.mi.(10,992 km²). **Population:** 2,650,000. **Capital:** Kingston, 770,000. **Government:** Constitutional monarchy. **Language:** English. **Religion:** Mostly Protestant. **Exports:** Bauxite, sugar, bananas, coffee, rum, tobacco. □ Once the most important sugar and slave center in the region, Jamaica is now one of the world's leading producers of bauxite (aluminum ore). Heavy rainfall is responsible for the dense forests, rushing rivers, and cascading waterfalls. Jamaica's scenic beauty and carefree lifestyle—which tourists find so attractive—mask the island's underlying poverty.

NETHERLANDS ANTILLES L

The Dutch possessions consist ot two groups of islands: (1) Close to Venezuela are *Aruba, Bonaire,* and *Curacao.* Aruba actually became an independent nation in 1996; 75 sq.mi.(193 km²). Population, 68,000; Capital, Oranjestad. *Bonaire:* 111 sq.mi. (288 km²). Population, 9,900; Capital, Kralendijk. *Curacao:* 171 sq.mi.(444 km²). Population, 171,000. Capital, Willemstad. Aruba and Curacao have huge refining facilities for Venezuelan oil. (2) East of Puerto Rico are *Saba, St. Eustatius,* and the southern part of *St. Martin*; all are very tiny islands.

PUERTO RICO M

Area: 3,435 sq.mi.(8,897 km²). **Population:** 3,900.000. **Capital:** San Juan, 470,000. **Government:** Self-governing commonwealth of the U.S. **Language:** Spanish; English. **Religion:** Roman Catholic. **Exports:** Pharmaceuticals, chemicals, light industry, sugar, bananas, coffee. □ The U.S. acquired Puerto Rico from Spain following the Spanish-American War of 1898. U.S. aid and investments have made Puerto Rico the region's most industrialized island. Most Puerto Ricans are of Spanish descent. They are U.S. citizens but cannot vote in presidential elections. Statehood is a controversial issue.

ST. KITTS N - NEVIS N'

Area: 101 sq.mi.(262 km²). **Population:** 42,300. **Capital:** Basseterre, 15,000. **Government:** Constitutional monarchy. **Language:** English. **Religion:** Protestant. **Exports:** Sugar, cotton, vegetables. □ St. Kitts-Nevis is a two-island nation lying east of Puerto Rico. The majority of the population lives on St. Kitts (also called St. Christopher), once the base for British operations in the Caribbean.

ST. LUCIA O

Area: 237 sq.mi.(614 km²). **Population:** 160,000. **Capital:** Castries, 46,000. **Government:** Constitutional monarchy. **Language:** English. **Religion:** Roman Catholic. **Exports:** Bananas, coconuts, light industry. □ The second language spoken here, French patois, is a reminder of the island's history; ownership alternated between the British and French prior to independence in 1978.

ST. VINCENT & GRENADINES P

Area: 150 sq.mi.(389 km²). **Population:** 120,000. **Capital:** Kingston, 25,000. **Government:** Constitutional monarchy. **Language:** English, French. **Religion:** Protestant; Roman Catholic. **Exports:** Bananas, coconuts, arrowroot. □ St. Vincent and 100 islands of the Grenadine chain make up this nation.

TRINIDAD Q & TOBAGO Q'

Area: 1,990 sq.mi.(5,154 km²). **Population:** 1,200,000. **Capital:** Port of Spain, 64,000. **Government:** Republic. **Language:** English. **Religion:** Roman Catholic; Protestant. **Exports:** Oil products, asphalt, sugar, rum. □ Trinidad has produced oil for over 100 years. It also refines oil from Venezuela, only 7 mi.(11 km) away. Pitch Lake is the world's largest deposit of natural asphalt. Calypso music, which began in Trinidad, is often played and sung to the beat of empty oil drums. Forty percent of the population are "real" Indians, descended from 19th century immigrants from India. Trinidad is heavily industrialized, whereas Tobago remains rural, with only 5% of the population.

VIRGIN ISLANDS (U.S.) M'

Area: 132 sq.mi.(342 km²). **Population:** 110,000. **Capital:** Charlotte Amalie, 12,500. **Government:** Self-governing territory. **Language:** English. **Religion:** Protestant. **Exports:** Refined petroleum products, light industry, rum. □ The Virgin Islands were so named by Columbus because of their pristine beauty. In 1917, the U.S. purchased the hilly Virgin Islands from Denmark. St. Croix (kroy) is the largest and most industrialized island. St. John is mostly a National Park. St. Thomas is a shopper's paradise and favored most by tourists.

14

ATLANTIC OCEAN*

20° N

MEXICO
GUAT.
BEL
HON.
ES
NIC.
CR
PAN

CUBA
JAMAICA HAI
DOM. REP.
PUERTO
RICO
GREATER ANTILLES
LESSER ANTILLES

CARIBBEAN SEA*

10° N

Panama
Canal
Cartagena
Maracaibo
TOBAGO
TRINIDAD

L

L

E

G
J
N

Galapagos Is.
(Ecuador)
F

F

0°
EQUATOR

Guayaquil

Fortaleza

Natal

I

C

10° S

Salvador

B

PACIFIC OCEAN*

1,500 miles
2,400 km

H

Rio de Janeiro
São Paulo

20° S

TROPIC OF CAPRICORN

1,000

D

ATLANTIC OCEAN*

30° S

Juan Fernández Is.
(Chile)
D
Valparaíso

500

K

A

Mar del Plata

40° S

South Georgia I.
(U.K.)

0

Strait of
Magellan

M

50° S

D
Cape Horn

A

SOUTH AMERICA: THE COUNTRIES

INDEPENDENT NATIONS

ARGENTINA A / BUENOS AIRES
BOLIVIA B / LA PAZ, SUCRE
BRAZIL C / BRASÍLIA
CHILE D / SANTIAGO
COLOMBIA E / BOGOTÁ
ECUADOR F / QUITO
GUYANA G / GEORGETOWN
PARAGUAY H / ASUNCIÓN
PERU I / LIMA
SURINAME J / PARAMARIBO
URUGUAY K / MONTEVIDEO
VENEZUELA L / CARACAS

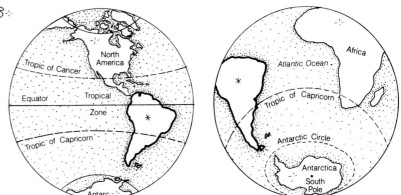

South America lies almost entirely east of North America, and most of it is within the tropics. Seasonal changes south of the Equator occur at opposite times from those in the Northern Hemisphere (e.g., January is the warmest month). South America is much closer to Antarctica than any other continent.

FOREIGN POSSESSIONS

FALKLAND ISLANDS M / STANLEY
FRENCH GUIANA N / CAYENNE

South America, the fourth-largest continent (6,900,000 sq.mi., 17,871,000 km²), has a population of 330 million. The continent is a study in extremes: affluent cities and wretched slums; Stone Age cultures and ultra-modern urban areas; densely populated coastal areas and virtually empty interiors; rich deposits of natural resources but no practical way to access them; and the greatest disparity of all—the gulf between the extraordinary wealth of the elite and the desperate poverty of the masses.

Most of these nations are struggling with stagnant economies and overburdened by massive foreign debt, a serious problem compounded by a population explosion that is in large measure due to the influence of the Roman Catholic Church. The Church arrived shortly after the conquistadors and converted the native Indian population to the religion and culture of the European invaders. Natives were forced to work in the mines, farms, and settlements; vast numbers died in the process. Just as they destroyed the Aztec civilization in Mexico, a handful of heavily armed Spaniards conquered the flourishing 500-year-old Inca civilization. Descendants of the Incas are still the majority in Peru, Ecuador, and Bolivia. The continent's earliest inhabitants are believed to have migrated from North America some 20,000 years ago. It may have taken another 10,000 years for them to reach the southern tip of the continent. Well over half of all South Americans are mestizos (of mixed Indian and European ancestry, who speak Spanish and are western oriented); mulattos (of mixed black and European backgrounds); and pure-blooded Indians and blacks. The remainder are of European descent. There were few Indians on the east coast, and they were nearly wiped out by the colonialists. Africans were imported as replacements for the decimated native population. Settlers in Argentina and Uruguay encouraged European immigration instead of resorting to slavery, and today the population in those countries is largely white. When slavery was abolished in the Guianas, and blacks left the plantations, the British and Dutch looked to India and southeast Asia for a new source of labor. Guyana and Suriname are now dominated by the descendants of those Asian indentured workers.

In the early 19th century, revolutionary fever swept the continent. European powers were too weakened by domestic wars to prevent their colonies from breaking away. New nations were formed under democratic constitutions, but the presence of actual democracies has been exceedingly rare. For nearly two centuries, extremely wealthy and influential families have been able to maintain enormous landholdings by encouraging dictatorial or military rule. Although the trend is now toward democratically elected governments, in the absence of meaningful land reform, these emerging democracies will have to contend with political unrest bred by the widespread poverty.

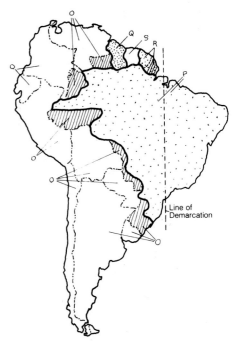

18TH CENTURY EUROPEAN COLONIES *

SPAIN o
PORTUGAL p
GREAT BRITAIN q
FRANCE r
NETHERLANDS s

In 1494, two years after Columbus discovered the New World, the Pope sought to avoid future conflict by drawing the "Line of Demarcation" down what was then believed to be the center of South America. He gave Portugal the right to all lands east of the line, and Spain received everything to the west. Later, Portugal was permitted to expand the Brazilian colony westward. Neither nation wanted the swampy coast and rugged forests of the northeast. That region was later claimed by the British (Guyana), Dutch (Suriname), and French (French Guiana), none of whom respected the Pope's territorial decrees in South America—or anywhere else. The shaded areas on the map below represent lands that Brazil would acquire from its neighbors.

SOUTH AMERICA: THE PHYSICAL LAND

MAJOR RIVER SYSTEMS
AMAZON A
TRIBUTARIES B
ORINOCO C
TRIBUTARIES D
RÍO DE LA PLATA
PARANÁ E
PARAGUAY F
URUGUAY G
TRIBUTARIES H

LAND REGIONS
ANDES MOUNTAINS I
GUIANA HIGHLANDS J
BRAZILIAN HIGHLANDS K
CENTRAL PLAINS
LLANOS L
SELVAS M
GRAN CHACO N
PAMPAS O
PATAGONIA P

Separating the highlands in the east from the Andes in the west are the *Central Plains*. The *Llanos* are grassy cattle-grazing regions of Colombia and Venezuela. The *Selves*, the rain forests of the Amazon River Basin, cover parts of Brazil, Peru, and Bolivia. The *Gran Chaco*, a generally dry and scrubby region of Bolivia, much of Paraguay, and part of Argentina, is known for the amazingly hard quebracho trees. The large, fertile, grassy, *Pampas* region is Argentina's breadbasket. *Patagonia* is a cold, wind-swept land of deserts in southern Argentina.

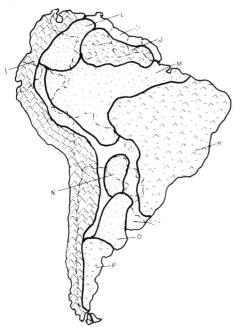

South America shares a common geology with Africa, from which it separated nearly 200 million years ago (p. 2). Like North America, ice cream cone-shaped South America is broad in the north and narrow in the south. Both continents have three regions: ancient highlands in the east; younger and taller mountains in the west; and a wide plain in the center. In South America, the ancient highlands are the rugged Guiana Highlands in the northeast and the densely populated Brazilian Highlands in the eastern bulge. The Andes are the world's longest and second-tallest mountain range (after the Himalayas). They extend 4,500 mi.(7,200 km) down the western edge of the continent, from the Caribbean to Cape Horn. The upper elevations have an eternal snow cover, even at the Equator. Mt. Aconcagua (22,831 ft., 6,960 m) is the tallest of nearly 40 Andean peaks that are higher than North America's Mt. McKinley (20,320 ft., 6,195 m). The Andes are still growing as the eastward-moving Nazca tectonic plate grinds under the South American plate (p. 2). This is a region of geologic volatility, part of the Pacific Rim's "Ring of Fire" (p. 2).

Dominating the Central Plains is the mighty Amazon River Basin of rivers, streams, and rainforests. The river is the world's largest, and its length of 4,000 mi.(6,400 km) is second only to the Nile. From headwaters in the Peruvian Andes, the Amazon wends its way across the continent, transporting 20% of the world's total river water. The flow is so great that fresh water can be detected in the Atlantic 50 mi.(80 km) from the river's mouth. Parts of the river basin remain unexplored. Primitive tribes live there in total isolation. The Amazon was named after the mythical Greek female warriors by a European explorer who claimed he saw women fighting among the native tribes. The rainforest (selva), which is the world's largest, is being destroyed at an alarming rate. This forest has been called the "earth's lungs," but trees actually perform the reverse function: they take in carbon dioxide and release oxygen. Scientists are concerned that the elimination of vast regions of oxygen-producing vegetation, along with the smoke from the burning brush, will add to the "greenhouse effect." The Amazon Basin has over 100,000 species of plants and animals; many are unique to the region and are in danger of extinction because of habitat destruction. South America, in general, is rich in distinctively different plant and animal life because of its relative isolation. The continent's native animals tend to be small; the tapir, which is the largest, is no bigger than a pony.

The continent's two other important river systems are the Orinoco and the Río de la Plata. The Orinoco begins in the Guiana Highlands and makes a wide arc as it drains the grasslands of Colombia and Venezuela on its way to the Atlantic. The Río de la Plata system includes many rivers that provide essential transportation routes for Argentina, Bolivia, Brazil, Paraguay, and Uruguay. The Paraná, Paraguay, and Uruguay are the principal rivers of the system. They empty into the Atlantic through an estuary (an ocean inlet) called the Río de la Plata. Buenos Aires and Montevideo are located on opposite banks of this estuary. Brazil and Paraguay have completed the world's largest hydroelectric plant at the Itaipú Dam on the Paraná River. In the same region, separating Brazil and Argentina, are the Iguaçu Falls, the world's largest. Depending on the seasonal flow of the Iguaçu River, there are as many as 275 separate falls in this spectacular complex.

The continent's two major lakes are Maracaibo and Titicaca. The latter, the largest, is in the Andes between Peru and Bolivia. At a height of over 12,000 ft.(3,659 m), it is the world's highest navigable lake. It is also known for the sailboats that are woven from reeds growing along its shore. Even more remarkable are the floating islands woven from the same reeds. Houses sit on these islands.

On a narrow strip of land along the coast of Peru and northern Chile are some of the world's driest deserts. Parts of the 600 mi.(960 km) Atacama Desert in Chile have never had recorded rainfall. This aridity, unusual on an ocean shore, is due to the interaction of the cold waters of the Humboldt (Peru) Current (p. 57) and the Andes Mountains. Tropical ocean storms are cooled as they pass over the current, and they release their moisture before reaching the coast. The Andes block storms from the east. Further up the coastline (beyond the range of the Humboldt Current) is one of the world's wettest regions, the Pacific coast of Colombia. South America does not have the temperature extremes of North America, although it does have a steamy equatorial region, Andean glaciers, hot deserts in Argentina, and the cold southern tip of Tierra del Fuego. Temperatures on the southeast coast are moderated by the warm Brazil Current (p. 57).

SOUTH AMERICA: NORTHERN COUNTRIES & BRAZIL

Although located in the heart of the tropics, the five northern countries of South America have temperatures ranging widely from the steamy coastal lowlands to the cold of snow-covered Andean peaks. In Colombia and Venezuela, most people live in the high valleys of three parallel extensions of the Andes. Residents of Guyana, Suriname, and French Guiana have to live in the coastal heat because the interior is covered by the nearly impenetrable, and partly unexplored, Guiana Highlands. The presence of these highlands may have been the reason why both Spain and Portugal avoided this part of South America. In the Venezuelan Highlands are the world's tallest waterfalls, Angel Falls (3,212 ft.,979 m), nearly 20 times higher than Niagara. The Falls were unknown until 1935, when they were spotted from the air by an American pilot. The populations of Colombia and Venezuela have mixed Indian, Spanish, and African roots. In both countries, 20% of the people are the white descendants of Spanish colonialists. As in other Latin American nations, the white minority owns the land and the industries and controls the government. In the three smaller countries, fewer than 3% of the overall population is white; Guyana and Suriname actually have Asian majorities—descendants of indentured workers from India and Indonesia who were brought to this region to replace emancipated slaves.

COLOMBIA A

Area: 439,750 sq.mi.(1,138,952 km²). **Population:** 38,600,000. **Capital:** Bogota, 5,750,000. **Government:** Republic. **Language:** Spanish. **Religion:** Roman Catholic. **Exports:** Coffee, textiles, bananas, emeralds, gold. **Climate:** Hot to freezing, depending on altitude. ☐ Colombia holds two geographical distinctions in South America: it is the only nation to have both Atlantic and Pacific coastlines, and it provides the only land entrance to the continent (from Panama). Colombia exports 90% of the world's supply of emeralds, but the nation's real treasure is its fine coffee—only Brazil grows more coffee. In recent years, cocaine and marijuana have become the largest cash crop. Drug lords have resorted to widespread urban terror in resisting government attempts to curtail their trade. The focus of drug activities is Medellín (1,750,000), one of three major cities located in the fertile Andean valleys. Bogota is the capital and cultural center; Medellín is the finance and industrial center, and the continent's leading textile producer. Colombia has the third-largest population in South America. The eastern 2/3 of Colombia is a sparsely populated, grassy plain (Llano) that merges with the Orinoco and Amazon Basins. Because land travel in the country is obstructed by natural barriers, Colombia was among the first nations to have a commercial airline.

FRENCH GUIANA B

Area: 35,200 sq.mi.(91,168 km²). **Population:** 163,000. **Capital:** Cayenne, 42,000. **Government:** French Overseas Dept. **Language:** French. **Religion:** Roman Catholic. **Exports:** Bananas, shrimp, sugar, hardwoods. **Climate:** Tropical. ☐ French Guiana (ghee ah' na) is the only foreign possession left on the continent. For nearly 100 years it was used as an overseas penal colony. Devil's Island, the most notorious of Guiana's three prisons, was a living hell for political prisoners sent from France. Another former prison colony, Kourou, is now the launching site for the European Space Agency's satellite program. Almost all of the residents are black or Creole (mixed black and French ancestry), and most live on the coast. The considerable mineral wealth in the extremely rugged Highlands is not readily accessible. This very poor possession of France relies on the mother country for continued support.

GUYANA C

Area: 83,100 sq.mi.(215,229 km²). **Population:** 710,000. **Capital:** Georgetown, 280,000. **Government:** Republic. **Language:** English; Creolese; Hindi. **Religion:** Hindu 35%; Christianity 55%; Islam 10%. **Exports:** Bauxite, sugar, rice. **Climate:** Tropical. ☐ Some aspects of Guyana (guy ah' na), formerly British Guiana, resemble an Asian country rather than a Latin American one. The Asian majority, who control commerce, have often been in conflict with the black minority (40%), who are descendants of slaves that were replaced by Asians. The blacks have held political power and have run a socialist government. A language called Creolese, which borrows from all ethnic groups, is spoken along with English. Many coastal areas, including Georgetown, the capital, are below sea level and are protected by a network of dikes and canals, many of which were built by the Dutch over 300 years ago. Georgetown is known for the white-painted wooden buildings common in many Caribbean island cities. Guyana is a world leader in the production of bauxite. In the 1970s, the nation attracted foreign religious cults, the most famous of which ended in the Jonestown Massacre, when an American cult committed mass suicide at the urging of its leader.

SURINAME D

Area: 63,240 sq.mi.(163,792 km²). **Population:** 430,000. **Capital:** Paramaribo, 210,000. **Government:** Republic. **Language:** Dutch. **Religion:** Christianity 35%; Hindu 30%; Islam 20%. **Exports:** Bauxite, aluminum, bananas, timber. **Climate:** Tropical and damp. ☐ In the 17th century, the Dutch made what was surely the worst land swap in history: they gave New Netherlands and New Amsterdam (the state and city of New York) to the British in exchange for what is now Suriname. Formerly Dutch Guiana, Suriname (or Surinam) is truly the melting pot of South America. Among the many ethnic groups living in relative harmony are Hindus, Indonesians, mulattoes, blacks, native Indians, Chinese, and Europeans, all speaking a common language. Black descendants of slaves, called "bush negroes," live a native African lifestyle in the densely wooded interior. Like Guyana, Suriname is a leading producer of bauxite.

VENEZUELA E

Area: 352,144 sq.mi.(912,052 km²). **Population:** 22,800,000. **Capital:** Caracas, 1,300,000. **Government:** Republic. **Language:** Spanish. **Religion:** Roman Catholic. **Exports:** Oil, iron and steel, coffee, sugar, cotton, food products. **Climate:** Varies according to altitude. ☐ Venezuela means "little Venice" in Spanish. Early explorers were reminded of the Italian city when they first saw Indian villages perched on stilts in the shallow waters along the shores of Lake Maracaibo. The black ooze that sullies the lake eventually made Venezuela, in the 1930s, the world's first oil exporting nation and the wealthiest country in South America. But oil profits have enriched only the ruling elite. The government is saddled with a huge foreign debt, accumulated from loans made against future oil revenues that never materialized, owing to a drop in prices. Venezuela has the continent's largest iron deposits and an expanding steel industry. Caracas, the capital, is modern and prosperous, but like other cities in Latin America it is surrounded by shantytowns. The government has enacted land reform and social programs to improve the quality of rural life and slow the migration to the overcrowded cities. Much of Venezuela's land is drained by the Orinoco River. The river basin is an alluvial plain subject to both floods and drought.

BRAZIL F

Area: 3,286,480 sq.mi.(8,511,983 km²). **Population:** 170,000,000. **Capital:** Brasilia, 2,500,000. **Government:** Republic. **Language:** Portuguese. **Religion:** Roman Catholic. **Exports:** Coffee, bananas, soybeans, cotton, beef, timber, autos, machinery, iron ore. **Climate:** Amazon region is tropical; northeast is subtropical; Brazilian Highlands region is temperate. ☐ The fifth-largest nation in the world is the superpower of South America Nearly half the size of the entire continent, and bordering all of the other countries except Ecuador and Chile, it has over half of South America's population. Ten metropolitan areas have over a million people each. Brazil is the leading agricultural, mining, and industrial nation in Latin America. With less than 5% of its land under cultivation, Brazil is still able to lead the world in the production of coffee, bananas, and sugarcane. Most of the cane is converted into alcohol to fuel the nation's automobiles. About 30% of the world's coffee is grown in the southern Highlands, where the climate for this crop is ideal: hot, wet summers and mild, dry winters. The large coffee estates are called "fazendas," the Portuguese version of the Spanish "haciendas" that are so common in Latin America. Brazil ranks among the leaders in soybeans, beef, cotton, and timber. The nation is mineral-rich, with huge reserves of iron ore and deposits of nearly every important mineral. Despite this natural wealth, Brazil has become the Third World's largest debtor nation. Inflation has been rampant, reaching 1,000% in the late 1980s. In 1990, the first democratically elected President in 30 years, Fernando Collor, instituted radical austerity measures. Brazil is a federal republic with 23 states, 3 remote territories, and a federal district, Brasília. In 1822, it gained independence from Portugal and became South America's only monarchy. Even as a republic (since 1889), it has endured a history of totalitarian regimes.

With most of its population, wealth, and industry located on or near the Atlantic coast, Brazil can be compared to 19th-century America. But unlike Americans, Brazilians have been reluctant to move west, even though their government has constructed roads, offered free land, and even moved its capital to the futuristic city of Brasília, 600 mi.(965 km) from the coast. Major changes are taking place in the Amazon rainforest as enormous areas of vegetation are being stripped away to provide timber and grazing land. In response to international concern, Brazil has pledged to regulate the cutting down of the rainforest. Brazil is the only country in Latin America that has a Portuguese culture and language, but people of Portuguese descent make up only a small part of the population. Many Brazilians immigrated from other European, Muslim, and Asian nations. More Roman Catholics live in Brazil than in any other country in the world. About 40% of the people are nonwhite: caboclos (of mixed Indian and white descent); mulattoes (mixed black and white); and native Indians, some of whom live a Stone Age existence deep in the Amazon Basin. Except for the mistreatment of its native Indian population (a practice common to every developing nation), race relations in Brazil are generally good. Lack of education, rather than race, is a barrier to economic advancement.

The country is divided into three regions: (1) The Amazon River Basin is a warm and extremely humid rainforest, sparsely populated by Indian tribes. (2) The northeast is a forested plain that covers the Atlantic bulge; 30% of the population lives here. The plain is often ravaged by drought. (3) The central and southern plateaus (Brazilian Highlands) contain 75% of the agriculture, 80% of the mining and manufacturing, and over 50% of the population. This area has the best climate in Brazil. Located here are São Paulo (16,500,000), the fastest-growing industrial center in Latin America, and Rio de Janeiro (10,500,000), a business and trade center. Rio sits in a spectacular harbor setting, surrounded by a wall of mountains that nearly seals it off from the mainland. The city is best known for the most extravagant of the four-day carnivals celebrated throughout the nation before the Christian holiday of Lent.

With a huge landmass, a favorable climate, and untapped natural resources, Brazil has the potential to become a world superpower in the next century. Industrial production will be greatly aided by a network of hydroelectric plants going up around the country. Brazil and Paraguay have built Itaipú Dam, the world's largest power station, on the Paraná River. One hundred and twenty-five miles to the south are the colossal Iguaçu Falls, which offer an even greater hydroelectric potential.

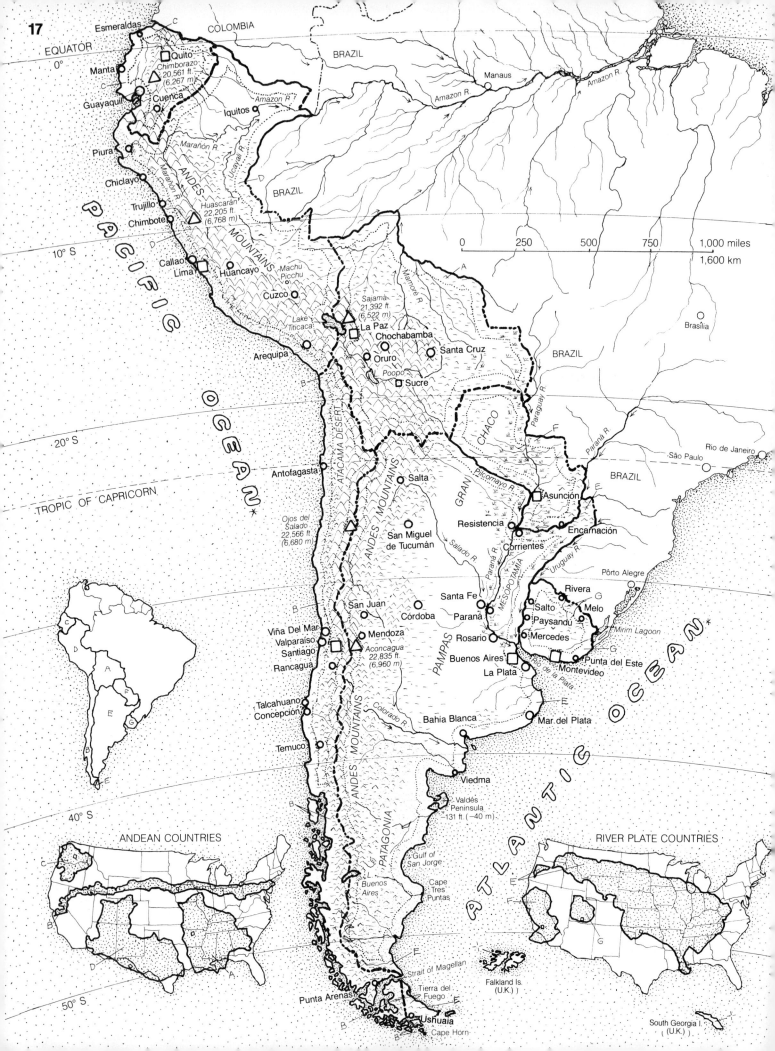

17

EQUATOR

COLOMBIA

Esmeraldas
C
Quito
Chimborazo
20,561 ft.
(6,267 m)
Manta
Guayaquil
Cuenca
Iquitos
Piura
Marañón R.
Chiclayo
ANDES
Trujillo
Marañón R.
Chimbote
Huascarán
22,205 ft.
(6,768 m)
Callao
Lima
Huancayo
Machu
Picchu
Cuzco
Lake
Titicaca
Arequipa
Sajamá
21,392 ft.
(6,522 m)
La Paz
Chochabamba
Oruro
Santa Cruz
Poopo
Sucre

PACIFIC OCEAN

ATACAMA DESERT

ANDES MOUNTAINS

Antofagasta
Salta
Ojos del
Salado
22,566 ft.
(6,680 m)
San Miguel
de Tucumán
Salado R.
San Juan
Córdoba
Santa Fe
Paraná
Viña Del Mar
Valparaíso
Mendoza
Aconcagua
22,835 ft.
(6,960 m)
Rosario
Santiago
Rancagua
Buenos Aires
La Plata
Talcahuano
Concepción
Colorado R.
Bahia Blanca
Temuco
Viedma
Valdés
Peninsula
-131 ft. (−40 m)

PAMPAS

PATAGONIA

Gulf of
San Jorge
L.
Buenos
Aires
Cape Tres
Puntas

Strait of Magellan
Punta Arenas
Tierra del
Fuego
Ushuaia
Cape Horn

BRAZIL

Manaus
Amazon R.
Amazon R.
Amazon R.

BRAZIL

0   250   500   750   1,000 miles
                              1,600 km

Marmoré R.

Brasília

BRAZIL

CHACO
Paraguay R.

GRAN
Pilcomayo R.
Asunción
Resistencia
Encarnación
Corrientes
Paraná R.
MESOPOTAMIA
Uruguay R.

São Paulo
Rio de Janeiro

BRAZIL

Pôrto Alegre
Rivera
Salto
Melo
Paysandú
Mirim Lagoon
Mercedes
Punta del Este
Montevideo
Rio de la Plata
Mar del Plata

ATLANTIC OCEAN

Falkland Is.
(U.K.)

South Georgia I.
(U.K.)

ANDEAN COUNTRIES

RIVER PLATE COUNTRIES

SOUTH AMERICA
ANDEAN COUNTRIES

The Andes divide Ecuador, Peru, and Bolivia into three regions: a western coastal desert, a central mountain range, and an eastern rain forest. In the Andes, people live at altitudes approaching 15,000 ft.(4,573 m). Life in the region is extremely hard. The mountainous terrain and the cold, oxygen-poor air have bred deep-chested people with short, stocky limbs. The staple of their meager diet is the potato, which may have originated here thousands of years ago. The llama, the "camel of the Andes," is a smaller, humpless relation of the camel. It provides transportation, wool, meat, and hides. The Andes nations have the largest Indian populations in the Western Hemisphere. Many of them are descendants of the highly evolved Inca civilization that flourished for 500 years before being destroyed by the Spanish.

BOLIVIA A

Area: 424,100 sq.mi.(1,098,420 km²). **Population:** 7,850,000. **Capital:** La Paz, 1,000,000; Sucre, 90,000. **Government:** Republic. **Language:** Spanish; two Indian dialects. **Religion:** Roman Catholic. **Exports:** Natural gas, tin, coffee, silver, cotton. **Climate:** Dry except in the northeast. ☐ With neither a coastline nor adequate road, rail, or water transportation, landlocked Bolivia is South America's poorest nation. On a plateau (Altiplano) in the Andes sits the world's highest capital, La Paz (12,000 ft., 3,659 m), the seat of all government functions except sessions of the Supreme Court, which are held in Sucre, the official capital. The population is 50% Indian and 35% mestizo. Most Bolivians speak their own native languages. Only the wealthy and middle classes speak Spanish, the official language. The country was named after Simon Bolívar, the "George Washington of South America," a Venezuelan who helped bring an end to 300 years of Spanish rule. Bolivia's long history of dictatorships and military coups was interrupted in 1952, when tinworkers led a revolution that brought in a reform-minded government. In less than 12 years it was overthrown by a military coup. Democracy did return in 1985. Bolivia has lost much mineral-rich land, including a vital Pacific coastline, in border wars with Chile and Paraguay.

CHILE B

Area: 292,250 sq.mi.(756,928 km²). **Population:** 14,700,000. **Capital:** Santiago, 4,800,000. **Government:** Republic. **Language:** Spanish. **Religion:** Roman Catholic. **Exports:** Copper, iron ore, chemicals, fruits, fish. **Climate:** Very dry to very wet. ☐ Picture a strip of land as long as the United States is wide, but averaging less than 150 mi.(240 km) in width, and you have the dimensions of the world's narrowest nation, Chile (chee' lay). Rainfall varies from nothing at all in the mineral-rich Atacama Desert in the north to more than 200 in.(508 cm) per year on the stormy islands in the south. The eastern third of the country is occupied by the Andes. Most of the population is confined to the central valley, a Mediterranean climate zone lying between the coastal range and the Andes. Punta Arenas, on the Strait of Magellan, is the world's southernmost major city; Ushuaia, Argentina, a town on Tierra del Fuego, is even farther south. Mestizos make up 70% of the population, but the European 20% is responsible for the continental flavor of Chilean cities. Until a democratically elected Marxist was overthrown by the military in 1973, Chile had a long tradition of political freedom, beginning with the liberal policies of its revolutionary leader and first president, Bernardo O'Higgins. Military rule ended in 1989; earlier gains in land reform were reversed, and confiscated property was returned to the large landowners. An important part of the economy is the world's largest production of copper. Chile's Easter Island (p. 36), famous for its huge stone figures of unknown origin, lies 2,300 mi.(3,680 km) west of the mainland.

ECUADOR C

Area: 108,000 sq.mi.(279,720 km²). **Population:** 12,400,000. **Capital:** Quito, 1,200,000. **Government:** Republic. **Language:** Spanish. **Religion:** Roman Catholic. **Exports:** Oil, bananas, coffee, fish products. **Climate:** Varies according to altitude. ☐ Ecuador (ek' wa dor) means "Equator" in Spanish. Quito, its charming capital, lies only 15 mi.(24 km) from the Equator, but with an altitude of 9,000 ft.(2,744 m), it enjoys being the "City of Eternal Spring." A hot, swampy coast produces the world's largest crop of bananas and feather-light balsa trees. Local palms supply the straw used in making Ecuador's famous "Panama hats." Almost all exports pass through the port of Guayaquil (1,500,000), the largest city and the commercial center. Politically conservative Quito is generally hostile to the progressive and liberal policies of Guayaquil. Mestizos make up 55% of the population, Indians 30%; most of the rest are the white elite. East of the Andes lies the "Oriente," a primitive part of the Amazon River Basin. Oil from deposits in this area is piped over the Andes to the port of Esmeraldas. Ecuador owns the Galapagos Islands (p. 14), 600 mi.(960 km) offshore, where plants and animals living in isolation have evolved into unique species. A visit by Charles Darwin in the 19th century helped formulate his theory of evolution.

PERU D

Area: 496,200 sq.mi.(1,285,158 km²). **Population:** 26,200,000. **Capital:** Lima, 6,500,000. **Government:** Republic. **Language:** Spanish. **Religion:** Roman Catholic. **Exports:** Copper, lead, silver, zinc, fish, coffee, guano. **Climate:** Coast is mild and dry; east of the Andes is tropical. ☐ What was once the heart of the Inca civilization is now the third-largest country in South America. The presence of Inca wealth induced the Spanish to found the coastal city of Lima (lee' muh) as the headquarters for their South American empire. The Andean city of Cuzco, the Western Hemisphere's oldest continuously inhabited city, was the Inca capital. Machu Picchu, at a higher elevation, was unknown

to the Spanish. Its walled ruins were discovered in the early 1900s. Over 9 million Inca descendants make up the largest Indian population of any country in the Western Hemisphere. Most of them live in poverty. Cities, farms, and factories are confined to the coastal deserts, irrigated by runoff from the Andes. The cold ocean currents that prevent rain (p. 15) also support an abundant marine life. Peru is a leading exporter of fish, much of it in the form of animal feed made from ground anchovies. The fish attract seabirds, whose droppings (guano) contribute to a major fertilizer industry. Over-fishing has led to a decline in the bird population, which in turn has reduced guano production. In the jungles of the northeast, at the headwaters of the Amazon, the port of Iquitos was built during the 19th century rubber boom. It still trades with vessels from the Atlantic Ocean that have sailed across an entire continent.

RIVER PLATE COUNTRIES

The River Plate nations are united by South America's second largest river system. The Río de la Plata (River Plate) is an estuary of the Atlantic; the Paraná and Uruguay Rivers are its tributaries. The capitals of these three nations are located on this vital transportation system. Argentina and Uruguay, which have literate populations of European descent, are two of the continent's most prosperous nations. Paraguay, a landlocked nation with a mostly mestizo population, has struggled under years of military repression. In addition to the Spanish language, the three nations share a common passion for maté, a locally grown herb tea. It is customarily served in a gourd and sipped through a straw (preferably made of silver).

ARGENTINA E

Area: 1,075,500 sq.mi.(2,785,545 km²). **Population:** 36,300,000. **Capital:** Buenos Aires, 3,700,000. **Government:** Republic. **Language:** Spanish. **Religion:** Roman Catholic. **Exports:** Beef, hides, sheep, wool, grains, cotton, and foods. **Climate:** Mild in the north, cooler in the south. ☐ Argentina, the world's leading beef exporter, is the continent's second-largest country in size and population. During the early part of the century, the sale of beef and cereals made it one of the world's richest nations. Under economic policies that began with Juan Perón in 1946, agriculture was heavily taxed to pay for social programs; foreign investment was deterred by nationalization of industry. Such policies led to Argentina's economic problems. The population reflects its multinational origin. Spanish is said to be spoken with an Italian accent. Buenos Aires is a distinctly European city. There are five regions in Argentina: (1) The western Andes contain the tallest peak in the Western Hemisphere, Aconcagua (22,835 ft.,6,950 m). (2) The scrub forests of the Gran Chaco are in the north. (3) Mesopotamia, a damp agricultural region, is bordered by the Paraná and Uruguay rivers. (4) The treeless, grassy plains of the Pampas contain fertile farmland, large ranches staffed by gauchos (cowboys), and major population and industrial centers. (5) Patagonia, a dry, inhospitable, windswept and thinly populated region, occupies the southern plateau. South of Patagonia is the island of Tierra del Fuego, which Argentina shares with Chile. Lying 300 mi.(480 km) to the east are the Falkland Islands, a British possession, which Argentina calls the Malvinas. In 1982, Argentina attempted to take the islands by force. Its defeat led to political turmoil at home and the overthrow of the military government.

PARAGUAY F

Area: 157,047 sq.mi.(406,752 km²). **Population:** 5,300,000 **Capital:** Asunción, 525,000. **Government:** Republic. **Language:** Spanish, Guaraní. **Religion:** Roman Catholic. **Exports:** Beef, hides, tannin, coffee, cotton. **Climate:** Warm and humid. ☐ The Paraguay (pa' ra gwy) River divides this landlocked country into two dissimilar regions: the Gran Chaco to the west and the Oriental to the east. The Paraguay and Paraná Rivers give the nation access to the Atlantic. Enormous hydroelectric dams, built with Brazil on the Paraná River, bring in vital revenues. Depending on the season, the Gran Chaco may be a dustbowl or a swampland. In the Chaco grows the remarkable quebracho ("axbreaker") tree. Its virtually indestructible hardwood—too heavy to float—is used as railroad ties, telephone poles, and road-paving material. The tree also produces tannin (used in curing animal hides). Believing that the Gran Chaco contained oil, Paraguay seized it from Bolivia in a costly war in 1932. Most Paraguayans live in the gentle environment of the Oriental. They speak a native language, Guaraní, as a second tongue. After suffering under an endless succession of dictators, Paraguay currently is a democracy, following 35 years of repression under General Alfredo Stroessner.

URUGUAY G

Area: 68,500 sq.mi.(177,415 km²). **Population:** 3,300,000. **Capital:** Montevideo, 1,500,000. **Government:** Republic. **Language:** Spanish. **Religion:** Roman Catholic. **Exports:** Meat products, hides, wool, textiles. **Climate:** Mild and humid. ☐ Until the 1950s, Uruguay (oo' roo gwy) was a model society that was called the "Switzerland of South America." Its agricultural economy was prosperous, and its political and social policies were the most advanced on the continent. But when world commodity prices declined, Uruguay could no longer pay for its social bureaucracy. A new conservative direction led to political unrest and the formation of an urban guerrilla movement. The upheaval brought the military to power, and in the process of restoring order, Uruguay became a nation with the most political prisoners per capita in the world. In 1985, a civilian government was elected. Uruguay's landscape is a gentle, grassy plain. The climate is always mild. A largely white population of Spanish and Italian extraction is the most urbanized (90%) in South America. Uruguay is one of the few nations in the world to have all of its land in use in one form or another.

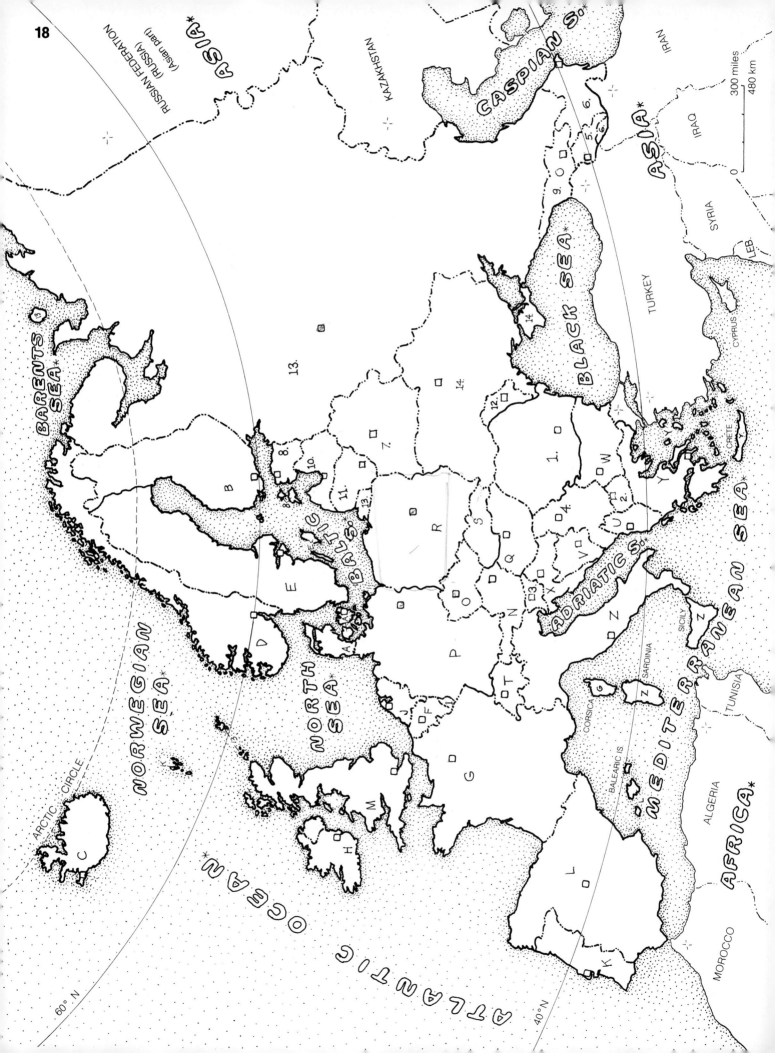

EUROPE: THE COUNTRIES

Europe is as far north as Canada; London is 650 mi.(1,040 km) farther north than New York City. But with the warming influence of the Gulf Stream (called the North Atlantic Drift in Europe), western Europe enjoys a milder climate than Canada and other regions of a similar latitude.

US
EUROPE*

Europe, with 3,810,000 sq.mi.(9,867,900 km²), is about 10% larger than the United States (Alaska and Hawaii included).

NORTHERN
DENMARK A COPENHAGEN
FINLAND B HELSINKI
ICELAND C REYKJAVIK
NORWAY D OSLO
SWEDEN E STOCKHOLM

WESTERN
BELGIUM F BRUSSELS
FRANCE G PARIS
IRELAND H DUBLIN
LUXEMBOURG I LUXEMBOURG
NETHERLANDS J AMSTERDAM
PORTUGAL K LISBON
SPAIN L MADRID
UNITED KINGDOM M LONDON

CENTRAL
AUSTRIA N VIENNA
CZECH REPUBLIC O PRAGUE
GERMANY P BERLIN
HUNGARY Q BUDAPEST
POLAND R WARSAW
SLOVAKIA S BRATISLAVA
SWITZERLAND T BERN

SOUTHEASTERN
ALBANIA U TIRANA
BOSNIA & HERZEGOVINA V SARAJEVO
BULGARIA W SOFIA
CROATIA X ZAGREB
GREECE Y ATHENS
ITALY Z ROME
ROMANIA 1. BUCHAREST
MACEDONIA 2. SKOPJE
SLOVENIA 3. LJUBLJANA
YUGOSLAVIA 4. BELGRADE

EASTERN
ARMENIA 5. YEREVAN
AZERBAIJAN 6. BAKU
BELARUS 7. MINSK
ESTONIA 8. TALLINN
GEORGIA 9. TBILISI
LATVIA 10. RIGA
LITHUANIA 11. VILNIUS
MOLDOVA 12. KISHINEV
RUSSIAN FEDERATION 13. MOSCOW
UKRAINE 14. KIEV

Europe is the second-smallest continent (after Australia), and its population of 720 million is second only to that of Asia. This makes Europe is the most densely populated of all continents. Most of the population resides in the industrial regions of the United Kingdom, the Netherlands, Belgium, northern France, Germany, southern Poland, the Czech Republic, northern Italy, Ukraine, and Russia. The fewest people are found in Iceland and northern Norway, Sweden, and Finland. Only the largest 40 nations are displayed on this plate. Those not included, but which are discussed on p. 21, 22, and 23, are Liechtenstein, Monaco, San Marino, and Vatican City; all are small enough to fit within the borders of a large city. The Ural Mountains define Europe's eastern border, splitting the Russian Federation (Russia) into European and Asian parts. Even though its Asian part (Siberia) is more than three times the size of European Russia, the nation is considered European because most Russians live west of the Urals and practice a European culture. Turkey is regarded as Asian because only 3% of its land mass is in Europe

and its culture is mostly Eastern.

Territorial barriers, in the form of mountains, rivers, lakes, gulfs, channels, and peninsulas, have preserved the continent's distinctly different cultures. Most people speak some form of four Indo-European languages: (1) Celtic (Breton, Irish, Scottish Gaelic, and Welsh); (2) Latin-Romance (French, Italian, Portuguese, Romanian, and Spanish); (3) Germanic (Dutch, English, German, and the Scandinavian languages—Danish, Icelandic, Norwegian, and Swedish); (4) Slavic (Bulgarian, Czech, Polish, Russian, Serbo-Croatian, and Slovak). Christianity is the principal religion, and Roman Catholicism is its most widely practiced form, especially in southwestern Europe. Protestant branches of Christianity dominate Great Britain, Scandinavia, and northern Europe. The Eastern Orthodox branch is the major religion in the eastern and southeastern parts. Judaism flourished until the Second World War, when the Germans annihilated 3/4 of Europe's Jewish population (approximately 6 million people).

Before the late 1980s, no one could foresee the end of the communist stranglehold on the Soviet Union and its captive nations of Eastern Europe. But by the 1990s, communism was gone, and with it the Soviet Union. All 15 republics became independent nations; Russia, the largest, is the Russian Federation. The Eastern Bloc of Soviet-controlled nations ousted their rulers. Yugoslavia disintegrated, and in its place five democratic countries emerged. The "Cold War" that lasted for over 40 years, between the communist nations of Eastern Europe and the democratic nations that made up the North Atlantic Treaty Organization (NATO) in the West, was finally over. In the late 1990s, NATO began to assume a larger role for maintaining the peace; former East Bloc nations sought admittance. The European Community (EC), an economic union of 15 nations (with others waiting to join), has broken down all trade barriers among member nations and is planning to introduce a single unit of currency (the Euro) in 1999. An even more difficult task facing it is the long-range goal of creating a "United States of Europe."

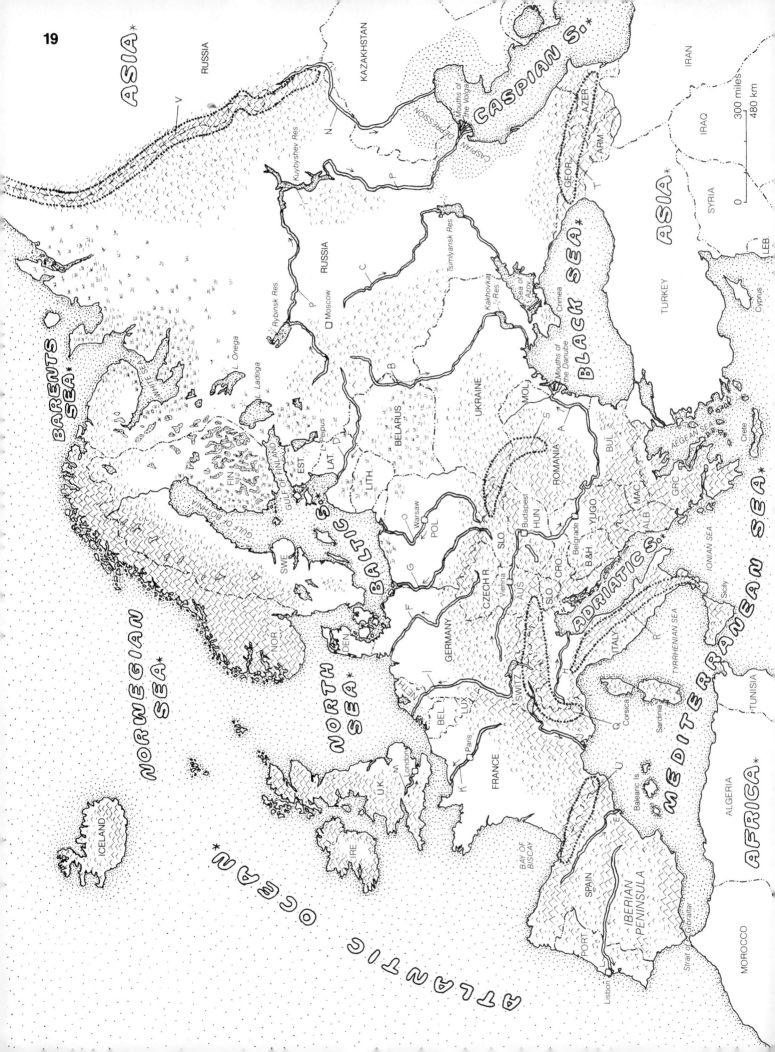

EUROPE: THE PHYSICAL LAND

Europe is a collection of peninsulas that together form the western peninsula of the Asian land mass. Some geographers regard the two continents as one: "Eurasia." The many peninsulas give Europe a longer coastline for its size than any other continent. With so many nations having access to the sea, Europeans have had a long history of shipbuilding, exploration, foreign trade, and fishing. Major fishing industries flourish on the Atlantic coast but not in the Mediterranean Sea, where the shallow Strait of Gibraltar bars the entry of the deep, cold Atlantic currents necessary to sustain large fish populations. There is very limited fishing in the highly polluted Black Sea.

Mountains play a major role in defining the landscape. The Urals in Russia form Europe's eastern boundary. Mountains in the northwest cover most of Norway, Sweden, and Great Britain and part of Ireland. It is believed that these low mountains were part of the Appalachians when North America and Europe were joined, 200 million years ago (p. 2). The larger, taller, and much younger Alpine System spans southern Europe from Spain to Russia. The Alps contain over 1,000 glaciers and almost all of the continent's tallest peaks except Europe's highest peak, Mt. Elbrus (18,480 ft., 5,634 m), in the Caucasus, close to the Caspian Sea. The Caspian is actually the world's largest saltwater lake. Its surface is the lowest point in Europe (-90 ft., -27 m). The land that surrounds the sea, the Caspian Depression, is also below sea level.

PRINCIPAL RIVERS :

DANUBE A
DNEPR B
DON C
DVINA (W) D
EBRO E
ELBE F
ODER G
PO H
RHINE I
RHÔNE J
SEINE K
TAGUS L
THAMES M
URAL N
VISTULA O
VOLGA P

PRINCIPAL MOUNTAIN RANGES :

ALPS Q
APENNINES R
CARPATHIANS S
CAUCASUS T
PYRENEES U
URALS V

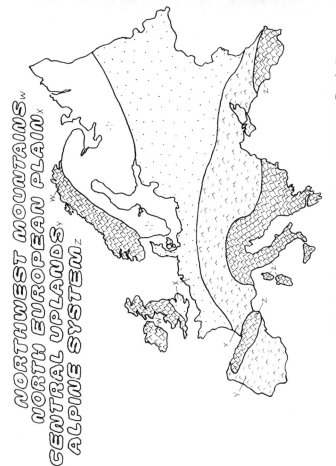

NORTHWEST MOUNTAINS W
NORTH EUROPEAN PLAIN X
CENTRAL UPLANDS Y
ALPINE SYSTEM Z

PRINCIPAL LAND REGIONS :

Europe's most productive agricultural and industrial region is the Northern European Plain, which includes most of European Russia, the Baltic states, Belarus, Poland, northern Germany, the Netherlands, Belgium, northern France, and southeastern Britain. This region has also been the site of Europe's bloodiest military battles; the flat lands and rolling hills are a natural avenue for invading armies. Spanning the continent, between the Northern Plain and the Alpine System, is the less densely populated Central Uplands region of plateaus and rocky highlands. Most of its inhabitants live on small farms nestled in fertile valleys.

The continent's extensive river systems have historically provided important transportation routes. The longest river, the Volga (2,194 mi, 3,510 km), is the nucleus of a much larger river and canal network that services populated areas of Russia and links the northern and southern coasts. The longest river in western Europe, the Danube (1,776 mi, 2,842 km), flows eastward through three capital cities. The Rhine (820 mi, 1,520 km) flows northward through Switzerland, Germany, and the Netherlands to the North Sea. The Rhine carries the most commercial traffic of any river in the world, by far.

European climate varies widely, from a damp, temperate northwestern coast, moderated by the North Atlantic Drift, to the continental temperature extremes of the interior regions. The pleasant Mediterranean climate of southern Europe (mild, damp winters and warm, dry summers) has given its name to similar climates in other parts of the world (California, central Chile, the cape of South Africa, and parts of Australia's south coast).

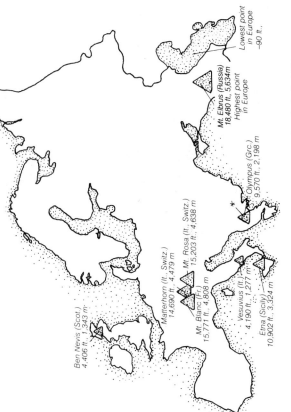

Ben Nevis (Scot.) 4,406 ft., 1,343 m
Matterhorn (It./Switz.) 14,690 ft., 4,479 m
Mt. Rosa (It./Switz.) 15,203 ft., 4,638 m
Mt. Blanc (Fr.) 15,771 ft., 4,808 m
Vesuvius (It.) 4,190 ft., 1,277 m
Etna (Sicily) 10,902 ft., 3,324 m
Mt. Olympus (Grc.) 9,570 ft., 2,198 m
Mt. Elbrus (Russia) 18,480 ft., 5,634 m Highest point in Europe
Lowest point in Europe -90 ft.

PRINCIPAL MOUNTAIN PEAKS :

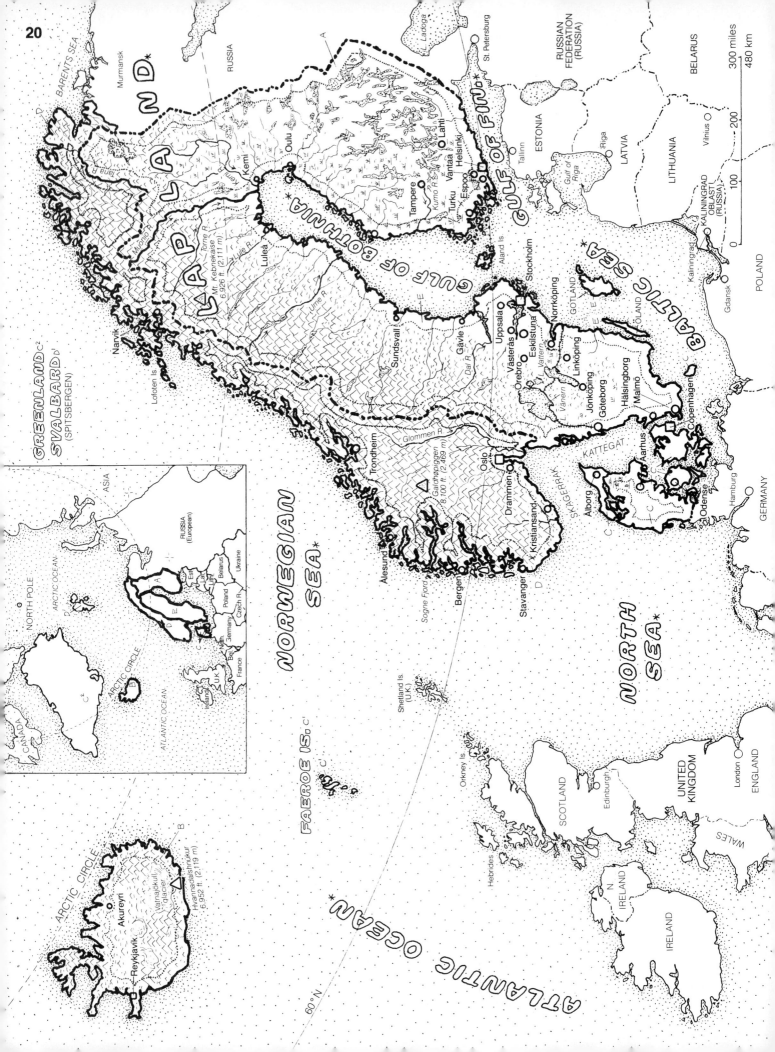

BARENTS SEA

Murmansk

RUSSIA

L. Ladoga

St. Petersburg

RUSSIAN FEDERATION (RUSSIA)

BELARUS

300 miles
480 km

GREENLAND C²

SVALBARD D¹
(SPITSBERGEN)

Tana R.

LAPLAND

Oulu

Kemi

Lahti

Vantaa

Helsinki

ESTONIA

Tallinn

Gulf of Riga

Riga

LATVIA

LITHUANIA

Vilnius

200

100

GULF OF FINLAND

Espoo

Turku

Tampere

Kumo R.

Mt. Kebnekaise
6,926 ft. (2,111 m)

Torne R.

Muonio R.

Lule R.

Luleå

Ume R.

KALININGRAD OBLAST (RUSSIA)

Kaliningrad

POLAND

Gdansk

BALTIC SEA

Narvik

Lofoten Is.

Åland Is.

Stockholm

GOTLAND

ÖLAND

Sundsvall

Gävle

Uppsala

Västerås

Eskilstuna

Norrköping

Dal R.

Örebro

L. Vättern

Linköping

L. Vänern

Jönköping

Göteborg

Hälsingborg

Malmö

Copenhagen

KATTEGAT

Ålborg

Aarhus

Odense

Hamburg

GERMANY

Glommen R.

Trondheim

Galdhøpiggen
8,100 ft. (2,469 m)

Oslo

Drammen

Kristiansand

SKAGERRAK

ÅLESUND

Bergen

Sogne Fjord

Stavanger

NORWEGIAN SEA *

FAEROE IS. D C¹

Shetland Is. (U.K.)

Orkney Is.

Hebrides

SCOTLAND

Edinburgh

NORTH SEA *

UNITED KINGDOM

London

ENGLAND

WALES

IRELAND

IRELAND

ATLANTIC OCEAN *

60° N

NORTH POLE

ARCTIC OCEAN

ASIA

RUSSIA (European)

Est.
Lat.
Lith.

Poland

Belarus

Ukraine

Ger.
Nor.

Czech R.

U.K.

France

Ireland

CANADA

ATLANTIC OCEAN

ARCTIC CIRCLE

ARCTIC CIRCLE

Akureyri

Vatnajökull Glacier

Hvannadalshnúkur
6,952 ft. (2,119 m)

Reykjavik

EUROPE: NORTHERN ::

California

CN: (1) Use a bright color for Denmark (C) and the Faeroe Islands (C') on the large map. (2) Color the word "Lapland" gray in the northern region. (3) On the small map, note Svalbard (D') in the Arctic Ocean.

Only Denmark, Norway, and Sweden are Scandinavian, but all five countries of Northern Europe share a common history and continue to maintain strong ties through membership in the Nordic Council. During the Middle Ages, they were all politically united under Danish rule. The Scandinavian languages have Germanic roots and are similar enough to be understood in any of the three countries. Finland's Asiatic language and ethnic origins are significantly different from those of its neighbors. The five nations are democratic, prosperous, socially progressive, highly literate, welfare states with free-market economies. Curiously, Denmark, Norway, and Sweden have kept their monarchies. Scandinavia was the home of the Vikings, a fierce tribe of seafarers who explored the northeastern coast of North America (they preceded Columbus by 500 years) and raided the British Isles, northern Europe, and European Russia. From the 9th through the 11th centuries, the Danish and Norwegian Vikings explored the North Atlantic and North America while the Swedish Vikings went eastward to Russia.

The landscape of the northern nations shows strong evidence of Ice Age activity. Glacial action left thousands of lakes, rivers, fjords, islands, and glacial moraines (debris). Ocean currents of the North Atlantic Drift (Gulf Stream) keep the climate temperate, except in the northern interiors. The entire region is subject to long, dark winters; Scandinavians call the part north of the Arctic Circle the "Land of the Midnight Sun," referring to two summer months of endless sunlight. Winter brings two months of darkness. This region, also called Lapland, is home to some 60,000 Lapps—short, colorfully dressed Asian-looking people. Lapps (Samis, as they prefer to be called) have resisted all government attempts to assimilate them. Some prefer the life of reindeer-herding nomads, roaming freely across international borders in the Arctic region they call Sapmi.

FINLAND A

Area: 130,120 sq.mi.(336,750 km²). **Population:** 5,150,000. **Capital:** Helsinki, 525,000. **Government:** Republic. **Language:** Finnish; Swedish (6%). **Religion:** Lutheran. **Exports:** Timber, paper products, plywood, manufacturing and engineering products, farmed furs, textiles. **Climate:** Cold and snowy in the north. ☐ Finland is a densely forested land of 60,000 lakes. In the southern lake and canal region, steamships service a 200 mi.(320 km) strip of countryside. Finland's prosperous economy emphasizes timber and paper products, including the world's largest plywood industry. Timbering is carefully monitored to protect the environment. Finland enjoys one of the world's highest rates of industrial growth. Its exports of technology include entire factories. In the south, farms are surprisingly productive despite the short growing season. Finland excels in the design of fabrics, housewares, and furniture. It has produced two of the 20th century's finest architects: Alvar Aalto and Eliel Saarinen. The latter's son, Eero, became famous in America for such works as the Gateway Arch in St Louis. Unlike Scandinavians, the Finns are descendants of Asian settlers. Their language is related to Estonian and Hungarian, which have similar Siberian origins. The Finns are hardy people, passionately devoted to the ritual of the sauna bath—baking in a room heated by hot rocks, followed by a plunge into icy water. Finland's capital was the site of the historic 1975 Helsinki Agree-

ment, a milestone in the promotion of world peace. Throughout history, Finland has been ruled by Sweden or Russia. Finland has lost many wars and much territory to Russia and the Soviets. Relations between Finland and its giant neighbor improved following World War II; even before the demise of the Soviet Union, it was Finland's chief trading partner.

ICELAND B

Area: 39,750 sq.mi.(102,953 km²). **Population:** 271,000. **Capital:** Reykjavik, 103,500. **Government:** Republic. **Language:** Icelandic. **Religion:** Lutheran. **Climate:** Moderate. **Exports:** Fish products, aluminum, wool, sheep products. ☐ A former colony of Norway and Denmark, prosperous and socially progressive Iceland is closer to North America than to Europe. The treeless island is not quite as cold as its name suggests, although it does boast Europe's largest glacier, Vatnajokull. With the North Atlantic Drift passing by, harbors rarely freeze, and along some coastal areas a green ground cover is present for most of the year. Iceland has been called the "Land of Ice and Fire." It is the most seismically active nation in the world and is constantly being enlarged as new islands are formed by volcanic eruptions (p. 2). Homes are heated by hot springs. The word "geyser" comes from the most famous of Iceland's many hot springs, "Geysir," which intermittently spouts steam nearly 200 ft.(61 m) high. Fishing and the processing of fish products constitute the principal industry, and the island's territorial waters are zealously guarded. The literacy rate in Iceland (99.9%) is the world's highest. Reykjavik, the most northerly of all capital cities, has more bookstores per capita than any other city. The Icelandic language is remarkably similar to the Old Norse spoken by the Viking settlers. Iceland's Parliament, the world's oldest form of representative government, was established by the Vikings in the 9th century.

SCANDINAVIAN COUNTRIES : *

DENMARK C

Area: 16,365 sq.mi.(42,352 km²). **Population:** 5,350,000. **Capital:** Copenhagen, 1,400,000. **Government:** Constitutional monarchy. **Language:** Danish. **Religion:** Lutheran. **Exports:** Meat, fish, dairy products, machinery, porcelain, pharmaceuticals, furniture. **Climate:** Mild but damp. ☐ With few natural resources other than a low, flat, and fertile landscape (enriched by glacial moraine), this tiny nation provides its citizens a very high standard of living. Danish foods such as butter, cheese, bacon, and ham are the profitable exports of a highly regulated agricultural industry. Cooperative farms are restricted in size and farmers must pass licensing examinations. Denmark is a major exporter of housewares and furniture—"Danish modern" has become an international style. Most Danes live and work on the many ferry-connected islands east of the Jutland Peninsula, which Denmark shares with Germany. Except for the 40 mi.(840 km) border across the peninsula, Denmark is totally surrounded by water. It has almost 500 islands. On the easternmost island, 10 mi.(16 km) from Sweden, is Copenhagen, the capital and cultural and industrial center. In its harbor is a statue of "the Little Mermaid," a character from a Hans Christian Andersen fairy tale. In the heart of Copenhagen is the famous Tivoli Gardens amusement park, the heart of Europe's entertainment center. Denmark owns Greenland and the Faeroe Islands, which lie north of the British Isles. Both of these self-governing possessions were founded by Viking explorers.

NORWAY D

Area: 125,051 sq.mi.(323,631 km²). **Population:** 4,420,000 **Capital:** Oslo, 485,000. **Government:** Constitutional monarchy. **Language:** Two forms of Norwegian, plus many dialects. **Religion:** Lutheran. **Exports:** Oil, fish, timber products, chemicals, aluminum, ships. **Climate:** Moderate along the coast and southern parts. ☐ Norway, the northernmost Scandinavian country, is a mountainous plateau with little farmland; only 3% of the country is cultivated. Farms are confined to the southern lowlands around Oslo, the nation's capital and industrial, cultural, and recreational center. Norway's scenic western coastline has over 100,000 islands; its special beauty comes from the many majestic fjords (fee ords)—long, narrow inlets of ocean that penetrate steep, coastal cliffs. The name "Viking" comes from the Old Norse word "vik," which means "inlet." Sogne Fjord, the longest, is 127 mi.(204 km) long. The 1,700 mi.(2,720 km) coastline is actually many times longer if one includes the coasts of the thousands of islands and fjords. Fishing is no longer Norway's main industry. It has been displaced by oil and natural gas from the North Sea. Norway has become quite wealthy; how to spend it all is a growing problem. In addition to oil and natural gas, Norway is blessed with an endless amount of cheap energy. In addition to oil and natural gas, it has the largest amount of hydroelectricity, per capita, in the world. The snowcapped mountain ranges that supply the waterpower also provide Norwegians with their favorite form of recreation: skiing.

SWEDEN E

Area: 173,231 sq.mi.(443,124 km²). **Population:** 8,890,000. **Government:** Constitutional monarchy. **Capital:** Stockholm, 700,000. **Religion:** Lutheran. **Exports:** Machinery, timber, autos, transportation equipment. **Climate:** Moderate in the south. ☐ Both capitalists and socialists can admire Sweden, sometimes called the "Land of the Middle Way." Public and private ownership share in the profit-driven industrial economy, which provides its citizens with a high standard of living plus a broad range of social benefits and services. The Swedes spend more on vacations, per capita, than any other nation. The majority of the people live in urban apartments, but many own second homes in the country. Except for the long, gray winters, Sweden could be the "perfect country." It is the largest of the northern countries and occupies a long, narrow plain that slopes from the mountains (which it shares with Norway) eastward to the Baltic Sea. Over half the nation is covered by forests, whose growth is monitored by a regulatory program as comprehensive as Finland's. Lake Vanern, the largest freshwater lake in western Europe, is the heart of a lake and canal linkage between Goteborg and the Baltic Sea. The northeastern coast remains frozen six months of the year. Most Swedes live in the southern lowlands region. Here, located on the Baltic, is Stockholm, the beautiful capital city built on 14 islands connected by bridges. Swedish industry is based on three abundant resources: iron ore, timber, and hydroelectric power. The nation is unique in the way it zealously guards the environment from industrial harm. Like Switzerland, Sweden maintains a large standing army, and like its neutral counterpart on the continent, it avoided the great wars of this century.

EUROPE: WESTERN

CN: (1) Color the British-owned Channel Islands off the coast of France. (2) Do not color Andorra, located between France and Spain. (3) Color the Mediterranean islands.

BELGIUM A

Area: 11,785 sq.mi.(30,499 km²). **Population:** 9,950,000. **Capital:** Brussels, 950,000. **Government:** Constitutional monarchy. **Language:** Flemish; French; German. **Religion:** Roman Catholic. **Exports:** Steel and engineering products, food, textiles, glassware. **Climate:** Moderate; moist. □ Brussels, the location for many economic, political, and military organizations, has been the de facto capital of Western Europe since the end of World War II. It is the headquarters for the European Community, an organization dedicated to the ultimate unification of the nations of Europe. Steel is Belgium's largest industry, and the nation's farms are the most productive (and smallest) in Europe. The nation is divided between the Flemings in the north, who speak a Dutch dialect, and the French-speaking Walloons in the south. It's as if each half of Belgium relates more to a nation on its border than to the other half. Over the years, the flat, strategically located nation has been the site of numerous battles fought by invading armies. Napoleon "met his Waterloo" (his final defeat) near the Belgian village of that name.

FRANCE B

Area: 212,467 sq.mi.(549,864 km²). **Population:** 58,000,000. **Capital:** Paris, 10,000,000. **Government:** Republic. **Language:** French; other dialects. **Religion:** Roman Catholic. **Exports:** Coal, iron ore, autos, foods, wine, textiles. **Climate:** Moderate in the west; continental in the interior; Mediterranean in the south. □ The largest country in western Europe is an unusual mixture of sophistication and provinciality. The French have traditionally excelled in diplomacy, science, art, architecture, music, literature, fashion, wine, and cooking; nevertheless, France is primarily a nation of farms, towns, and villages—the largest agricultural country in Europe. Picturesque regions are identified with famous wines (Champagne, Bordeaux, Burgundy, etc.). It is often said that there is France and there is Paris. The French capital is the premier tourist city, with its magnificent architecture, museums (the Louvre), cathedrals (Notre Dame), parks, boulevards, restaurants, high fashion, and the 19th-century engineering marvel, the Eiffel Tower. France has become a European leader in high technology. It is the world's largest producer of nuclear power plants and transportation systems (TGV high-speed trains) and is a collaborator in commercial aviation design (Concorde and Airbus jets) and space exploration (Ariane rocket). A diverse countryside includes coastlines facing three seas (North Sea, Atlantic Ocean, and the Mediterranean), farmland and plateaus in the interior, and the towering Alps (including Mt. Blanc) in the southeast.

ANDORRA. Located high in the Pyrenees with an area of 190 sq.mi.(465 km²) and a population of 60,000, this Catalan-speaking republic is governed by both France and Spain, which results in the duplication of services such as the mail, schools, and currency. Tourists enjoy duty-free shopping and the fine ski resorts.

MONACO. With an area of 0.75 sq.mi.(1.9 km²), this principality of 30,000 on the French Riviera is aptly described by the word "tiny." Home to many millionaires attracted by elegant, tax-free living, Monaco made headlines when the actress Grace Kelly left Hollywood to marry Prince Rainier. His palace is in the capital, Monaco. This nation is home to the city of Monte Carlo, with its famed gambling casino.

IRELAND C

Area: 27,120 sq.mi.(70,186 km²). **Population:** 3,600,000. **Capital:** Dublin, 1,000,000. **Government:** Republic. **Language:** Gaelic and English. **Religion:** Roman Catholic. **Exports:** Meat and dairy products, textiles, technology, whiskey. **Climate:** Mild and moist. □ A damp climate and limestone-rich soil are responsible for the intensely green countryside that gives Ireland (Eire) the name "Emerald Isle." Although tourism is the chief industry, a boom has been created by foreign high-tech investments. This has halted 150 years of emigration (mostly to America), which began with the potato famine of the 1840s. Peat, a combustible soil, is the nation's primary fuel. It is dug out of damp, spongy bogs that cover a sixth of the country. The people have a strong oral tradition, and though Gaelic is the traditional language, Ireland has produced some of the finest writers of the English language: Shaw, Swift, Joyce, Wilde, Yeats, and Beckett.

LUXEMBOURG D

Area: 997 sq.mi.(2,580 km²). **Population:** 410,000. **Capital:** Luxembourg, 76,000. **Government:** Constitutional monarchy. **Language:** French; German; Letzeburgesch. **Religion:** Roman Catholic. **Exports:** Steel and chemical products. **Climate:** Continental. □ Higher than its "low" neighbors, Luxembourg, with its hills and valleys, forests, castles, and quaint villages, is the more scenic. Luxembourg, the capital, is a center for banking and finance. A huge steel industry is at the heart of Luxembourg's prosperity. Residents have to learn three languages: French for government affairs, German for writing, and Letzeburgesch (a German dialect) for conversation.

NETHERLANDS E

Area: 16,040 sq.mi.(41,544 km²). **Population:** 15,800,000. **Capital:** Amsterdam, 750,000. **Government:** Constitutional monarchy. **Language:** Dutch. **Religion:** Roman Catholic 40%; Protestant 25%. **Exports:** Engineering and dairy products, natural gas, cut diamonds, flower bulbs. **Climate:** Moderate and damp. □ Before Belgium and Luxembourg gained their independence in 1830, they were part of the Netherlands, a name that means "low countries." During the past 600 years, the Netherlands (Holland) has increased its size by 40% by pumping out the North Sea. The reclaimed lands, called "polders," are protected by a network of dikes, ditches, and canals. Amsterdam, the capital, is built on a polder. Steam and electricity have replaced windpower as the energy source for the pumps. Windmills are still a familiar sight on the Dutch landscape, along with fields of tulip bulbs. Bicycles fill the streets of Amsterdam, the center for trade, finance, and manufacturing. Rotterdam (660,000), on the Rhine River and with access to the Atlantic Ocean, is the world's busiest seaport.

PORTUGAL F

Area: 35,500 sq.mi.(91,945 km²) **Population:** 9,900,000. **Capital:** Lisbon, 700,000. **Government:** Republic. **Language:** Portuguese. **Religion:** Roman Catholic. **Exports:** Cork, olive, and wood products; wines; sardines. **Climate:** Mild; Mediterranean on the south coast. □ The exports of western Europe's poorest nation are agricultural: cork, from the world's largest cork forests; olive oil; and fine wines (principally port and Madeira). Port is produced in the region surrounding Porto (340,000), the second-largest city. Most of the population lives in the wide, fertile coastal areas. Portugal's fine weather, church architecture, castles, and quaint villages draw a growing tourist trade. Lisbon, located on an inland sea formed by an estuary of the Tagus River, is one of Europe's most beautiful capitals. The Portuguese-owned Azores and Madeira Islands (p. 39) are popular vacation resorts as well as agricultural regions.

SPAIN G

Area: 194,884 sq.mi.(504,750 km²). **Population:** 38,900,000. **Capital:** Madrid, 3,8100,000. **Government:** Constitutional monarchy. **Language:** Spanish; Basque; Catalan; Galician. **Religion:** Roman Catholic. **Exports:** Autos, wine, olive products, cork. **Climate:** West is mild; interior is continental; southeast is Mediterranean. □ It isn't just good weather and low prices that make "sunny Spain" the most popular tourist attraction in Europe. Cut off from the rest of the continent by the Pyrenees, and only 10 mi. (16 km) from North Africa, Spain developed a unique blend of Western and Moorish cultures (the Moors occupied the country for over 700 years). Most tourists head for the Mediterranean coast, but many venture inland to see bullfights, great art, medieval castles, and dramatic scenery. Spain shares the rugged Iberian peninsula with Portugal. Most of it is a craggy, high, dry, treeless plateau called the Meseta. All roads and rails lead to Madrid, the centrally located capital and the cultural, industrial, and commercial center of Spain. Spanish traditions, such as the siesta (a midday rest period) and the evening paseo (a walk before a late supper), are gradually giving way to the demands of a rapidly industrializing society. People in the distinctive regions of Galicia, the Basque provinces, and Catalonia speak their own languages (the ancient Basque language is related to no other) and have long sought independence from Spain. The exciting city of Barcelona (1,650,000) is Spain's second-largest and the capital of Catalonia. At the mouth of the Mediterranean, the Rock of Gibraltar (1,398 ft., 426 m) occupies most of a narrow, 4 mi.(6.4 km) peninsula, controlled by the British for nearly three centuries. Britain refuses to leave and maintains a naval base at the Rock.

UNITED KINGDOM H

Area: 94,250 sq.mi.(243,919 km²). **Population:** 59,000,000. **Capital:** London, 7,000,000 **Government:** Constitutional monarchy. **Language:** English; some Welsh and Gaelic. **Religion:** Protestant. **Exports:** Engineering products, autos, chemicals, food, textiles. **Climate:** Temperate and moist. □ The British Isles include the United Kingdom (Great Britain and Northern Ireland) and the Republic of Ireland. The island of Great Britain includes the countries of England, Scotland, and Wales. The capital of the Kingdom, London, is also the capital of England. The names "United Kingdom," "Great Britain," "Britain," and "England" are used interchangeably, and not always accurately. As recently as the early 20th century, the British Empire, the largest in history, governed a quarter of the globe. The Empire, over which "the sun never set," is nearly gone, but its influence remains. Britain still presides over the Commonwealth of Nations, an organization of 75 former colonies, current dependencies, and territories. The British Isles have been protected from continental aggression by the English Channel, which was formed by rising seas following the last Ice Age. The last invasion was the Norman Conquest of 1066. The rugged hills of Scotland in the north and Wales in the west are well suited for the grazing of livestock. Southeastern England has a significant amount of highly productive farmland. Wales has the richest coal reserves on the island. Its capital, Cardiff (280,000), and most of the population are located on a narrow coastal plain. Belfast (310,000) is the capital of Northern Ireland. This strife-torn region was called Ulster when it was the northern province of Ireland, before the southern part won its independence from Britain in 1921. The continuing violence stems from Protestant rejection of demands made by the downtrodden Catholic minority, who seek unification with Ireland. The proud, independent people who live in scenic Scotland share the Gaelic language and a common heritage with the Protestants who migrated to Northern Ireland hundreds of years ago. The stone buildings of Edinburgh (440,000) make the Scottish capital an exceptionally beautiful city. Glasgow (750,000), on the western shore, is Scotland's industrial center. Offshore are three sparsely populated archipelagos: the Hebrides, the Orkney Islands, and the Shetland Islands (p. 20). South of England, and close to the French coast, are the Channel Islands, the largest of which, Jersey and Guernsey (homes of the famous breeds of cattle), are popular vacation spots as well as producers of specialty export crops. Densely populated England has 80% of the Kingdom's people. The Midlands region is the country's industrial heartland, and Birmingham (1,050,000) is the largest industrial city. The industrial revolution, fueled by large coal and iron reserves, had its beginnings in 17th-century England. London, located on the historic Thames River, is the cultural heart of the United Kingdom and a center for world finance. The discovery of North Sea oil in the 1970s reversed a decline in the economy that began with the breakup of the Empire after World War II.

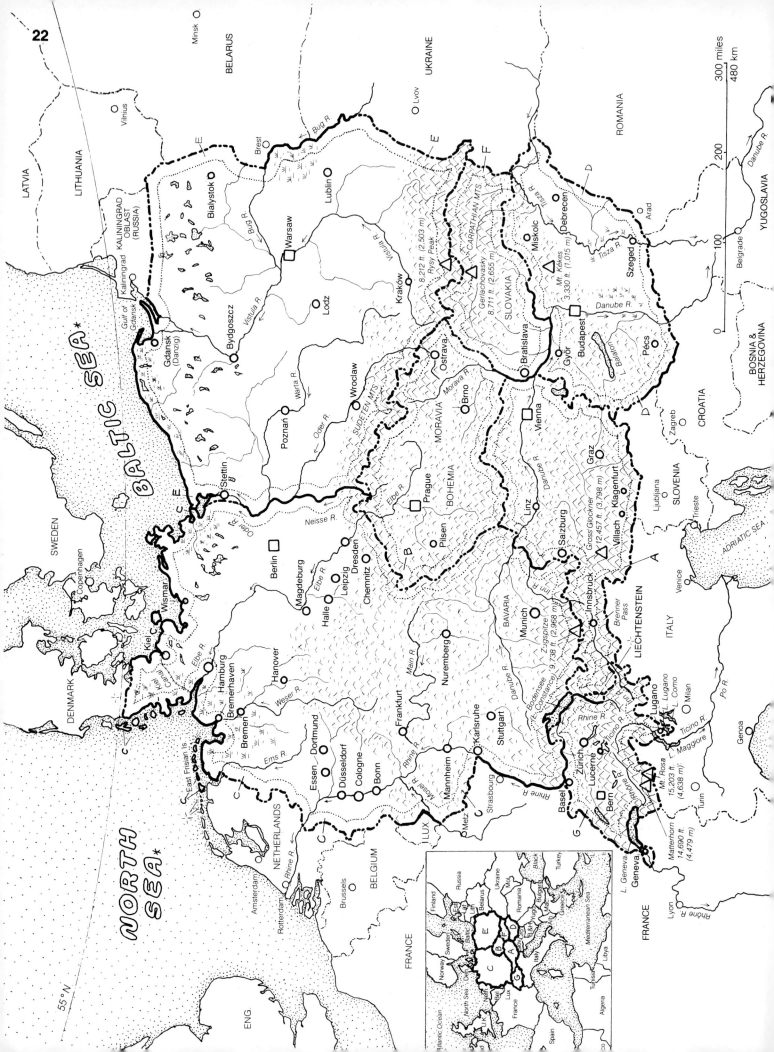

EUROPE: CENTRAL

CN: Do not color Liechtenstein, wedged between Switzerland and Austria.

After its defeat in World War II, Germany was divided into two nations: East and West Germany. Europe itself was split into two armed camps: the free-market democracies of the West and the communist nations of the East (most of which were under Soviet domination). Austria, Liechtenstein, and Switzerland were in the western group, and Czechoslovakia, Hungary, and Poland were part of the Eastern Bloc. During the "Cold War" period, the nations of the West prospered while the communist economies were barely able to provide necessities. In 1989, without opposition from the Soviet Union, the people of the Eastern Bloc rose up and threw their communist rulers out of power, paving the way for multiparty elections, the institution of free-market economies, and the restoration of religious freedom. The completely unanticipated and swift collapse of communism in Europe culminated with the reunification of Germany in 1990.

AUSTRIA A

Area: 32,375 sq.mi.(83,850 km²). **Population:** 8,130,000. **Capital:** Vienna, 1,725,000. **Government:** Republic. **Language:** German (many dialects). **Religion:** Roman Catholic. **Exports:** Engineering, chemicals, forest products. **Climate:** Moist summers; cold, dry winters. ☐ Austria, even more than Switzerland, is dominated by the Alps; the towering peaks occupy 70% of the nation (versus 60% of Switzerland). Austria is famous for its ski resorts, music festivals, and health spas. Especially popular is the Mozart festival in Salzburg, the composer's birthplace. South of Innsbruck, on the Italian border, is the Brenner Pass, the main route across the Alps in central Europe and part of an amazing network of rail and road tunnels linking Austria, Switzerland, and Italy. Austria's central location gave it the title "Crossroads of Europe." Before World War II, Vienna, the home of Sigmund Freud, was the cultural, educational, scientific, and medical center of Europe. During the postwar period, it remained politically neutral, serving as a buffer between East and West. Many international organizations are headquartered in Austria. In the fertile Danube Valley, small, highly efficient farms supply most of the nation's needs. Austria's industry is also efficient and benefits from hydroelectric power supplied by Alpine runoff. Austria is especially concerned about preserving both its urban and rural environments by strictly enforcing laws designed to protect them.

CZECH REPUBLIC B

Area: 30,387 sq.mi.(78,703 km²). **Population:** 10,300,000. **Capital:** Prague, 1,225,000. **Government:** Parliamentary democracy. **Language:** Czech. **Religion:** Roman Catholic. **Exports:** Manufactured goods, machinery, transportation equipment, chemicals, metals, foods. **Climate:** Continental. ☐ For 75 years, the Czech (Chehk) Republic was the western half of Czechoslovakia, a nation created after World War II by combining the Czech regions of Bohemia and Moravia in the west with Slovakia in the east. The two halves were complete opposites: the Czechs were industrialized, prosperous, and cultured people, influenced by Austria and Germany; the Slovaks were mostly illiterate peasants, ruled for centuries by Hungary. Nearly 50 years of Soviet domination following World War II narrowed the gap between the two halves: illiteracy was wiped out in Slovakia and industrialization raised their standard of living while Czech quality of life declined under communism. In 1993, four years after gaining independence, Czechoslovakia formally split apart. Prague, the capital, is one of Europe's most historic cosmopolitan centers.

GERMANY C

Area: 137,748 sq.mi.(356,866 km²). **Population:** 82,100,000 **Capital:** Berlin, 3,000,000. **Government:** Republic. **Language:** German. **Religion:** Roman Catholic in the south, Lutheran in the north. **Exports:** Autos, engineering and chemicals, lignite, foods, textiles. **Climate:** Coast is moderate; inland is continental. ☐ Can anyone explain how a nation that has contributed so much to the world in the fields of literature, philosophy, science, medicine, and, most notably, music (Bach, Brahms, Beethoven, Schubert, and Wagner) could have inflicted the magnitude of suffering and destruction that Germany created in World War II? After its defeat, the nation was divided into East and West Germany. East Berlin was made the capital of East Germany, and West Berlin was a detached part of West Germany within the East. With Western aid, West Berlin became a showcase of capitalism. As a result, East Berlin was forced to erect the infamous "Berlin Wall" to prevent its citizens trom fleeing to the West. The wall was torn down in 1990, and the two Germanys were reunified after over 40 years of division. From the ashes of World War II, West Germany had become the world's fourth-largest industrial economy, East Germany the tenth largest—an even greater achievement, since it was mostly farmland prior to the War, with only a quarter of Germany's population. Germany's main industrial centers and mineral deposits are located in the western Ruhr and Rhine valleys, the largest industrial region in continental Europe. Germany's population is the largest in Europe outside of Russia. Most major cities are located in the west, on the bustling Rhine River. The southern Rhine Valley is known for its vineyards and castles. Hamburg (1,700,000), on the Elbe River close to North Sea, is the second largest city, busiest port, and the cultural center of northern Germany. Munich (1,275,000), the capital of Bavaria in the south, best known for beer festivals, has become a major industrial center. East Germany has been the world's leading producer of lignite, a low grade of soft coal. The marshy Baltic coastal regions Germany shares with Poland are not suitable for farming.

HUNGARY D

Area: 35,911 sq.mi.(93,032 km²). **Population:** 10,210,000. **Capital:** Budapest, 2,000,000. **Government:** Republic. **Language:** Hungarian. **Religion:** Roman Catholic. **Exports:** Transportation vehicles, pharmaceuticals, poultry, bauxite, steel. **Climate:** Mild summers, continental winters. ☐ Hungarians, unlike their Slavic neighbors, are descendants of the Magyars, who came from central Russia in the ninth century. The Hungarian language, called Magyar (modjar), is related to Finnish and Estonian. The country is vertically bisected by the Danube River. To the east lies a low and flat agricultural plain (the site of an ancient sea). To the west is a hilly region including the popular, recreational Lake Balaton, the largest lake in central Europe. Two cities divided by the Danube, Buda and Pest, make up the nation's capital and industrial center. A quarter of the population lives in metropolitan Budapest. Hungary's feudalistic agricultural industry was improved by restructuring and the modern techniques introduced by the communists, but economic growth in the 1980s was due to Hungary's leadership among Eastern Bloc nations in pushing through the economic and political reforms that led to the revolution of 1989.

POLAND E

Area: 120,732 sq.mi.(312,677 km²). **Population:** 38,600,000. **Capital:** Warsaw, 1,650,000. **Government:** Republic. **Language:** Polish. **Religion:** Roman Catholic. **Exports:** Engineering, steel, timber, foods, coal, sulphur. **Climate:** Continental away trom the coast. ☐ Poland is a mostly flat country in the Northern European Plain. The land gradually rises from the Baltic coast until it reaches the mountainous border the nation shares with the Czech Republic and Slovakia. Throughout its history, Poland has had its boundaries changed by invading armies. World War II began when Germany crossed Poland's western frontier. The Germans were especially murderous in Poland: they killed 6 million Poles and 3 million Jews (Europe's largest Jewish community). Warsaw, the capital, was leveled; it has been completely rebuilt, with the "old town" faithfully

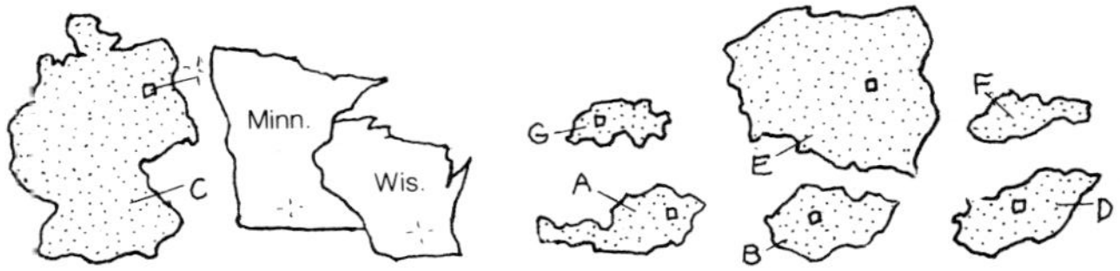

restored. After the War, Poland was essentially moved westward because of border adjustments. It lost its eastern third to the Soviet Union (now Belarus and Ukraine), but gained the southern half of Germany's East Prussia, a Baltic region detached from pre-war Germany; the Soviets annexed the northern half of it, calling it Kaliningrad Oblast (now a detached part of Russia). Poland also acquired the port city of Danzig (now called Gdansk) and a large part of Germany east of the Oder-Neisse Rivers (the city of Wroclaw was formerly Breslau). As a result of these border changes, Poland gave up lands containing oil potash, forests, and farms, but received an expanded Baltic coastline with two established ports, rich farmland, coal deposits, and a modern industrial region. Aided by enormous coal reserves, the economy changed from an agricultural to an industrial base. In the early 1980s, shipyard workers from Gdansk, led by Lech Walesa, formed Solidarity, an anticommunist union. They were supported by a strong Catholic Church. These actions were among the significant early efforts by citizens of Eastern Bloc nations to rid themselves of communism and Soviet domination.

SLOVAKIA F

Area: 18,860 sq.mi.(48,845 km²). **Population:** 5,400,000. **Capital:** Bratislava, 447,000. **Government:** Parliamentary democracy. **Language:** Slovak. Religion: Roman Catholic. **Exports:** Machinery, chemicals, foods. **Climate:** Continental. ☐ See Czech Republic for a description on how Slovakia came to be a nation in 1993. Slovakians, who have cultural links to Hungary and Russia, are enjoying true independence for the first time in their history. Bratislava, Europe's newest capital, is the only large city in this previously rural nation.

SWITZERLAND G

Area: 15,941 sq.mi.(41,287 km²). **Population:** 7,260,000. **Capital:** Bern, 200,000. **Government:** Republic. **Language:** German 70%; French 20%; Italian 9%; many dialects. **Religion:** Protestant 52%; Roman Catholic 45%. **Exports:** Engineering and chemical products, scientific instruments, pharmaceuticals, chocolate, watches, cheese. **Climate:** Temperature varies according to altitude. ☐ "Switzerland" is synonymous with "scenic," "skillful," "stable," and a standard of living that is the highest in Europe. The majestic Alps, which cover the southern half of the country, are a year-round tourist attraction. The pyramid-shaped Matterhorn (14,690 ft., 4,479 m) is one of the world's most striking peaks. The Alps are Europe's watershed: the Rhine flows to the North Sea, the Rhone to the Mediterranean, the Ticino (via the Po) to the Adriatic, and the Inn (via the Danube) to the Black Sea. Basel, on the Rhine, is the major port city. The offical language is German, but a majority speak the dialect called Swiss German. Political stability (185 years of neutrality) has made Swiss banks a favorite depository for foreigners. In the 1990s, the banking system was rocked by disclosures of World War II deposits, made by the Nazis, of the wealth taken from their victims. Swiss reluctance to acknowledge this collaboration or admit to the existence of substantial deposits made by Jews during that period only intensified the attack on the Swiss reputation of strict political and financial neutrality. The Swiss obsession with neutrality has prevented the country from wanting to join the United Nations, for fear of having to take political positions.

LIECHTENSTEIN. South of Lake Constance, on the east bank of the Rhine and wedged between Switzerland and Austria, is a tiny (61 sq.mi, 158 km²) country of 31,200 German-speaking citizens. Liechtenstein relies on Switzerland to perform many services: the issuance of currency, mail delivery, telephone service, etc. The nation depends on tourists, light industry, the sale of postage stamps, and the residency of hundreds of corporations seeking a tax haven.

UKRAINE

MOLDOVA

Odessa

Prut R.

Iaşi

Siret R.

Mouths of the Danube

Galaţi

Constanţa

BLACK SEA *

Danube R.

CARPATHIAN MTS.

Moldoveanu ▲ 8,343 ft (2,543 m)

Ploieşti

□ Bucharest

Ialomiţa R.

Ruse

Varna

Burgas

Bosporus

Istanbul

Sea of Marmara

TURKEY

ASIA *

Cluj

TRANSYLVANIAN ALPS

Danube R.

Olt R.

BALKAN MTS.

Stara Zagora

Plovdiv

Marisa R.

RHODOPE MTS.

Musalla Mt. ▲ 9,596 ft (2,925 m)

Kavalla

Dardanelles

Aegean Islands

Timişoara

Iskâr R.

□ Sofia

Struma R.

Strimon R.

Salonika

AEGEAN SEA *

Athens

Piraeus

Tisza R.

Danube R.

Morava R.

VOJVODINA

YUGOSLAVIA

Belgrade

SERBIA

Sava R.

KOSOVO

Skopje

MACEDONIA

Vardar R.

Olympus ▲ 9,570 ft (2,198 m)

PINDUS MTS.

Volos

Lárisa

Gulf of Corinth

Patras

PELOPONNESUS

Shkodër

Tirana

Durrës

Vlorë

CRETE

Iráklion

SEA OF CRETE

Rhodes

SLOVAKIA

HUNGARY

Budapest

Danube R.

MONTENEGRO

BOSNIA & HERZEGOVINA

Sarajevo

Strait of Otranto

Ionian Is.

IONIAN SEA *

SEA

CROATIA

Zagreb

Ljubljana

SLOVENIA

Triglav 9,393 ft (2,865 m) ▲

Split

Dubrovnik

DALMATIAN COAST

ADRIATIC SEA *

Bari

Gulf of Taranto

Reggio di Calabria

Messina

Strait of Messina

Catania

AUSTRIA

LIECHTENSTEIN

Venice

Trieste

Gulf of Venice

SAN MARINO

APENNINES

Foggia

Naples

Vesuvius ▲ 4,190 ft (1,277 m)

TYRRHENIAN SEA *

Palermo

SICILY

Etna ▲ 10,902 ft (3,324 m)

MALTA

GERMANY

SWITZERLAND

ALPS

Verona

L. Garda

PO VALLEY

Bologna

Florence

Arno R.

Tiber R.

Rome

□ VATICAN CITY

Inn R.

Rhine R.

Ticino R.

L. Lugano

L. Maggiore

L. Como

Milan

Ticino R.

Po R.

Genoa

Gulf of Genoa

LIGURIAN SEA

Turin

Rhône R.

FRANCE

Nice

MONACO

Elba

CORSICA (France)

SARDINIA

Cagliari

MEDITERRANEAN

SEA *

TUNISIA

Tunis

AFRICA *

0   100   200   300 miles
0   480 km

45° N

35° N

Russia

Belarus

Poland

Germany

Czech R.

Austria

Slov.

Hun.

Ukraine

Mol.

Turkey

Black Sea

Syria

Jordan

Egypt

Libya

Algeria

Tunisia

Mediterranean Sea

Spain

France

Switz.

Bel. Lux.

UK

Neth.

EUROPE: SOUTHEASTERN I

The coastal areas of southern Europe enjoy a Mediterranean climate of dry, hot summers and mild, moderately wet winters. These lands were once thickly forested, but centuries of logging, fires, overgrazing, and limited rainfall have stripped the landscape of almost all native trees. Though the land is only marginally fertile, with irrigation it can grow grapes, olives, figs, dates, citrus fruits, and chestnuts.

ITALY

Area: 116,237 sq.mi.(301,054 km²). **Population:** 56,800,000. **Capital:** Rome, 2,850,000. **Government:** Republic. **Language:** Italian; many dialects. **Religion:** Roman Catholic. **Exports:** Engineering and food products, autos, plastics, wine, silk. **Climate:** Mediterranean except in the north. □ Like Greece, Italy offers the visitor the antiquity of a great classical civilization (the Roman Empire) in a scenic setting. In addition, the landscape, climate, and waters around southern Italy very much resemble Greece. In the north one encounters further attractions: great art, architecture, museums, opera, and music festivals, and in the far north there is the scenery of the Alps and Alpine lakes. The Po Valley, in the north, is the richest industrial and agricultural region in all of southern Europe; Italy has become the world's fifth-largest industrial power. Milan is Italy's industrial and financial center. Turin is the automobile capital, and Genoa is the busiest seaport. The wealthy northern region stands in stark contrast to the widespread poverty of the south. Venice, on the northern shore of the Adriatic, is like no other city; canals substitute for streets, great art and sumptuous architecture abound. Centrally located Florence is the nation's premier city of magnificent museums and Renaissance art. Close to the Swiss border are the beautiful lake regions of the Italian Alps. Near the Tyrrhenian coast is the capital, Rome, the "Eternal City," the site of the ruins of the capital of the Roman Empire. Most of Rome's great art and architecture was commissioned by the Catholic Church. The world headquarters for the Church is Vatican City, an independent nation located within Rome's city limits. The only major city and industrial center in southern Italy is Naples. Located in an especially beautiful setting, the city is famous for the foot of Mount Vesuvius. This still-active volcano is famous for the eruption that buried Pompeii and two other Roman cities. Only 7 mi.(11.2 km) from the northern shore of the Adriatic, is the site of a 2 mi.(3.2 km) from the toe of the Italian boot is Sicily, the largest island in the Mediterranean. Sicily is home to the Mafia, a organization connected with criminal activities in Italy and abroad. In the mountains of Sicily is Europe's tallest active volcano, Mount Etna (10,902 ft., 3,324 m). The mountains are an extension of the Apennines, the chain that runs the length of the Italian peninsula. The Strait of Messina, which separates the island from the mainland, is probably a mountain pass that has sunk below sea level. Westward across the Tyrrhenian Sea is Sardinia, the Mediterranean's second-largest island. Its culture bears the imprint of numerous invasions. To the north, lying between the island of Corsica (France) and the Italian mainland, is the tiny island of Elba, the site of Napoleon's exile.

VATICAN CITY. Located within Rome, is the smallest nation in the world. It is no larger than a few city blocks and has a population of 1,000. The Vatican holds the enormous church of St. Peter's and the Vatican Palace, the home of the Pope. It is the site of a glorious art collection that includes the recently restored ceiling of the Sistine Chapel, painted by Michelangelo.

SAN MARINO. Once a city-state in medieval Italy, San Marino, with a population of 24,900, is the world's oldest and smallest (24 sq.mi., 61 km²) republic. It lies nestled in the Apennines within sight of the Adriatic Sea. Like other postage stamp-sized countries, it derives revenue from selling stamps.

MALTA. About 375,000 people live on the three small islands (120 sq.mi., 310 km²) south of Sicily that form this nation. Through the centuries, Malta, the largest island, has been occupied by numerous invaders. Britain, the last nation to control Malta, made it a naval base, which is now a busy shipyard.

BULGARIA c

Area: 42,823 sq.mi.(110,912 km²). **Population:** 8,250,000. **Capital:** Sofia, 1,120,000. **Government:** Republic. **Language:** Bulgarian. **Religion:** Eastern Orthodox. **Exports:** Food products, grains, metals, textiles, high technology. **Climate:** From continental to Mediterranean. □ Even as it shifts to an industrial economy, Bulgaria continues to be the Balkan Peninsula's leading agricultural nation and eastern Europe's winter breadbasket. Bulgaria is the world's largest supplier of attar of roses, an oil used in the manufacture of perfume. Byzantine-style architecture, cultural artifacts, and the consumption of yogurt (solidified, fermented milk) are reminders of 500 years of Turkish rule. In Bulgaria, unlike Russians to drive the Turks out, as they did in Bulgaria in 1878. Consequently, the majority of Bulgarians are Christians, followers of the Eastern Orthodox Church. Memory of the hated Turkish rule is so deeply embedded in the national conciousness that in 1984 the government forced the Turkish and Muslim minorities (1.5 million people) to Bulgarize their names and refrain from speakng Turkish in public. This policy was reversed in 1990.

GREECE d

Area: 50,943 sq.mi.(131,840 km²). **Population:** 10,700,000. **Capital:** Athens, 3,000,000. **Government:** Republic. **Language:** Greek. **Religion:** Eastern Orthodox. **Exports:** Food products, olives, wine, citrus fruits. **Climate:** Mediterranean. □ Over 400 sun-drenched islands, deep-blue skies, an unrivaled historic past, good food, music, and dance all are part of Greece's attraction. Almost all the islands in the Aegean archipelago (including most of those off the Turkish coast) belong to Greece. Crete, the largest island, was the site of a highly advanced Minoan civilization that flourished 5,000 years ago. Most of Greece is a ragged peninsula whose southern tip, the Peloponnesus, is separated from the mainland by the Gulf of Corinth and a canal. The Peloponnesus was the site of the Persian Wars, and the battles between Athens and Sparta. The Greek maritime tradition (the oldest in Europe) dates back to that period. Today, Greece operates the supertankers of the world's largest merchant fleet. The Acropolis ("high city"), the heart of ancient Greece, is located on a hill in the middle of Athens. These ruins of structures built 2,500 years ago represent the birthplace of western culture and its democratic ideals. Although modern Greece is rapidly becoming industrialized, agriculture remains a significant part of the economy. Wheat, cotton, and tobacco are grown in addition to the usual Mediterranean crops (olives, grapes, dates, and citrus).

BALKAN COUNTRIES f

Albania, Bulgaria, Greece, Romania, and the five nations of the former Yugoslavia (shown on the next plate) occupy the Balkan Peninsula. The word "balkan" means "mountain" in Turkish. Agriculture is not very productive because of the rugged environment and outmoded methods of farming; in many rural villages, donkeys are still the principal form of transportation. Since World War I, Greece has been the only non-communist country in the region, but in 1989, the revolutionary fervor of Eastern Europe reached the Balkans and communism was under attack in every state but Albania. The generally depressed economies, following the transition to capitalism, have only reinforced the Balkan Peninsula's reputation for being "Europe's poorest corner." The Balkans have also been an area of intense ethnic and religious hatred, both within and between nations. These attitudes were exacerbated during the post-World War I upheaval, when national boundaries were significantly altered and minority populations were trapped within hostile countries. Years of communist control only concealed the underlying problems, which erupted when the nations were liberated. Nowhere were the consequences more tragic than among the nations of the former Yugoslavia.

ALBANIA b

Area: 11,100 sq.mi.(28,748 km²). **Population:** 3,330,000. **Capital:** Tirana, 290,000. **Government:** Republic. **Language:** Albanian. **Religion:** Muslim 70%. **Exports:** Food products, wine, minerals. **Climate:** Mediterranean along the coast; continental inland. □ Albania is the smallest Balkan country in size, population, and per capita income. For over 40 years, this small mountainous nation has had the most repressive of all communist governments; it had severed relations with other communist nations because of their "liberal" policies, thus isolating itself from both the democratic and the communist worlds. In 1967, all mosques and churches were closed and the country became the world's first officially atheistic state. Until then Albania had been the only European nation with a Muslim majority—the result of 500 years of Turkish rule. In the 1990s, Albania became the last European country to throw out its communist government. But conditions remain severe, and there has been considerable emigration to Italy, across the Strait of Otranto.

ROMANIA e

Area: 91,699 sq.mi.(237,500 km²). **Population:** 22,400,000. **Capital:** Bucharest, 2,350,000. **Government:** Republic. **Language:** Romanian. **Religion:** Eastern Orthodox. **Exports:** Engineering, chemical, food products, oil. **Climate:** Continental. □ Romanians trace their ancestry and their Latin-derived language back nearly 2,000 years to the Roman occupation. The nation's oil reserves were once the largest in Europe—they fueled the Nazi war machine, but now appear to be nearing depletion. The economy began to decline in the 1980s because of the repressive and destructive policies of its communist leader, Nicolae Ceausescu. In 1989, the government was overthrown and the hated dictator was executed. Before World War II (when life was brighter) the capital, Bucharest, was referred to as the "Paris of the Balkans"—many of its neighborhoods were designed after the French capital. Life in modern Romania has been hardest for over 2 million persecuted Hungarian Magyars, the largest ethnic minority in Europe. Their ancestors were trapped by boundary changes after World War I. Many live in the agricultural and recreational region of Transylvania (home of the fictitious Count Dracula) in the western part of the country. Much of Romania is dominated by the Carpathian Mountains, which trace a wide arc running north to south. To the east, where the Danube reaches the Black Sea, is a huge delta, home to 300 bird species.

EUROPE: SOUTHEASTERN II

NATIONS OF THE FORMER YUGOSLAVIA

When so many diverse and antagonistic cultures agreed to be united in the creation of the former Yugoslavia, in 1918, one can only wonder what was their motivation. Consider that this new nation would bring together five major nationalities (including at least twenty minorities), speaking three "official" languages, using two different alphabets , and practicing three separate religions. What chance of success would any country have, facing those obstacles? The answer, obviously, is "none," but it took over 70 years to undo the mistake. Even though five separate nations now stand in Yugoslavia's place, the error of its creation in 1918 continues to have tragic consequences.

Yugoslavia means "land of the southern Slavs," and most of the ethnically diverse citizens of the former country share the same Slavic heritage. Around the 5th century A.D., Slavs from central and eastern Europe began migrating south into territory controlled by the declining Roman Empire. Slavs settling in the western region would eventually establish the states of Croatia, Slovenia, and parts of Bosnia-Herzegovina. They used the Roman (Latin) alphabet and adopted Roman Catholicism as their religion. Slavs migrating to the eastern portion would eventually become today's Serbians, Montenegrans, and Macedonians; they used the Cyrillic alphabet and practiced the Eastern Orthodox religion (a branch of Christianity that split from the Roman Catholic Church in the 11th century).

Over succeeding centuries, these lands were subject to foreign domination: the Austro-Hungarian empire controlled Croatia, Slovenia, and part of Bosnia, while the eastern region and the rest of Bosnia were under the domination of the Ottoman Turkish Empire, whose nearly 500-year-old rule lasted until the second half of the 19th century. In 1914, World War I broke out when Archduke Franz Ferdinand, heir to the Austro-Hungarian Empire, was assassinated by a Bosnian-Serb nationalist in Sarajevo, the capital of Bosnia. The collapse of both empires by the end of the war liberated the remaining Balkan nations under their control (Serbs and Montenegrans had won their freedom 40 years earlier). In the year of the armistice, 1918, the peoples of these free nations, filled with Slavic pride and believing in strength through unity, formed a constitutional monarchy called the "Kingdom of the Serbs, Croats, and Slovenes." In 1929, the Serbian monarch abolished the constitution, changed the name of the country to Yugoslavia, and began an unpopular dictatorial rule. During this period leading up to World War II, tensions grew as the Croats and Slovenes, resentful of Serbian rule, pressed for autonomy.

The Croatian hatred of Serbia paved the way for the invasion of Yugoslavia by Germany and its allies (the Axis nations) in 1941. The Nazis allowed Croatia to remain independent because of its friendly, fascist government. They also placed Bosnia and Herzegovina under Croatian control. During the war, Croats committed the first instance of "ethnic cleansing" in Yugoslavia: Rounding up thousands of Jews and Serbs for execution. Recent Serbian atrocities (at least the ones committed against the Croats) may have been in retaliation for events of that period. Elsewhere, bands of resistance fighters formed to fight the Axis invaders. The most successful group, the Partisans, was commanded by Marshal Tito, a Croatian communist. In 1945, Tito, a national hero, took control of the entire country and created a communist government.

Tito kept a tight lid on ethnic strife. His brand of communism allowed workers to run the collective farms and factories (small farmers were entirely exempted from government control), and the individual republics were permitted self-rule. The country was opened to world trade and tourism.Tito remained independent of Moscow—unlike the nations of the eastern European Communist Bloc. Tito's independence enabled him to receive Western economic assistance, making Yugoslavia the industrial leader of the communist world. After Tito's death in 1980, ancient ethnic rivalries and animosities, held in check under his rule, erupted, making effective government impossible.

In 1991, Croatia and Slovenia declared independence from Serbia. Fighting broke out as Serbia, which still considers itself to be the former Yugoslavia, sent troops to aid Serbian enclaves within Croatia and Slovenia. Fighting in Slovenia lasted only 10 days, but the UN had to monitor a ceasefire in Croatia, which had a larger Serbian population. In 1997, the Eastern Slavonian region of Croatia, held by Serbia, was returned to Croatia.

When Bosnia and Herzegovina declared independence in 1992, a full-scale war erupted between Bosnia's three ethnic groups: Roman Catholic Croats, Eastern Orthodox Serbs, and Muslims. The Bosnian Serbs, with military aid from Serbian Yugoslavia, embarked on a genocidal program of "ethnic cleansing," in which they attempted to kill or drive Muslims out of Bosnia. Western nations, outraged by Serbian aggression, imposed severe economic sanctions against Yugoslavia to discourage further military aid to Bosnian Serbs. It is estimated that 250,000 Bosnians had been killed and over two million rendered homeless by the time a ceasefire was put in place in 1995 by a NATO-led force. No longer having the military support of Yugoslavia, the president of the Bosnian Serbs was forced to join with the Croatian and Muslim presidents in Dayton, Ohio (USA), to sign a peace treaty that divided Bosnia and Herzegovina into two autonomous regions: a Muslim-Croat federation encompassing 51% of the country and a Serbian republic with 49%. (see map insert). NATO deployed 60,000 troops (20,000 of which were American) to keep the peace. After two years, the joint government is still unable to govern, and it remains dangerous for refugees to return to homes in enemy-controlled areas.

With the four other republics declaring independence and moving toward democracy and free-market economies, Serbia and Montenegro are all that remain of the original Communist Yugoslavian Federation. These former republics regard themselves as a continuation of the old Yugoslavia and have kept its name. Hence the names "Yugoslavia," "Serbia," and "Serbia and Montenegro" all refer to the same country.

The breakup of the original Yugoslavia and the ensuing strife have had a disastrous effect on the economies of the new nations. Under the prior communist government, each republic was economically dependent on the others: Raw materials, manufactured goods, and agricultural products came from different republics.

BOSNIA & HERZEGOVINA A

Area: 19,734 sq.mi. (51,233 km^2). **Population:** 3,365,000. **Capital:** Sarajevo, 525,000. **Government:** Republic. **Language:** Serbo-Croatian. **Religion:** Muslim (40%), Orthodox (32%), Roman Catholic (15%). **Exports:** Machinery, manufactured goods, food products. **Climate:** Hot summers and cold winters; Mediterranean on the coast. ☐ The Turkish invaders of the 15th century joined the mountainous and forested Bosnia (Bahz' nee uh) in the north with the rocky landscape of Herzegovina (Hurt suh go' vee nuh) in the south. Throughout the country one can see minarets, domes, mosques, and covered bazaars—evidence of over 400 years of Turkish rule. Sarajevo, the nation's capital, is a blend of East and West, while Mostar, the capital of the Herzegovina region, is more Turkish in appearance. The Turks introduced Islam, accounting for a large Muslim population which has been severely reduced by Serbian atrocities and refugee flight. Bosnia was one of the most backward and poorest republics under the old federation.

CROATIA B

Area: 21,829 sq.mi. (56,538 km^2). **Population:** 4,675,000. **Capital:** Zagreb, 725,000. **Government:** Parliamentary Democracy. **Language:** Serbo-Croatian. **Religion:** Roman Catholic (77%), Eastern Orthodox (11%). **Exports:** Machinery, transportation equipment, manufactured goods, chemicals, textiles, and food stuffs. **Climate:** Continental inland, Mediterranean on the coast. ☐ Prior to the dissolution of Yugoslavia, Croatia (Kroh ay' shuh) was the second most prosperous and industrialized of the nation's republics, but recent fighting has discouraged modernization of industry and foreign investment. Only recently was a third of Croatian territory (in the East), returned to Croatian control by Serb separatists. Croatia is also burdened with thousands of Croatian and Muslim refugees fleeing Bosnia. Croatia's Dalmatia region is a world-famous, 1,100-mi. (1,760-km.) stretch of recreational coastline and islands. The other major tourist coastal attraction is the Venetian-influenced Istria region in the northwestern tip of Croatia. Croatia is one of Europe's oldest states, established as a Catholic nation in 924.

MACEDONIA C

Area: 9,928 sq.mi. (24,856 km^2). **Population:** 2,210,000. **Capital:** Skopje, 450,000. **Government:** Democracy. **Language:** Macedonian (70%), Albanian (21%). **Religion:** Eastern Orthodox (67%), Muslim (30%). **Exports:** Manufactured goods, machinery, textiles, tobacco, food products. **Climate:** Continental. ☐ Macedonia (Mass uh doh' nee uh), the "magical country of archaeology," has been the home of numerous civilizations since the Stone Age. During the 4th century B.C., it was the northern part of greater Macedonia (which included what is now northern Greece and part of Bulgaria) and home of Alexander the Great. Before the Turks arrived in 1371, Macedonia had been occupied many times. In 1913, after the Balkan Wars, Macedonia was divided nearly equally between Serbia and Greece—with Bulgaria receiving a tiny share. Macedonia and Serbia later became part of Yugoslavia, and when Tito took control in 1945, he recognized the Macedonians as a separate people (with their own language) and designated their country a Yugoslavian republic. Macedonia holds the fortunate distinction of being the only former Yugoslavian republic to have achieved a non-violent independence. Greece has challenged Macedonia's use of the same name as Greece's famous northern province. Until a final ruling, the interim name is "The Former Yugoslav Republic of Macedonia."

SLOVENIA D

Area: 7,836 sq. mi. (20,296 km^2). **Population:** 1,975,000. **Capital:** Ljubljana. **Government:** Democracy. **Language:** Slovenian. **Religion:** Roman Catholic. **Exports:** Machinery and transport equipment, manufactured goods, chemicals. **Climate:** Mediterranean on coast, continental in the eastern half. ☐ By far the most prosperous republic of the former Yugoslavia, Slovenia (Sloh vee' nee uh) has strong economic ties with Western Europe. The absence of ethnic diversity (most citizens are Slovenes) made Slovenia a lesser target for Yugoslavian (Serbian) intervention and aided the peaceful move toward independence. The most European of the former republics, Slovenia has applied for admission into the European Union. Its Adriatic port of Koper is an important conduit for trade from the landlocked nations of Austria and Hungary on Slovenia's northern border. The nation's rugged terrain contains rich deposits of coal, mercury, lead, and zinc. The Julian Alps, in the far north, is the former Yugoslavia's most popular skiing area.

YUGOSLAVIA E

Area: 39,518 sq.mi. (102,350 km^2). **Population:** 11,200,000. **Capital:** Belgrade,1,300,000. **Government:** Republic. **Language:** Serbo-Croatian. **Religion:** Eastern Orthodox (65%), Muslim (19%). **Exports:** Machinery and transport equipment, manufactured goods, chemicals, and food products. **Climate:** Continental inland, Mediterranean near coast. ☐ Serbia (Sur' bee uh) and Montenegro (Mahn' tuh nay groh) are all that remain from the original Yugoslavian Federation. The breakup created a severe strain on Serbia, whose economy was interconnected with the seceding republics—Croatia and Slovenia were the industrial centers for the old Yugoslavia, and Serbia was the nation's "breadbasket." Former communists, who run the country, refuse to allow dissent from traditional socialistic policies. Yugoslavia must also support over 200,000 Serbian refugees expelled from Croatia. Economic conditions are even worse in Montenegro, which blames Serbia for its problems, prompting rumors of secession. Montenegro, Yugoslavia's window to the Adriatic, is part of the popular Dalmatian Coast. In the interior region of Montenegro are the Black Mountains, whose name in Italian (a Venetian influence) is "Montenegro." Another stress point is Kosovo, a province along Yugoslavia's Albanian border. It has an Albanian Muslim majority seeking unification with Albania.

25

SWEDEN

FINLAND

Lake Onega

Lake Ladoga

BALTIC SEA

Åland (Fin)

GULF OF FINLAND

St. Petersburg

A
Tallinn

Pärnu

Gulf of Riga

Tartu

Lake Peipus

D

Gotland (Swe)

B

B

Riga

Daugava R.

Pskov

Liepāja

C

Daugavpils

B

E

D

Šiauliai

C

Vitsyebsk

Dnepr R.

D

Kaliningrad

Vilnius

C

Smolensk

Minsk

Orsha

POLAND

E

Dnyapro R.

Baranarichy

PINSK MARSHES

Brest

Prypyat R.

Mazyr

Homyel'

Bryansk

Vistula R.

Warsaw

Desna R.

G

Rivne

Chernobyl

Kiev Reservoir

Sumy

D

Kraków

Kiev

Kharkiv

L'viv

Dnipro R.

SLOVAKIA

Dnister R.

Vinnytsya

Kramanchug Reservoir

Luhans'k

CARPATHIAN MOUNTAINS

Miskol

HUNGARY

Dnipropetrovs'k

Donets'k

Kryvyy Rih

Zaporizhzhya

F

Bălți

Dniester R.

Mykolayiv

Mariupol'

Rostov

ROMANIA

Prut R.

Chișinău

F

G

Odesa

G

SEA OF AZOV

Kerch

CRIMEA

Simferopol'

G

Danube R.

Mouths of the Danube

BLACK SEA

Sevastopol'

Yalta

D

0   100   200   300 miles
480 km

Atlantic Ocean

Norway

Sweden

Finland

North Sea

Den

Balt

Ireland

U.K.

Neth

Germany

Poland

Russia

France

Bel

Lux

Czech R.

Slo

Austria

Hun

Ukraine

Switz

Slov

Cro

Romania

Portugal

Spain

Italy

B &H

Yugo

Bulgaria

Mac

Black Sea

Georgia

Greece

Turkey

Arm.

Mediterranean Sea

Alb

Leb

Syria

Iraq

Morocco

Algeria

Tunisia

Libya

Egypt

Jordan

Saudi Arabia

Ist

Washington

A

B

C

Kentucky

Tennessee

E

North Dakota

Minnesota

G

South Dakota

Massachusetts

F

Connecticut

EUROPE: EASTERN I.

FORMER SOVIET BALTIC REPUBLICS

In 1940, at the outset of World War II, the Soviet Union, under a pact signed with Germany, seized and annexed the three Baltic Sea nations of Estonia, Latvia, and Lithuania. They were the last three republics to become part of the Soviet Union. Estonia, Latvia, and Lithuania, which share a common history, had become independent nations only 20 years earlier, after the close of World War I. Within a year, the Nazis violated their agreement by occupying the Baltic States as the first step in the invasion of the Soviet Union. In 1944, the Soviets faced fierce local resistance when they recaptured the three republics. Thousands of patriots were killed, and many leading politicians, businessmen, and intellectuals were forcibly relocated to remote parts of the Soviet Union. They were replaced by a wave of Russian immigrants assigned to high positions in government, industry, and education. The Soviets prized these states for their access to the Baltic Sea, for the security they offered as a buffer between Russia and the rest of Europe, and for the industrial economies of Estonia and Latvia, which enjoyed the highest standard of living in the Soviet Union.

When the Soviets began to loosen controls in the late 1980's, Estonia, Latvia, and Lithuania were the first republics to bolt the Union. Since gaining independence in 1991, they have been seeking an economic and military alliance with the West—a goal consistent with their refusal to join the Commonwealth of Independent States, the alliance of former Soviet republics. The final withdrawal of Russian troops from Baltic territories left behind many Russian settlers and their descendants—about a third of the population in Estonia and Latvia, but less than 9% in Lithuania. No longer in positions of power, and generally unfamiliar with the local languages, the Russians in Estonia and Latvia are facing severe discrimination from hostile populations. Many of them have had to leave the only home they have ever known.

The terrain of these nations is fairly similar: flat and marshy near the coast, and gradually becoming hilly and wooded along the Russian borders. The coastline has been a popular recreation area, but rampant industrial pollution has forced the closure of some resorts. The proximity to the Baltic keeps the climate cool in the summer and moderately cold in the winter, with average rainfall.

ESTONIA A
Area: 17,410 sq. mi. (45,100 km²). **Population:** 1,425,000. **Capital:** Tallinn, 450,000. **Government:** Republic. **Language:** Estonian, Russian. **Religion:** Lutheran, Russian Orthodox. **Exports:** Textiles, vehicles, machinery, chemicals. **Climate:** Temperate. □ The most northerly of the Baltic nations, Estonia (Eh stoh' nee uh), is virtually a peninsula, bounded by the Gulf of Finland to the north, the Baltic Sea to the west, the Gulf of Riga to the southwest, and Lake Peipus along most of its eastern border with Russia. Hundreds of islands dot its coastline. Although the nation shares a common history with Latvia and Lithuania—at various times, they were ruled by Denmark, Germany, Sweden, Poland, and Russia—the highly literate people of Estonia have a culture and language that have much more in common with Finland. The Finns are especially interested in Estonia's coastal and historic tourist attractions. Estonia's post-communist, free-market economy, powered by a rich, native supply of shale oil, has made it the most prosperous and most economically stable of the former Soviet republics.

LATVIA B
Area: 24,750 sq.mi. (64,100 km²). **Population:** 2,400,000. **Capital:** Riga, 875,000. **Government:** Republic. **Language:** Lettish, Russian. **Religion:** Lutheran, Russian Orthodox. **Exports:** Oil products, timber, food, textiles. **Climate:** Moderate. □ Latvians are also called "Letts," which comes from "Lettish," the name of the Latvian language. Lettish, one of Europe's oldest tongues, can trace its roots to Sanscrit, the language of ancient India. Latvia's industrialized economy ships most of its exports from its capital, Riga, located at the mouth of the Dvina River, and one of the busiest ports on the Baltic Sea. The popular coastal areas of Latvia have been hit the hardest by industrial pollution.

LITHUANIA C
Area: 25,170 sq. mi. (65,200 km²). **Population:** 3,600,000. **Capital:** Vilnius, 600,000. **Government:** Republic. **Religion:** Roman Catholic. **Exports:** Electronics, oil products, chemicals, food. **Climate:** Moderate. □ Even though remaining Russians constitute less than 10% of the population, former communist leaders have managed to hold on to key positions in government by winning important elections. This has resulted in better relations with Russia than either Estonia or Latvia has. Lithuania was mostly an agricultural country before the Soviets began industrializing it, and agriculture remains an important part of the economy. In spite of their communist backgrounds, Lithuanian leaders have taken the nation in the direction of privatization and the free market. Vilnius, the capital, has a particularly impressive historic district, which survived World War II. The same can't be said of one of Europe's largest prewar Jewish communities (nearly 8% of the nation's population), which perished under the Nazi occupation.

RUSSIA D
KALININGRAD OBLAST (RUSSIA). This tiny region, separated from the Russian Federation motherland, was formerly German East Prussia. Following its destruction and capture by the Soviets at the close of World War II, the oblast's surviving Germans were almost entirely replaced by Russians. There is currently little evidence of the historic German culture—Prussia was the seat of German militarism—that prevailed for centuries. Kaliningrad, the capital, is home to half of the oblast's one million residents, most of whom observe neither a Russian nor a German heritage. Trade and tourism have been attracted to Kaliningrad Oblast's beautiful and fertile landscape located on the clement Baltic coast.

FORMER SOVIET EASTERN EUROPEAN REPUBLICS

Belarus, Moldova, and Ukraine have little in common (past or present) with the Baltic republics, and very little—other than religion—with each other. But even their religion represents separate branches of the Eastern Orthodox Church—itself a branch of Christianity. Unlike the Baltic States, they have maintained ties with Russia and the other former republics by forming the Commonwealth of Independent States. Since Russian is no longer the official language in these nations (or anywhere outside of Russia), only Russian minorities speak it.

BELARUS E
Area: 80,150 sq. mi. (207,600 km²). **Population:** 10,410,000. **Capital:** Minsk, 1,700,000. **Government:** Constitutional Republic. **Language:** Belarussian, Russian. **Religion:** Eastern Orthodox. **Exports:** Machinery, transportation equipment, chemicals, foods. **Climate:** Continental extremes. □ Belarus (Behl uh roos') has been previously called Belorussia, Byelorussia, and White Russia. The western region (historically part of Belarus) was taken from Poland and added to the Soviet republic following World War II. Belarus was slowly privatizing its economy, after gaining independence, when Alexander Lukashenko was elected President in 1994. Since then, he has made every effort to turn the clock back to Soviet-style, repressive socialism. In 1997, Belarus and Russia signed an agreement providing for economic and military unification, but short of creating a single state. Belarus has been devastated not only by failed socialist policies but by the cost of an almost endless environmental cleanup and the treatment of a growing number of cancer patients following the 1986 nuclear explosion in Chernobyl. Prevailing winds blew the bulk of the radioactive debris into the southern region of Belarus, close to the Ukrainian power plant.

MOLDOVA F
Area: 13,010 sq. mi. (33,700 km²). **Population:** 4,460,000. **Capital:** Chisinau (Kishinev), 750,000. **Government:** Republic. **Language:** Romanian. **Religion:** Eastern Orthodox. **Exports:** Foodstuffs, wine, tobacco, and textiles. **Climate:** Warm summers and moderate winters. □ Moldova (Mahl doh' vah) was the second smallest Soviet republic (after Armenia), the most densely populated, and the most agriculturally productive (for its size). The land is ideally suited for agriculture: it has a fertile, rolling terrain, mild weather, and adequate rainfall. Lacking energy resources, Moldova's modest industrial programs were initially hurt when energy supplies from Russia were cut off after the breakup of the Soviet Union. A recent agreement with Russia provides for renewed shipments of natural gas. It is only because of twenty-five miles (40 km) of Ukrainian territory standing between Moldova and the Black Sea that the nation cannot deal directly with Middle Eastern suppliers of energy resources. There are plans to build a port on the Prut River (the boundary with Romania), which flows into the Black Sea. Moldova, which shares a common language and culture with Romania, was actually Romania's eastern region before it was taken into the Soviet Union as part of the 1940 pact with Nazi Germany. In 1991, sentiment among the Moldovan majority was running high for reunification with Romania. Fearing this possibility, the Russian and Ukrainian minorities carved out the Trans-Dniester republic from a strip of land on the eastern banks of the Dniester River and declared independence. When Moldova renewed its ties with the former Soviet Republics by joining the Commonwealth of Independent States, and in a 1994 plebiscite voted down unification with Romania, Trans-Dniester was asked to return.

UKRAINE G
Area: 233,090 sq.mi. (603,700 km²). **Population:** 50,130,000. **Capital:** Kiev, 2,700,000. **Government:** Republic. **Language:** Ukrainian, Russian. **Religion:** Ukrainian Orthodox. **Exports:** Coal, metals, chemicals, machinery, transportation equipment, and food products. **Climate:** Continental in the North and Mediterranean along the southern coastline. □ When the Soviet Union broke up, it was the loss of Ukraine that hurt Russia the most. Texas-sized Ukraine accounted for over 25% of the former Soviet Union's agricultural and industrial production; it was referred to as the "breadbasket" of the Soviet Union. Ukrainians bitterly resisted the forced collectivization of their farms after the nation was annexed by the Soviet Union in the early 1920's. An estimated 7-10 million Ukrainians died when Josef Stalin, the Soviet dictator, starved them into submission. As a result, in 1940 they welcomed the invading Nazis, but by the end of the war, millions had lost their lives. Following the hostilities, the Soviets expanded Ukrainian borders to include lands from Poland and Romania. In recent years, Ukraine and Russia have been wrangling over control of the Crimean Peninsula, on the Black Sea. Many Russians living in the Crimea and in Ukraine's eastern region have campaigned for unification with Russia. Besides being the most popular resort area in the former Soviet Union, the Crimea is strategically important to Russia. Sevastopol, its major seaport, is home base to Russia's Mediterranean fleet, ownership of which also is in dispute. In 1996, Ukraine transferred the last of 2,000 nuclear warheads to Russia. In return, Ukraine is receiving economic assistance from the International Monetary Fund, and is shifting to a free market economy. There is pressure on Ukraine to shut down the remaining nuclear reactors at Chernobyl, but it lacks the funds needed to entomb the plant, and it desperately needs the energy.

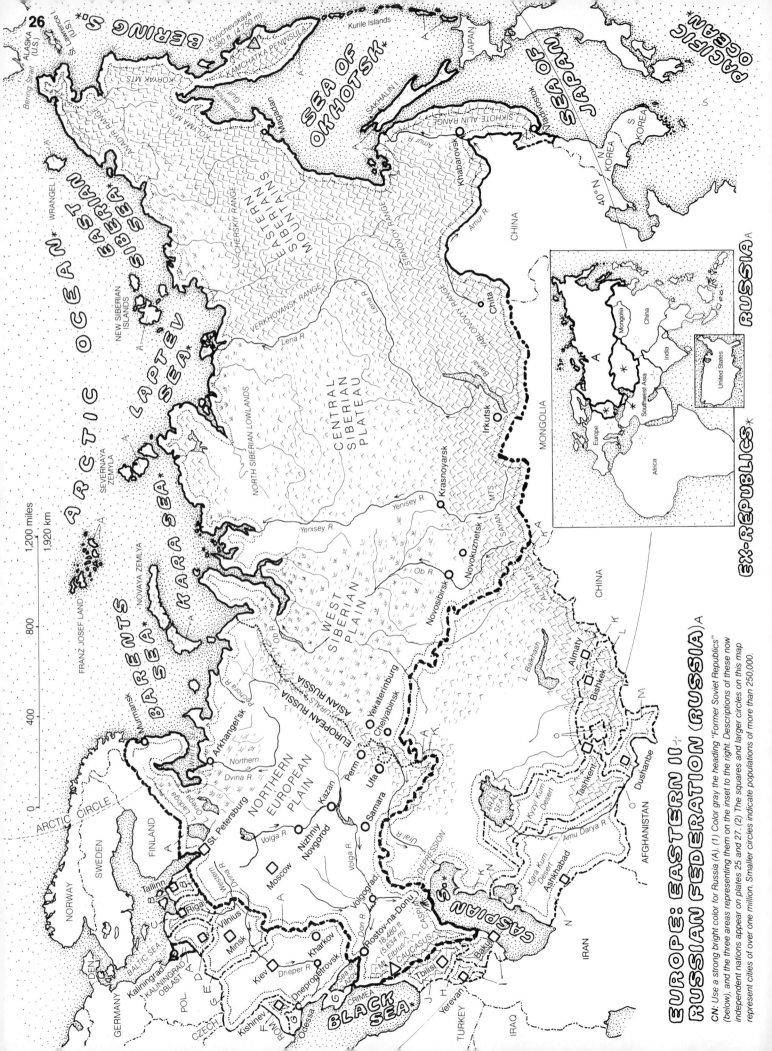

26

EUROPE: EASTERN II ÷
RUSSIAN FEDERATION (RUSSIA) A

CN: Use a strong bright color for Russia (A). (1) Color gray the heading "Former Soviet Republics" (below), and the three areas representing them on the inset to the right. Descriptions of these now independent nations appear on plates 25 and 27. (2) The squares and larger circles on this map represent cities of more than 250,000. Smaller circles indicate populations of over one million.

EX-REPUBLICS *

RUSSIA A

ALASKA (U.S.)

BERING Sr.

Bering Strait

Kluchevskaya
15,580 ft (4,750 m)

KAMCHATKA PENINSULA *

Kurile Islands

KORYAK MTS

KOLYMA MTS

CHERSKY RANGE

EASTERN SIBERIAN SUM

SEA OF OKHOTSK *

SAKHALIN I.

JAPAN

SEA OF JAPAN *

PACIFIC OCEAN *

KORES S.

KOREA N.

SIKHOTE ALIN RANGE

Vladivostok

40° N

ARCTIC OCEAN *

WRANGEL I.

NEW SIBERIAN ISLANDS

LAPTEV SEA *

VERKHOYANSK RANGE

Lena R.

STANOVOY RANGE

Amur R.

Khabarovsk

Amur R.

CHINA

MONGOLIA

SEVERNAYA ZEMLYA

NORTH SIBERIAN LOWLANDS

CENTRAL SIBERIAN PLATEAU

YABLONOVY RANGE

Chita

Baikal

Irkutsk

FRANZ JOSEF LAND *

KARA SEA *

NOVAYA ZEMLYA

Yenisey R.

Yenisey R.

Krasnoyarsk

SAYAN MTS

ARCTIC CIRCLE

BARENTS SEA *

WEST SIBERIAN PLAIN

Ob R.

Novokuznetsk

Novosibirsk

ALTAI

K

CHINA

Murmansk

WHITE SEA

Arkhangel'sk

Northern Dvina R.

Pechora R.

URALS

ASIAN RUSSIA
EUROPEAN RUSSIA

Yekaterinburg

Chelyabinsk

Perm

Ufa

L. Balkhash

Almaty

Bishkek

NORWAY

SWEDEN

FINLAND

St. Petersburg

Ladoga

Onega

NORTHERN EUROPEAN PLAIN

Moscow

Volga R.

Nizhniy Novgorod

Kazan

Samara

Volga R.

Ural R.

ARAL SEA

Tashkent

Kyzyl Kum Desert

Dushanbe

AFGHANISTAN

Kara Kum Desert

Amu Darya R.

Ashkhabad

CASPIAN SEA *

IRAN

DEN

GERMANY

Tallinn

Riga

BALTIC SEA

Kaliningrad
KALININGRAD OBLAST

Vilnius

Minsk

POL.

CZECH

Kishinev

ROM.

Kiev

Odesa

Dnieper R.

Dnepropetrovsk

Kharkov

Don R.

Volgograd

Sea of Azov

CRIMEA

Rostov-na-Donu

CAUCASUS

Elbrus *
18,480 ft (5,634 m)

BLACK SEA *

TURKEY

Tbilisi

Baku

Yerevan

IRAQ

1,200 miles
1,920 km

0   400   800

MONGOLIA

China

India

Southwest Asia

Europe

Africa

United States

FORMER SOVIET REPUBLICS*

ESTONIA B
LATVIA C
LITHUANIA D
BELARUS E
MOLDOVA F
UKRAINE G
ARMENIA H
AZERBAIJAN I
GEORGIA J
KAZAKHSTAN K
KYRGYZSTAN L
TAJIKISTAN M
TURKMENISTAN N
UZBEKISTAN O

Area: 6,592,800 sq. mi. (17,075,400 km^2). **Population:** 148,000,000. **Capital:** Moscow, 8,820,000. **Government:** Federation (Republic). **Language:** Russian (official), many ethnic dialects. **Religion:** Russian Orthodox. **Exports:** Oil products, natural gas, timber, metals, and chemicals. **Climate:** Except for a small, almost subtropical region adjacent to the Caspian Sea, most of the country is exposed to continental extremes. Limited rainfall is confined to summer months. Much of northern Russia is subarctic, whereas the far north (polar) is tundra.

☐ Russia (Rush' ah) was the largest of the fifteen republics that made up the former Soviet Union (also known as the USSR, or the Union of Soviet Socialist Republics). The other fourteen republics were once independent nations, or parts of other nations, adjacent to Russia. Most had been seized in the 19th century by the Russian Empire and later annexed by the Soviet Union after the Russian Revolution of 1917. The peoples of these republics are of different ethnic, cultural, and religious backgrounds. The only thing they previously held in common was their hatred of Soviet/communist control and its officially imposed Russian language. In 1990 and 1991, the declining Soviet Union appeared incapable of preventing the secession of one republic after another. By early 1992, the Soviet Union ceased to exist, and in its place were 15 independent nations. The world's largest, and second most powerful nation had disappeared, virtually overnight.

The former Soviet republic of Russia (officially called the Russian Federation) is now itself a federation of 21 republics. Moscow, the largest city and capital of the former Soviet Union, remains the capital. Even within the Russian Federation there are autonomous regions with separate cultures. One such republic, seeking to secede from Russia, is Chechnya, with an area the size of Connecticut, in the Caucasus Mountains of southern Russia. In 1991, it declared its independence, and for two years following the invasion of Russian troops in 1994, it waged a bloody war of resistance. A cease-fire in 1996 was followed by a 1997 peace treaty that failed to decide the issue of Chechnya's independence.

Even without the inclusion of the fourteen former republics, Russia remains the largest nation in the world (almost twice the size of the US). Its population ranks sixth behind China, India, the US, Indonesia, and Brazil. Roughly 25% of the nation's land mass is in Europe; the remainder, Siberia, lies east of the Urals, in Asia. The Urals, constituting the eastern border of Europe, are a chain of low mountains (average height: 3500 ft., 1,000 m) running for 1,250 miles (2,000 km) from the Kara Sea in the north to the Aral Sea in the south.

Some of Russia's leading natural features are the world's largest forest (the coniferous forests of Siberia); the largest flat plain (the northern European and western Siberian plains); the deepest freshwater lake (Lake Baikal, in Siberia, about a mile deep and containing as much water as all the Great Lakes in North America combined); Europe's largest lake (Lake Ladoga, near St. Petersburg); its tallest peak (Mt. Elbrus (18,480 ft., 5,634 m, in the Caucasus Mountains); and its longest river (the Volga River). Because of Russia's northerly location and distance from the warming influence of the Atlantic and Pacific Oceans, most of the nation experiences long and frigid winters. The earth's coldest temperatures have occurred in northeastern Siberia (the record is -90° F, -68° C). Permafrost (permanently frozen ground) covers most of the north, and much of the nation's vast river and canal network remains frozen during winter.

The sheer size of Russia and its abundant mineral deposits enable it to be the world's largest producer of oil, natural gas, iron ore, timber, coal, lead, manganese, titanium, mercury, and potash. Production would even be greater but for the remote locations of deposits, the adverse climate, difficult terrains, the long distances to major shipping routes as well as communist mismanagement.

The Soviet Union had been the world's largest producer of many varieties of fruits, grains, and vegetables, but most of its agriculture came from Soviet republics warmer than Russia—in particular, the fertile lands of European Ukraine, Belarus, and Moldova and the irrigated deserts of Soviet republics in Asia. These regions have growing environments superior to the less fertile, drier, and much colder Russia. This dependence on "foreign" sources for the bulk of its food supply was one of the major factors in the 1991 creation of The Commonwealth of Independent States. Russia, together with the former republics (but not including the Baltic states), is expecting this organization to facilitate trade, communication, security, and the preservation of ethnic rights among its members.

The loss of major agricultural regions is just one of many serious problems facing Russia. The nation has to deal with an unstable government run by a ruling elite; declining industrial and agricultural production; high inflation and low wages; widespread environmental pollution; corruption and lawlessness; a huge army with low morale; declining health and welfare services; and a growing gap between the very wealthy and the rest of the population. Former communist leaders (wearing new allegiances) are still in positions of power. Not many Russians know how to play the capitalist game, and the few who do are making fortunes. Traditional Russian law has proved inadequate to regulate the new commerce. Until more appropriate laws are written and properly enforced, organized crime and corruption can only increase.

Russians, who comprise many ethnic groups, are mostly Slavic people whose culture dates back many centuries. It is believed that Viking explorers, arriving from Scandinavia in the 9th century A.D., conquered a large region surrounding the center of Kiev (current capital of the Ukraine). They called the natives "Rus." These invaders were absorbed into the Russian population and by the end of the 10th century, a local prince introduced Christianity from the Church of Constantinople, establishing the Russian Orthodox Church. Several centuries later, Russia was overrun by Genghis Khan and was ruled by Mongols until the emergence of the first Russian Czar, Ivan IV (the Terrible), in the 16th century.

Peter the Great, in the early 18th century, adopted Western reforms and began the expansion of the Russian Empire by enlarging the western boundaries. He also turned the Russian capital at the time, the port city of St. Petersburg, into the nation's "Window to the West." He imported Western architects to design magnificent buildings, museums, boulevards, and a network of canals and bridges. Close to five million people live in the "Venice of the North"—Russia's center of culture. In World War II, St. Petersburg (called Leningrad under the Soviet regime) lost over one million citizens to starvation during a three-year siege by the Nazis.

In the second half of the 18th century, Catherine the Great accelerated both westernization and Russian expansionism. It should be noted that at the time of the Russian Revolution of 1917, a succession of Czars had brought into the Russian Empire most of the adjacent lands that later became republics of the Soviet Union. At the turn of the 20th century, a series of military defeats weakened the monarchy of Czar Nicholas II and set the stage for the 1917 Russian Revolution. The first revolutionary government was headed by moderate socialists ("Mensheviks") under Aleksandr Kerensky. They were defeated in a four-year civil war by the radical "Bolsheviks," led by Vladimir I. Lenin. In 1921, the Soviet Union was established, and until the death of Lenin in 1924, it pursued a fairly tolerant policy toward private enterprise, science, and free expression in the arts. But this "liberal" period ceased when Joseph Stalin won the power struggle that followed Lenin's death. Historians are now viewing Lenin less favorably, considering that he may have put in place policies that set the stage for the reign of terror that followed his death. Stalin liquidated thousands of suspected rivals or traitors. Millions died during the forced socialization of industry and agriculture. Stalin ordered the widespread murder and starvation of peasants who resisted the forced collectivization of farmlands. The remarkable progress made in both restructuring and industrializing the Soviet Union during the period leading up to World War II was accompanied by great hardships for the Russian people. Under Stalin's police state, the vast and forbidding region of Siberia was transformed into a slave labor camp (the Gulag) for political dissidents as well as common criminals.

World War II brought even greater suffering to the Russian People; Stalin's nonaggression pact with Hitler in 1939 left Russia unprepared for the German attack in 1941. Until the Germans were driven out, four years later, the cost to Russia in loss of life and destruction was enormous. The brutal Russian winters (which previously defeated Napoleon) and the heroic efforts of the Russian people—along with massive American aid—combined to destroy the Nazi offensive. Russian troops that chased the retreating Germans across countries of eastern Europe continued to occupy those lands after the war's end. With the Soviet military present to enforce the power grabs of local communists, those nations became satellites of the Soviet Union. An "Iron Curtain" descended across Europe, and the Cold War began between the United States and its democratic allies in the West, and the Soviet Union and its satellite nations of the eastern European bloc. Stalin's death in 1953 lessened the continuing threat of nuclear annihilation as succeeding Soviet leaders proved more flexible and entered into positive agreements with the West. But no leader, prior to Mikhail Gorbachev, had the courage or willingness to change his nation's repressive political system.

The costs of a forty-year nuclear arms race, ambitious space and missile programs, and the support of communist-led "wars of liberation" around the globe led directly to the disintegration of the Soviet Union. Although hunger, disease, unemployment, and illiteracy had been virtually eradicated, the standard of living in the Soviet Union was comparable to most third-world nations, and considerably below that of the industrialized West. Scant resources were available for housing and consumer goods. What was offered for purchase was generally dated in design and poorly constructed—long lines of hopeful consumers were commonplace. The scarcity of food couldn't always be blamed on the weather: Without personal financial incentives, collective farmers were reluctant producers.

In the 1980s, Gorbachev acknowledged that the nation, with its secretive, controlled economy, was falling farther behind the dynamic, capitalist nations of the West, and he undertook a massive restructuring ("perestroika") of Soviet society. Part of the program was an openness ("glasnost") in the conduct of government, elections, and personal expression. The first attempts to introduce free-market techniques were met with resistance—70 years of communism bred a society cared for "from cradle to grave" and fearful of a system that could bring rising prices, unemployment, competition, and a reduction in government benefits.

The loosening of the reins of repression and the radical changes occurring in Moscow signaled to the satellite nations of Eastern Europe—as well as to other republics of the Soviet Union—that the time had come to free themselves of the Russian/communist yoke. One of the earliest and most symbolic acts of the freedom movement was the tearing down of the Berlin Wall dividing East and West in Germany's capital city. The 1990s witnessed the fall of communism in Europe and the dissolution of the Soviet Union. In Russia, free elections were held, and the Russian people experienced true democracy for the first time in the nation's history. But whether this new democracy will survive is uncertain. Those currently in power have to defend the free-market policies—often referred to as "shock therapy"—that have brought much pain to the masses. The growing rate of crime, especially in the business community, is of national concern. Despite the repression and hardships of communism, there appears to be a growing desire on the part of the Russian people to regain the security of the past.

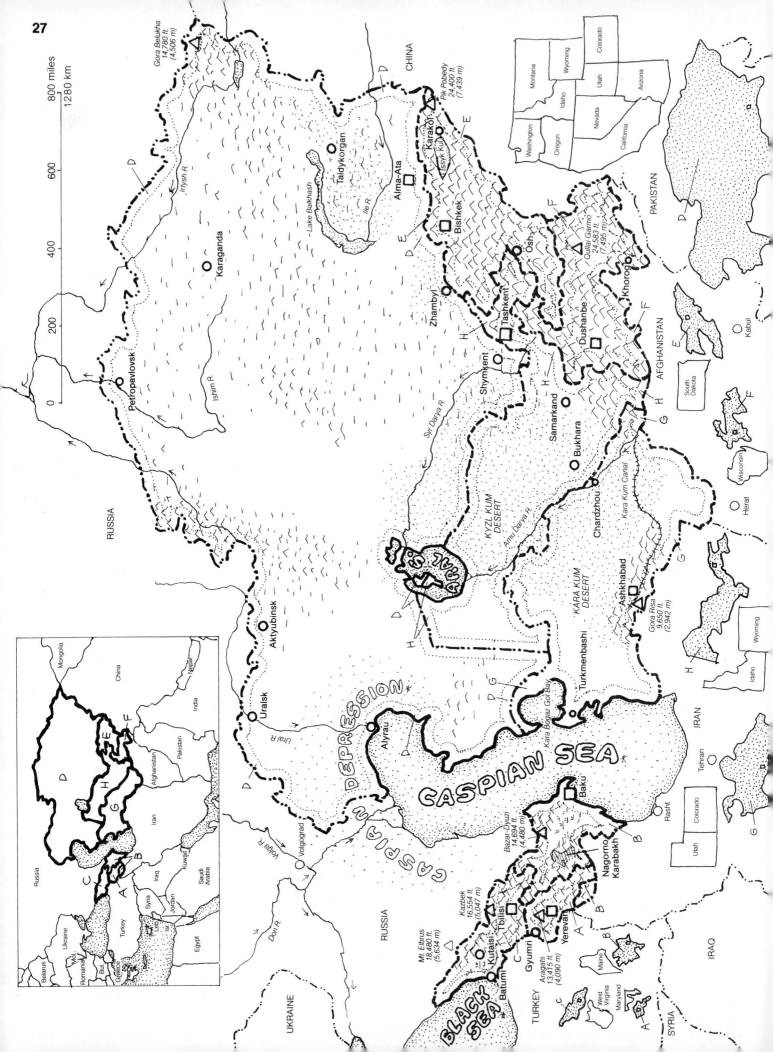

27

800 miles

1280 km

Gora Belukha
14,780 ft
(4,506 m)

Irtysh R.

CHINA

Pik Pobedy
24,400 ft
(7,439 m)

Karakol

Taldykorgan

Lake Balkhash

Ile R.

Alma-Ata

Issyk Kul

Bishkek

Montana

Wyoming

Colorado

Washington

Oregon

Idaho

Nevada

Utah

Arizona

California

Karaganda

Osh

Qullai Garmo
24,583 ft
(7,495 m)

PAKISTAN

Ishim R.

Zhambyl

Tashkent

Dushanbe

Khorog

AFGHANISTAN

Kabul

Petropavlovsk

Shymkent

Samarkand

South Dakota

Syr Darya R.

Bukhara

KYZL KUM
DESERT

Amu Darya R.

Chardzhou

Kara Kum Canal

Wisconsin

RUSSIA

ARAL SEA

Herat

KARA KUM
DESERT

Gora Risa
9,650 ft
(2,942 m)

Aktyubinsk

Ashkhabad

IRAN

Wyoming

Turkmenbashi

Kara Bogaz Gol Bay

Tehran

Idaho

Uralsk

Ural R.

Alyrau

Rasht

Volgograd

Volga R.

DEPRESSION

CASPIAN SEA

Baku

Bazar-Dyuzi
14,694 ft
(4,480 m)

Colorado

RUSSIA

CASPIAN

Nagorno
Karabakh

Don R.

Kazbek
16,554 ft
(5,047 m)

Tbilisi

Yerevan

Utah

Mt. Elbrus
18,480 ft
(5,634 m)

Kutaisi

Gyumri

Maine

Aragats
13,415 ft
(4,090 m)

West Virginia

Maryland

UKRAINE

Batumi

BLACK
SEA

TURKEY

SYRIA

IRAQ

Mongolia

China

Nepal

Russia

Ukraine

Kazakhstan

India

Pakistan

Afghanistan

Belarus

Turkey

Iran

Iraq

Kuwait

Saudi Arabia

Greece

Romania

Bul.

Syria

Jordan

Isr

Leb

Egypt

FORMER SOVIET: CAUCASUS REPUBLICS

The three nations of the Caucasus Mountains region—Georgia, Armenia, and Azerbaijan—are ancient societies dating back to prebiblical times. Because of their strategic locations, they have experienced centuries of invasions and occupations by Romans, Byzantines, Persians, Arabs, Mongols, and the Russians. Since gaining independence from the former Soviet Union in 1991, Georgia and Azerbaijan have been wracked by conflicts involving ethnic rivalries and separatist movements. Such violence has been particularly destructive to their fragile economies. Separatist movements in Georgia became so powerful that the Georgian government was forced to request a Russian military presence in order to stabilize the nation. The price of this assistance was an agreement with Russia that reduced Georgian sovereignty. Since 1988, Armenia and Azerbaijan have been engaged in violent conflicts over Nagorno-Karabakh, a region populated by Armenian Christians, but within the Islamic nation of Azerbaijan. Although a ceasefire ended hostilities in 1995, the situation remains precarious as the Armenian population of the disputed region seeks to unite with Armenia.

ARMENIA [A]

Area: 11,500 sq. mi. (29,800 km²). **Population:** 3,625,000. **Capital:** Yerevan, 1,200,000. **Government:** Republic. **Language:** Armenian. **Religion:** Armenian Orthodox. **Exports:** Industrial products, chemicals, and textiles. **Climate:** Continental extremes. □ Armenia (Ahr mee' nee uh), a land-locked country with a rugged terrain, lies within the Little Caucasus mountain range, an area subject to devastating earthquakes. In 1988, 10% of the nation's industrial and housing capacity was destroyed by one such quake. Azerbaijan has impeded Armenian recovery from the quake by its interference with vital trade routes. The people of Armenia are one of the world's oldest societies whose ancient empire stretched from Black to Caspian Seas. In 300 A.D., Armenia became the first nation in the world to designate Christianity as its official religion. Although the religion was to endure, it offered no protection against the numerous invasions and occupations that were to follow. By the early 19th century (1828), Armenia was part of the Russian Empire. Less than 100 years later, a sizable Armenian population, living in neighboring Turkey, became the victims of the 20th century's first holocaust, when an estimated one million Armenians were slaughtered by the Turks. Most of the survivors fled to what was then Russian Armenia. Armenians have been noted for their contributions to the fields of engineering, science, and literature.

AZERBAIJAN [B]

Area: 33,436 sq. mi. (86,000 km²). **Population:** 7,950,000. **Capital:** Baku, 1,800,000. **Government:** Republic. **Language:** Azeri. **Religion:** Muslim (Shite). **Exports:** Oil, natural gas, cotton, and caviar. **Climate:** Continental, semi-arid. □ Recent exploitation of oil reserves along the Caspian coast has been a boon for Azerbaijan (Ah zur by jahn'). This traditionally poor nation had been further burdened by the political turmoil and prolonged conflict with Armenia over the independence of Nagorno-Karabakh, an Armenian enclave within Azerbaijan. While negotiations following a ceasefire continue to determine the fate of the enclave, the heavily armed residents have seized additional surrounding territory from Azerbaijan to use as a buffer and through which to build a permanent road to Armenia. Azerbaijan shares its other major border with Iran. A large number of Azeris (equal to their population in Azerbaijan) reside in northern Iran, which was once part of an early Azerbaijani empire. Azerbaijan maintains

close relations with Turkey, which was deterred by Russia from taking part in the Nagorno-Karabakh dispute.

GEORGIA [C]

Area: 26,912 sq. mi. (69,700 km²). **Population:** 5,800,000. **Capital:** Tbilisi, 1,350,000. **Government:** Republic. **Language:** Georgian. **Religion:** Georgian Orthodox. **Exports:** Citrus, tea, food products, and industrial products. **Climate:** Mild with Mediterranean conditions in coastal areas. □ Except for lowlands adjacent to the Black Sea (a very popular resort area), the landscape of Georgia (Johr' juh) is dominated by the Great Caucasus and Little Caucasus mountain ranges. The former Soviet republic of Georgia has been the victim of violent ethnic rivalries and separatist movements since gaining independence in 1991. Ongoing strife has prevented the nation from exploiting its mineral resources and realizing the agricultural potential provided by the Mediterranean climate and fertile lowlands adjacent to the Black Sea. Georgian people and their culture are noted for their longevity. The culture and language date back to the early 4th century, and in Georgia there probably are more individuals per capita (unofficially, as records are scant) over the age of 100 than anywhere in the world.

FORMER SOVIET ASIAN REPUBLICS

These five Islamic nations of western Asia were the poorest republics of the former Soviet Union. Many inhabitants of the region are descendants of early Mongol invaders. For centuries, they lived as nomads until their lands were taken into the Russian Empire in the 19th century and by the Soviet Union in the 20th century. Forced industrialization and collective farming forever changed the life of those wandering tribes. Industrialization resulted in the exploitation of the region's considerable mineral resources, the building of local industries, and the creation of serious environmental pollution. Agricultural expansion was made possible by massive irrigation projects, most of which have been draining the Amu Darya River, which feeds what was the world's fourth-largest lake, the Aral Sea (normally larger than North America's Lake Huron). Since the sea is completely surrounded by land (Kazakhstan and Uzbekistan), it is technically a lake (the Caspian Sea is the world's largest lake). Diversion of its river source, together with evaporation, has shrunk the size of the sea to 1/3 of what it was just 40 years ago. A once thriving fish industry is gone. Pesticide contamination of the sea and surrounding groundwater has created a major ecological disaster. Health problems abound and life expectancy in the region is falling. Blowing sand from exposed lake beds is ruining farmland as far as hundreds of miles away.

KAZAKHSTAN [D]

Area: 1,049,150 sq. mi. (2,717,300 km²). **Population:** 17,590,000. **Capital:** Alma Ata,1,250,000. **Government:** Republic. **Language:** Kazakh, Russian. **Religion:** Muslim (Sunni), Russian Orthodox. **Exports:** Oil, natural gas, coal, metals, and cotton. **Climate:** Dry with continental extremes. □ In size, this huge nation was second only to Russia among the former Soviet Republics. The combination of enormous mineral wealth, agricultural potential, and suitable locations for nuclear plants, Soviet space programs, missile installations, and weapons factories brought millions of Russians to Kazakhstan (Kah zahk stahn') in the 20th century. Nearly as many Russians as Kazakhs live there. The nation still maintains very close military ties with Russia. Industrialization, mineral extraction, and the careless use of agricultural pesticides have inflicted a terrible toll on the environment.

Massive irrigation projects have dried up huge portions of the Aral Sea. The nation's terrain varies widely, from the Tien Shan Mountains in the east, and westward across the huge steppe and desert lands, down to 1,500 miles of low-lying coastline along the Caspian Sea.

KYRGYZSTAN [E]

Area: 76,640 sq. mi. (198,500 km²). **Population:** 4,900,000. **Capital:** Bishkek (formerly Frunze), 650,000. **Government:** Republic. **Language:** Kyrgyzian, Russian, Uzbek. **Religion:** Sunni Muslim 70%, Russian Orthodox. **Exports:** Wool, cotton, tobacco, chemicals. **Climate:** Extremely continental. □ This mountainous region, wedged between Kazakhstan and China, is the home of the nomadic Kyrgyz people, who violently resisted Soviet-enforced agricultural collectivization and industrialization. Ethnic violence between the Kyrgyz and the smaller Uzbek population has occurred since independence from the Soviet Union was achieved. The Uzbeks have been seeking to create an independent state within the country. Islamic Kyrgyzstan (Kihr gee stahn') has sought to establish closer relations with distant Turkey, which lies thousands of miles to the west; there is evidence of an early cultural connection between the two. Islam came to Kyrgyzstan in the seventh century.

TAJIKISTAN [F]

Area: 55,250 sq. mi. (143,100 km²). **Population:** 6,475,000. **Capital:** Dushanbe, 600,000. **Government:** Republic. **Language:** Tajik, Uzbek, Russian. **Religion:** Sunni Muslim 80%, Shiite Muslim 5%. **Exports:** Cotton, fruit, vegetable oils, and aluminum. **Climate:** Continental. □ Because of its extremely mountainous terrain, the lands that make up Tajikistan (Tah jhik ih stahn') have had a history of fewer foreign invasions than have the other nations of this group. The terrain also has made the Tajiks less nomadic than the other ethnic groups. Although they physically resemble the neighboring Uzbeks, Tajiks use a language similar to the Farsi spoken in non-neighboring Iran. Russia has given significant economic aid to the post-independence, pro-communist government. Rival groups, including communists, fundamentalists, nationalists, and pro-Western parties, have been battling for power ever since independence in the early 1990s. This political instability has further weakened an already poor economy.

TURKMENISTAN [G]

Area: 188,500 sq. mi. (488,100 km²). **Population:** 4,235,000. **Capital:** Ashkabad, 425,000. **Government:** Republic. **Language:** Turkmen. **Religion:** Muslim (Sunni). **Exports:** Cotton, grapes, oil, natural gas, and chemicals. **Climate:** Very dry and continental in the east. □ Although Turkey lies far to the west, Turkmen, the language of Turkmenistan, is a Turkish dialect. Much of the nation is covered by the Kara Kum desert, through which flows the Amu Darya river. Irrigation projects in the southern region are fed by the Kara Kum Canal (680 mi., 1100 km.), which diverts river water flowing north to the Aral Sea. The nation recently began to exploit its significant oil and natural gas reserves along the Caspian coastline. Plans are under way to construct a pipeline across Iran that will enable Turkmenistan to ship its natural gas to Turkey.

UZBEKISTAN [H]

Area: 172,740 sq. mi. (447,400 km²). **Population:** 23,950,000. **Capital:** Tashkent, 2,250,000. **Government:** Republic. **Language:** Uzbek. **Religion:** Muslim (Sunni). **Exports:** Cotton, textiles, fertilizers, and foodstuffs. **Climate:** Long, hot summers, but mild winters. □ The desert landscape of Uzbekistan (Ooz behk ih stahn') is sufficiently irrigated by rivers to make the nation the world's third-largest producer of cotton. Human habitation in this region dates back to prehistoric times. Uzbeks represent 70% of the population of this Asian group, the largest ethnic group. Of the many ethnic groups in the most populous nation of this Asian group, Uzbeks represent 70% of the population. Uzbekistan is also the least democratic of these former communist republics: former communist leaders have won power, reinstating censorship and restrictions on individual liberties.

28

ATLANTIC OCEAN *

GREENLAND

CANADA

NORTH AMERICA +

ALASKA (U.S.)

Iceland

ARCTIC

North Pole

OCEAN *

EAST SIBERIAN SEA *

BERING SEA

Ireland

U.K.

Norway

LAPTEV SEA *

KARA SEA *

Portugal

Spain

France

Bel.
Lux.
Neth.

Den.

Sweden

Finland

Germany

Switz.

Italy

Aust.

Slov.

Cro.

Czech R.

Poland

Lith.
Lat.
Est.

Belarus

Ukraine

ARCTIC CIRCLE

EUROPE *

RUSSIA

SEA OF OKHOTSK *

B&H

Yugo.
Alb.
Mac.

Hungary

Romania

Mol.

Kurile Islands

Greece

Bulgaria

A

BLACK S.

Georg.

Arm.
Azer.

CASPIAN S.

SEA OF JAPAN *

M'D. SEA *

G

15

1

Egypt

RED SEA *

E
D
C

F

J

B

18

19

16

17

Q

East China Sea

TROPIC OF CANCER

Sudan

Eritrea

K

H

Persian Gulf

M

I

W

Y

2

3

Ethiopia

Djib.

N

O

L

Gulf of Oman

Gulf of Aden

P

Somalia

V

S

PACIFIC OCEAN *

AFRICA *

ARABIAN SEA *

T

R

BAY OF BENGAL *

10

8

SOUTH CHINA SEA *

13

Y

6

14

EQUATOR 0°

Andaman Sea

Gulf of Thailand

U

X

5

Celebes Sea

9

9

11

12

7

INDIAN OCEAN *

7

7

Java Sea

7

7

Timor Sea

MADAGASCAR

0   500   1,000   1,500 miles

2,400 km

AUSTRALIA

RUSSIA A (ASIAN PART)
MIDDLE EAST I
CYPRUS B / NICOSIA
ISRAEL C / JERUSALEM
JORDAN D / AMMAN
LEBANON E / BEIRUT
SYRIA F / DAMASCUS
TURKEY G / ANKARA

MIDDLE EAST II
BAHRAIN H / MANAMA
IRAN I / TEHRAN
IRAQ J / BAGHDAD
KUWAIT K / KUWAIT CITY
OMAN L / MUSCAT
QATAR M / DOHA
SAUDI ARABIA N / RIYADH
UNITED ARAB EMIRATES O
YEMEN P / SANA   ABU DHABI

SOUTHERN
AFGHANISTAN Q / KABUL
BANGLADESH R / DHAKA
BHUTAN S / THIMPHU
INDIA T / NEW DEHLI
MALDIVES U / MALE
NEPAL V / KATHMANDU
PAKISTAN W / ISLAMABAD
SRI LANKA X / COLOMBO

FAR EASTERN
CHINA Y / BEIJING
JAPAN Z / TOKYO
MONGOLIA 1 / ULAN BATOR
NORTH KOREA 2 / PYONGYANG
SOUTH KOREA 3 / SEOUL
TAIWAN 4 / TAIPEI

SOUTHEASTERN
BRUNEI 5 / BANDAR SERI BEGAWAN
CAMBODIA 6 / PHNOM PENH
INDONESIA 7 / JAKARTA
LAOS 8 / VIENTIANE
MALAYSIA 9 / KUALA LUMPUR
MYANMAR (BURMA) 10 / RANGOON
PHILIPPINES 11 / MANILA
SINGAPORE 12 / SINGAPORE
THAILAND 13 / BANGKOK
VIETNAM 14 / HANOI

WESTERN
KAZAKSTAN 15 / ALMATY
KYRGYZSTAN 16 / BISHKEK
TAJIKISTAN 17 / DUSHANBE
TURKMENISTAN 18 / ASHGABAT
UZBEKISTAN 19 / TASHKENT

ASIA: THE COUNTRIES

CN: (1) On the small map, use gray for Asia in the dark outline (including the islands). (2) On the large map, color the two arrows representing the locations of the Maldives (U) and Singapore (12). (3) Note that the maps and text for the former Soviet republics (Western Asia, 15-19) are on the preceding plate.

Asia, the largest continent (17,230,000 sq.mi., 44,625,700 km²), contains 30% of the earth's landmass. It is both the widest and the deepest continent, stretching 6,000 mi.(9,600 km) from Turkey's Aegean coastline to the Pacific shores of Japan, and covers a similar distance from Siberia's Arctic tundra to the tropical Indonesian islands south of the Equator. Asia has the most people (3,500,000,000)—nearly 60% of the world's population. One out of every three human beings lives in either China or India. Because so much of Asia is extremely dry or mountainous, the world's most crowded population centers are generally located along the continent's coastlines or in its river valleys.

Asia is separated from Europe by a fictitious line that passes southward along the Ural Mountains to the Caspian Sea, then westward across the Caucasus to the Black Sea. The Sinai Peninsula is part of Asia, although it belongs to the African nation of Egypt—the two continents were connected until the Suez Canal was built. Asia was once linked to North America, but when ocean levels rose as glaciers melted, following the last Ice Age, the 50 mi.(80 km) Bering Strait was created, separating Siberia (Russia) and Alaska.

Northern Asia (Siberia, Mongolia, and northern China) is the most sparsely populated part of the continent. In the latter two nations, animal herding is the main industry. Most of Asia is rural, and the standard of living is very low, but in the Far Eastern region, Japan and the "Little Dragons" (Hong Kong, Singapore, South Korea, and Taiwan) had become extraordinarily productive industrial powers until the economic crash of the late 1990s. The wealthiest nations (per capita) in the world are located on the Arabian Peninsula. These desert monarchies are sitting on well over half of the world's known oil reserves.

Asia was the birthplace of many of the world's oldest civilizations. The Tigris-Euphrates Valley of the Middle East, the Indus River Valley of Pakistan, and the Huang He River Valley of China were the locations of flourishing, advanced societies. All of the world's major religions originated in Asia: Judaism, Christianity, and Islam came from the Middle East; Hinduism, Buddhism, Confucianism, Taoism, and Shintoism originated in South Asia and the Far East. Christianity, with the most adherents worldwide, actually plays a minor role in Asia. Hinduism, centered in India and Nepal, has the most followers. Islam, the second-largest faith, is dominant in Pakistan, Bangladesh, Malaysia, Indonesia, Turkey, Iraq, Iran, and the Arab nations of the Middle East.

Asians are generally placed in the Caucasoid and Mongoloid racial groups (p. 59). Caucasoids include the people of the Middle East (the Arab countries, plus Israel, Turkey, Iraq, Iran, and Afghanistan) and the Indians of South Asia (India, Pakistan, and Bangladesh). The Mongoloid race includes all of the people of the Far East and Southeast Asia.

29

ATLANTIC OCEAN *

ARCTIC OCEAN *

NORTH AMERICA

GREENLAND

CANADA

ALASKA (US)

Iceland

North Pole

Bering Strait

Ireland

U.K.

Portugal

Spain

France

Bel.

Neth

Lux.

Switz

Germany

Den

Norway

Sweden

Finland

KARA SEA *

LAPTEV SEA *

EAST SIBERIAN SEA *

BERING SEA

Italy

Czech R.

Poland

Slov.

Aust.

Cro.

B.&H.

Hungary

Slovakia

Lith.

Lat.

Est.

Belarus

ARCTIC CIRCLE

SIBERIA (Asian Russia)

SEA OF OKHOTSK

Alb

Mac.

Yugo

Romania

Greece

Bulgaria

Mol.

Ukraine

RUSSIA

EUROPE

Caspian Depression

Y

I

L

Q

J

MED. SEA

BLACK S. *

TURKEY

CYP

ISR

LEB

Georg

Arm

Azer

Aral Sea

KAZAKHSTAN

L. Balkhash

L. Baikal

B

Caspian S. *

UZBEKISTAN

KYRGYZSTAN

R

MONGOLIA

SEA OF JAPAN

SYRIA

JOR

IRAQ

TURKMENISTAN

TAJIKISTAN

A

S

X

W

GOBI DESERT

KOR

KOR

JAPAN

Egypt

D

N

AFGHANISTAN

U

V

PLATEAU OF TIBET

G

F

P

CHINA

Yellow Sea

EAST CHINA SEA *

RED SEA *

TROPIC OF CANCER

SAUDI ARABIA

IRAN

Persian Gulf

PAKISTAN

E

C

T

M

K

Sudan

U.A.R.

Gulf of Oman

NEPAL

O

TAIWAN

Eritrea

YEMEN

OMAN

INDIA

BANG

PACIFIC OCEAN

Ethiopia

Djib

Gulf of Aden

ARABIAN SEA *

INDIAN PENINSULA

MYANMAR

H

LAOS

THAI

INDO-CHINA PENINSULA

Hainan I.

VIETNAM

Somalia

AFRICA

BAY OF BENGAL

CAM.

SOUTH CHINA SEA *

PHILIPPINES

EQUATOR · 0°

MALDIVES

SRI LANKA

Andaman Sea

Gulf of Thailand

BRU.

Celebes Sea

MALAYSIA

MADAGASCAR

INDIAN OCEAN *

MALAYSIA

MALAY ARCHIPELAGO

INDONESIA

Java Sea

Timor Sea

INDONESIA

Java

AUSTRALIA

0   500   1,000   1,500 miles

2,400 km

PRINCIPAL RIVERS:

AMU DARYA A
AMUR B
BRAHMAPUTRA C
EUPHRATES D
GANGES E
HUANG HE (YELLOW) F
INDUS G
IRRAWADDY H
IRTYSH I
LENA J
MEKONG K
OB L
SALWEEN M
TIGRIS N
XI JIANG (PEARL) O
YANGTZE P
YENISEY Q

PRINCIPAL MOUNTAIN RANGES:

ALTAI R
ELBURZ S
HIMALAYAS T
HINDU KUSH U
KUNLUN V
PAMIRS W
TIAN SHAN X
URALS Y

LAND REGIONS:

DESERT Z
RAIN FOREST 1
LOWLANDS 2
MOUNTAINS 3
PLATEAU 4
HIGHLANDS 5
TUNDRA 6

The highest and lowest places on the planet (Mt. Everest and the Dead Sea) are in Asia, the continent with the tallest mountains, the largest and highest plateaus, the largest deserts, most of the longest rivers, the deepest lake, the largest forest region, the largest flat plains, the most active volcanoes, the most earthquakes, and the hottest, coldest, driest, and wettest climates. These geographical extremes characterize most of Asia—a vast, inhospitable environment where relatively few people live.

The most northerly part of Asia is the Arctic portion of Siberia (Asian Russia). South of this tundra is a huge highlands region containing the world's largest coniferous forest. To the west is the world's largest, flattest lowlands, the West Siberian Plain. South of these regions, and spanning the mid-section of Asia from Saudi Arabia to southern Mongolia, is a succession of large deserts, barren plateaus, and rugged mountains. Unlike the mountains of North and South America and Europe, Asia's massive ranges are located in the center of the continent. In this region, surrounded by some of the tallest peaks, is the Plateau of Tibet, the "Roof of the World" (15,000 ft., 4,573 m) and the world's highest inhabited plateau. Along Tibet's southern border are the Himalayas, the world's tallest mountains, including the tallest peak, Mt. Everest (29,028 ft., 8,848 m). These geologically young mountains continue to grow as the tectonic plate carrying the subcontinent of India grinds under the Eurasian plate (p. 2). The Asian subcontinent (the Indian Peninsula) is a triangular plateau that is subject to heavy monsoonal rainfall. Southeast Asia includes the rainforests of the Indochina Peninsula, the Malay Peninsula and Archipelago, and the Philippine Islands. Earthquake-prone regions (the Malay Archipelago, the Philippines, and the islands of Japan) are the peaks of seismically active, sunken mountain ranges. The Indonesian island of Java has the greatest concentration of active volcanoes in the world. The mountains of Central Asia contain the headwaters for most of the world's longest rivers. The Yangtze (3,915 mi., 6,265 km) is the third longest (after the Nile and Amazon) and is China's most important river for transportation, commerce, irrigation, and hydroelectric power.

Because Asia is so vast, the interior is far from the moderating influence of the Atlantic and Pacific Oceans, allowing the region to experience the earth's greatest temperature fluctuations—although the Polar regions are consistently colder, they are never warm. Central Asia has the kind of weather that only nomadic herders can endure. Southwest Asia (the Middle East) is parched and scorched—summer temperatures exceed 120° F (49° C). Southern and Southeast Asia are not nearly as hot, but high humidity and torrential rainfall (summer monsoons) create another form of discomfort. Heat and rain sustain the principal agriculture of the region: rice production.

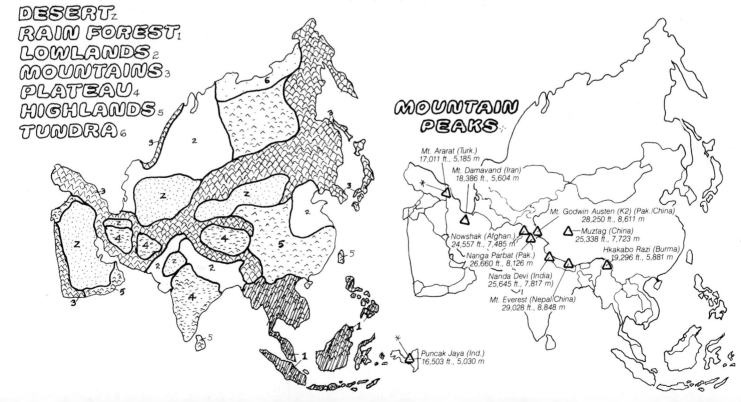

MOUNTAIN PEAKS:

Mt. Ararat (Turk.) 17,011 ft., 5,185 m

Mt. Damavand (Iran) 18,386 ft., 5,604 m

Mt. Godwin Austen (K2) (Pak./China) 28,250 ft., 8,611 m

Muztag (China) 25,338 ft., 7,723 m

Nowshak (Afghan.) 24,557 ft., 7,485 m

Nanga Parbat (Pak.) 26,660 ft., 8,126 m

Hkakabo Razi (Burma) 19,296 ft., 5,881 m

Nanda Devi (India) 25,645 ft., 7,817 m

Mt. Everest (Nepal/China) 29,028 ft., 8,848 m

Puncak Jaya (Ind.) 16,503 ft., 5,030 m

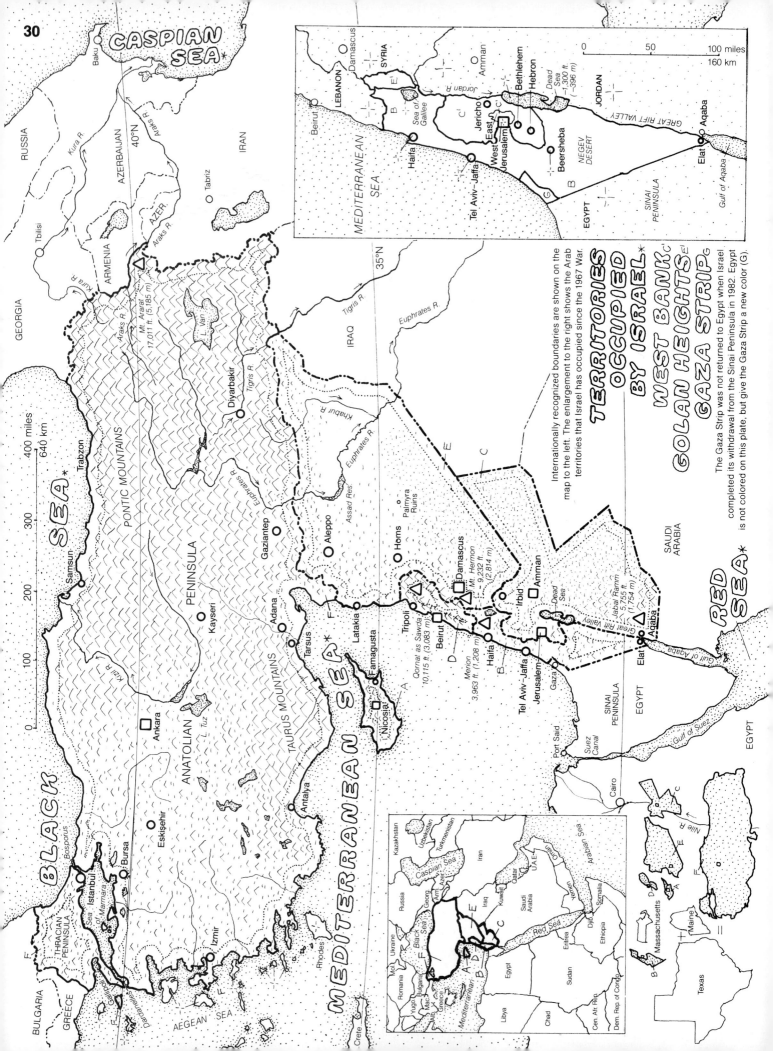

30

CASPIAN SEA*

TERRITORIES
OCCUPIED
BY ISRAEL*

WEST BANK C
GOLAN HEIGHTS E
GAZA STRIP G

Internationally recognized boundaries are shown on the map to the left. The enlargement to the right shows the Arab territories that Israel has occupied since the 1967 War.

The Gaza Strip was not returned to Egypt when Israel completed its withdrawal from the Sinai Peninsula in 1982. Egypt is not colored on this plate, but give the Gaza Strip a new color (G).

RUSSIA

Baku

AZERBAIJAN

40°N

Kura R.

Araks R.

AZER.

ARMENIA

GEORGIA

Tbilisi

Mt. Ararat
17,011 ft. (5,185 m)

Araks R.

IRAN

Tabriz

L. Van

Tigris R.

IRAQ

35°N

Euphrates R.

Khabur R.

Euphrates R.

Assad Res.

Palmyra Ruins

SYRIA

Damascus

Homs

Aleppo

Gaziantep

Diyarbakir

Tigris R.

Euphrates R.

PONTIC MOUNTAINS

Trabzon

Samsun

SEA*

BLACK

400 miles
640 km

300

200

100

0

ANATOLIAN

PENINSULA

Kayseri

Ankara

L. Tuz

Kızıl R.

Eskişehir

Bursa

Istanbul

Bosporus

Sea of Marmara

THRACIAN PENINSULA

Gallipoli

Dardanelles

BULGARIA

GREECE

Izmir

AEGEAN SEA

Crete

Rhodes

MEDITERRANEAN SEA*

Antalya

TAURUS MOUNTAINS

Adana

Tarsus

Latakia

Famagusta

Nicosia

Tripoli

Qornat as Sawda
10,115 ft. (3,083 m)

Beirut

Mt. Hermon
9,232 ft. (2,814 m)

Damascus

Meiron
3,963 ft. (1,208 m)

Haifa

Tel Aviv–Jaffa

Jerusalem

Gaza

Amman

Irbid

Dead Sea

Great Rift Valley

Jabal Ramm
5,755 ft. (1,754 m)

Elat

Aqaba

Gulf of Aqaba

RED
SEA*

SAUDI ARABIA

SINAI PENINSULA

EGYPT

Gulf of Suez

Suez Canal

Port Said

Cairo

Nile R.

EGYPT

Enlargement inset (top right):

SYRIA

Damascus

LEBANON

Beirut

Sea of Galilee

Jordan R.

Amman

JORDAN

Bethlehem

Hebron

Dead Sea
−1,300 ft. (−396 m)

Jericho

East Jerusalem

West Jerusalem

Haifa

Tel Aviv–Jaffa

Beersheba

NEGEV DESERT

EGYPT

SINAI PENINSULA

Gulf of Aqaba

Elat

Aqaba

GREAT RIFT VALLEY

MEDITERRANEAN SEA

0   50   100 miles
160 km

World locator inset (bottom center):

Russia

Kazakhstan

Uzbekistan

Turkmenistan

Iran

Caspian Sea

Arabian Sea

Qatar

U.A.E.

Oman

Yemen

Saudi Arabia

Kuwait

Iraq

Jordan

Black Sea

Ukraine

Mol.

Romania

Bulgaria

Yugo.

Greece

Mediterranean

Libya

Egypt

Red Sea

Sudan

Eritrea

Ethiopia

Djibouti

Somalia

Chad

Cen. Afr. Rep.

Dem. Rep. of Congo

Arm.

Azer.

Georg.

Size-comparison inset (bottom right):

Massachusetts

Maine

Texas

Nile R.

ASIA: MIDDLE EAST I

With the exception of Cyprus, Iran, Israel, and Turkey, the countries of the Middle East are Arab—their people speak Arabic and follow the Islamic religion. Citizens of Iran and Turkey are not Arabs, but they are Muslims (followers of Islam). Most Cypriots are Greek Orthodox Christians, and most Israelis are Jewish.

Although oil reserves are scant in this part of the Middle East, there are historic and religious riches for Jews, Christians, and Muslims. Not only was this region the birthplace of Judaism and Christianity, but the city of Jerusalem is holy for Muslims as well. Unfortunately, religious differences appear to play a major role in the discord of the area: Muslim versus Jew (Arab nations and Israel are in a technical state of war); Christian versus Muslim (in Lebanon and Cyprus); and Muslim versus Muslim (various Islamic factions in Lebanon).

Even the earth is divided in the Middle East: the Great Rift Valley, a crack in the earth's crust, runs south from Syria, under the Red Sea (p. 38). It defines the border of Israel and Jordan and contains the Jordan River, the Sea of Galilee, and the Dead Sea. The latter is the saltiest body of water on earth (28% salt vs. 3.5% in seawater), and is the lowest point on earth (-1,300 ft., -396 m below sea level). The absence of sea life and its absolute silence suggest no other name.

CYPRUS A

Area: 3,572 sq.mi.(9,251 km²). **Population:** 760,000 **Capital:** Nicosia, 187,000. **Government:** Republic. **Language:** Greek; Turkish. **Religion:** Greek Orthodox; Islam. **Exports:** Citrus, copper, wine, food products. **Climate:** Mediterranean. ☐ The words "Cyprus" and "copper" come from the Greek "kypros." Today, agriculture, not copper, is the main industry on this ruggedly beautiful island with an ancient past. The British Empire, the last of its many rulers, granted independence in 1960. A new constitution provided for joint rule between the Greek Orthodox majority (77%) and the Turkish Muslim minority, but unresolved differences led to the violence that brought in soldiers from Turkey, in 1974. The U.N. mediated a ceasefire, and a line was drawn across northern Cyprus and through Nicosia, the 5,000-year-old capital. The Turkish minority were given control of the northern third of the island, which they and Turkey consider to be an independent nation. The rest of the world regards Cyprus as a single nation under the Greek Cypriot government.

ISRAEL B

Area: 8,018 sq.mi.(20,767 km²). **Population:** 5,700,000 **Capital:** Jerusalem. 475,000. **Government:** Republic. **Language:** Hebrew; Arabic. **Religion:** Judaism; Islam. **Exports:** Oranges, produce, polished gems, electronics. **Climate:** Mediterranean on the coast, hot inland. ☐ A fifth of the world's Jews live in this land, from which their ancestors were driven by repeated invasions 2,000 years ago. For centuries, Jews scattered around the world dreamed of returning. Hebrew, the language of the Bible, is Israel's official language. In the late 19th century, European Jews formed the Zionist movement to create a homeland. After World War I, Britain, which received a mandate from the League of Nations to govern Palestine, permitted the gradual immigration of European Jews. Palestinian Arabs, whose ancestors had lived there for 1,500 years, grew alarmed at the growing number of immigrants and persuaded Britain to restrict immigration. After World War II, pressure to immigrate mounted as survivors of the Holocaust demanded entry. The United Nations divided Palestine into Jewish and Arab states, with Jerusalem to be under international control. The Arab nations rejected this plan and attacked the new nation of Israel in 1948. The Jews won the war and seized half of what was intended to be the Arab state, absorbing many Palestinians in the process. The thousands of Palestinians who had fled to neighboring Arab countries in anticipation of an Arab victory were denied re-entry by the Israelis. Those Palestinians and their descendants make up the 2,500,000 permanently displaced refugees living in Lebanon, Jordan, and Syria. The ones who did not leave Israel during the war have grown to 750,000 (15% of the population) and pursue their own culture. They have a higher standard of living than most Arabs in the Middle East and are accorded almost all of the rights held by Israelis, but they are victims of discrimination and live as a segregated minority. Forty years and three wars after its founding, Israel has acquired additional Arab territories: the Gaza Strip from Egypt, the Golan Heights from Syria, the West Bank and East Jerusalem from Jordan (see map in right-hand corner). After eight years of years of violently resisting the Israeli occupation of the West Bank, the Palestinian Liberation Organization (PLO)—with the aid of international pressure—was granted limited self-rule of the West Bank in 1994. There was also a vague plan for an eventual Israeli withdrawal, but the peace process has been repeatedly undermined by the expansion of Jewish settlements in the area, and by acts of extremists on both sides. Israel is technically at war with all its Arab neighbors except Egypt; the two countries signed a peace treaty in 1979 in which Israel returned to Egypt the Sinai Peninsula, but not the Gaza Strip. The capital of Israel is the modern city of West Jerusalem. When Israel captured and formally annexed East Jerusalem, it included the Old City, the site of ancient Jerusalem. Jews were once again able to pray at the "Wailing Wall," believed to be part of the ancient Temple of Solomon. The Old City is holy to Christians and Muslims as well. It was the setting for many events in the life of Christ, and Muslims believe that it was there that Muhammad ascended to Heaven, and Jerusalem is nearly as sacred to them as Mecca and Medina. The modern cities of Tel Aviv-Jaffa and Haifa are centers of industry and commerce. Out of the desert, Israel has created the most productive farmland in the Middle East. This tiny nation of a little more than 5 million is also one of the world's most militarily powerful and technologically advanced societies. Much of its strength is the result of aid from the United States, reparations from Germany (for crimes committed against Jews during the Holocaust), and financial support from Jews around the world.

JORDAN C

Area: 37,600 sq.mi.(97,384 km²). **Population:** 4,600,000. **Capital:** Amman, 1,150,000. **Government:** Monarchy. **Language:** Arabic. **Exports:** Produce, phosphates, tobacco, oil products. **Climate:** Desert. ☐ Jordan is landlocked, except for 15 mi.(24 km) of coastline on the Gulf of Aqaba. About 80% of its land is desert. The nation has no historic basis: the British made Jordan and other former colonies in the Middle East independent countries as part of the breakup of the Ottoman Empire after World War I. Jordan's only fertile land is in the Jordan River Valley. More than half of this area (the West Bank) and a quarter of Jordan's population were lost to Israel in the Six-Day War of 1967. Jordan wants the West Bank to become the new Palestinian homeland, and although Israel has agreed to move in that direction, it still argues that since Palestinians are the majority in Jordan, it already is a Palestinian state. Many Jordanians are members of formerly nomadic Bedouin tribes. Amman, the capital, has replaced war-torn Beirut, Lebanon, as the area's financial center.

LEBANON D

Area: 4,024 sq.mi.(10,422 km²). **Population:** 3,825,000. **Capital:** Beirut, 750,000. **Government:** Republic. **Language:** Arabic. **Religion:** Islam; Roman Catholic. **Exports:** Tobacco, fruits and vegetables, small industrial products. **Climate:** Mild, but varies according to terrain. ☐ Prior to the 1970s, beautiful Lebanon was the most stable and prosperous nation in this region. Its capital, Beirut, was called the "Paris of the Middle East." It was the regional center for finance and commerce and it provided shipping facilities for Jordan, Iraq, and Syria. During the 1980s, the country was in an almost continual state of civil war: Christians fighting Muslims, and Muslims fighting among themselves. The Roman Catholic Maronites constitute the highest Christian percentage (40%) of any Arab country. The Arab majority of Druse, Shiite, and Sunni Muslims have been contesting the traditional power wielded by the Maronites. The fighting stopped in the 1990s, but the peace is fragile. Before the war, a visitor traveling the 40 mi.(64 km) width of the country could see Mediterranean beaches, a green coastal plain, snowcapped mountains (in the winter), the Beqaa Valley, and the edge of the Syrian desert. The famed cedars of Lebanon are nearly extinct, but reforestation programs are intended to bring them back.

SYRIA E

Area: 71,510 sq.mi.(185,211 km²). **Population:** 16,000,000. **Capital:** Damascus, 1,600,000. **Government:** Socialist republic. **Language:** Arabic. **Religion:** Islam. **Exports:** Cotton, textiles, tobacco, fruits, oil. **Climate:** The coast is mild; interior is dry. ☐ Syria is a land filled with the history of ancient civilizations. It was the trade center of the caravan routes that linked the continents of Europe, Asia, and Africa. The eastern half of the country is part of a river-irrigated region that has sustained life in the Middle East for thousands of years. Called the "Fertile Crescent" because of its shape, it is bounded by the Tigris and Euphrates Rivers. It begins at the Persian Gulf and arcs northwestward through most of Syria and down into Israel. Damascus, the 5,000-year-old capital of Syria, may be the world's oldest continuously inhabited city. Archaeologists have been excavating the ruins of Palmyra for over 50 years but are far from completely uncovering the ancient trade center. During this century, Syria (like Lebanon) was part of the Ottoman Empire and became a French mandate following World War I. Since gaining independence in 1946, it has been ruled by a succession of military regimes. Modest oil deposits help pay for basic government expenses. National animosity focuses on Israel, which continues to occupy Syria's strategic Golan Heights, which loom over Israel's northeastern border.

TURKEY F

Area: 300,960 sq.mi.(779,486 km²). **Population:** 62,700,000. **Capital:** Ankara, 2,750,000. **Government:** Republic. **Language:** Turkish. **Exports:** Cotton, carpets, fruit and nuts, tobacco, chromium. **Climate:** Thrace and the coast are Mediterranean; interior is continental. ☐ Turkey sits on two peninsulas, each in a different continent. The Thracian Peninsula (Thrace), on the southern tip of the Balkans, represents only 3% of Turkey. It is separated from mainland Turkey by the Sea of Marmara and two narrow straits: the Dardanelles and the Bosporus. The remainder of Turkey occupies the Anatolian Peninsula (also called Anatolia or Asia Minor), a rugged plateau rimmed by mountains. Turkey's largest city, Istanbul (7,400,000), straddles the Bosporus and sits on both continents. The city was inhabited long before the ancient Greeks named it Byzantium. The Romans made it the capital of their Empire in 324 AD and changed its name to Constantinople. When it became the capital of the Ottoman Empire in 1453, the Turks gave it its present name. The tallest peak in the Middle East, Mount Ararat (17,011 ft., 5,186 m), is on the eastern border with Armenia and Iran. It is believed by some to be the resting place of Noah's Ark. Both the Euphrates and the Tigris Rivers originate in Turkey. In Syria and Iraq there is concern that Turkey's hydroelectric and irrigation projects will reduce the flow of those vital rivers. For over 500 years, Turkey was the heart of the Ottoman Empire (p. 47), which ruled the Middle East, North Africa, and southeastern Europe. The Empire's collapse came when it fought on the losing (German) side in World War I. Turkey's present borders were defined in 1923, the year it became a republic. The first president, Kemal Ataturk, westernized the nation: he eliminated traditional dress (including the veil and fez), the Arabic alphabet, Islamic schools and laws, polygamy, and the subjugation of women. Despite the rise of Islamic fundamentalism, a woman was elected prime minister in 1993.

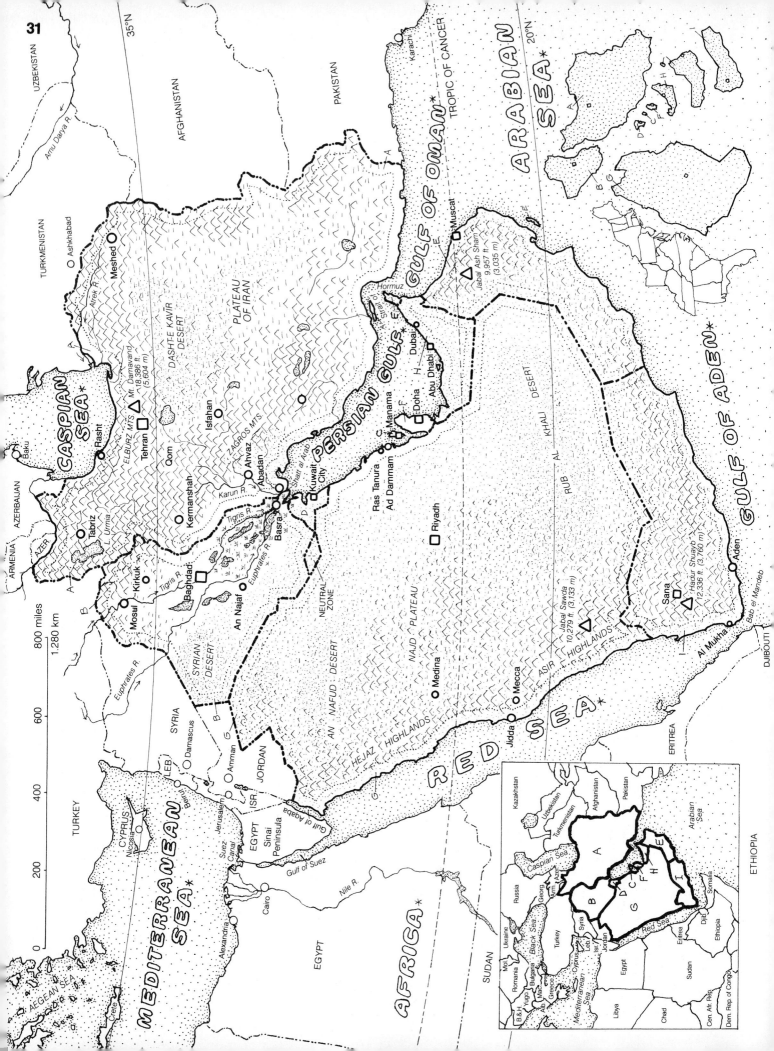

ASIA: MIDDLE EAST ☷ ∴

Oil became the major source of revenue in this region when, in the 1970s, the newly formed OPEC (Organization of Petroleum Exporting Countries) curtailed production and prices skyrocketed. Almost overnight these countries went from poor desert nations to rich, urban welfare states—no taxes and free social and educational services. During the 1980s, Iran and Iraq were locked in a devastating war that left them out of the oil bonanza. Iraq was to precipitate another disastrous war in 1990 (see Iraq for a description). Even though oil prices have dropped, they are much higher than they were before OPEC took control of its oil reserves. Most Arabian nations earn additional income from worldwide and domestic investments.

The Arabian Peninsula is one of the world's hottest and driest regions. There are virtually no rivers or streams, but underground water surfaces at oases and private wells. The wealthier nations rely on distilled seawater. The peninsula's barren land slopes eastward from highlands along the Red Sea to the oil-rich lowlands of the Persian Gulf. Most Arab nations became independent of Great Britain in the mid-20th century, but English is still learned as a second language and is used in commercial world.

Kurdistan, a region in eastern Turkey, northern Iraq, northwestern Iran, and the tip of Syria, is a "borderless nation" seeking a political identity. It is the home of the Kurds, a fiercely independent mountain group of Muslims whose population is estimated at 20 million (the largest ethnic minority in the Middle East). They have resisted every attempt by a national government to assimilate them. They have been brutalized by both Turkey and Iraq. The latter gassed Kurdish villages for siding with Iran during the Iraq/Iran war of the 1980s.

IRAN ₐ

Area: 636,259 sq mi(1,648,004 km²). **Population:** 69,000,000. **Capital:** Tehran, 6,500,000. **Government:** Islamic republic. **Language:** Persian (Farsi). **Religion:** Islam. **Exports:** Oil and oil products. **Climate:** Dry and continental. ☐ The last Shah tried to modernize Iran (ih ran') with the profits from its enormous oil reserves; this led to his downfall in 1979. Religious conservatives led by the Ayatollah Khomeini came to power and governed under Islamic law. Iran isolated itself from the world community and received little help when it was invaded by Iraq. An eight-year, bloody war ended in a stalemate in 1988. Iran (formerly Persia) rests on a huge plateau with an average elevation of 4,000 ft (1,220 m), bordered by northern and western mountain ranges that form a giant V pointing at Turkey. The northern slope of the Elburz Mountains, facing the Caspian Sea, is the only region with enough rainfall to support agriculture. The Caspian is a rich provider of the sturgeon eggs that support an important caviar industry. On the southern slopes of the Elburz Mountains is the capital, Tehran, the Middle East's largest city and the cultural, religious, and business center of Iran.

IRAQ ₈

Area: 169,930 sq mi(438,419 km²). **Population:** 23,100,000. **Capital:** Baghdad, 4,700,000. **Government:** Republic (socialist). **Language:** Arabic; Kurdish. **Religion:** Islam. **Exports:** Oil, dates, copper, wool, hides. **Climate:** Hot summers and mild winters. ☐ Iraq (ih rak') produces 80% of the world's supply of dates, but oil is by far its major source of revenue. The fertile land between the Euphrates and Tigris rivers, which in ancient times was called Mesopotamia (Greek for "between the rivers"), supported the earliest recorded civilizations (9,000 years old). Early Sumerian, Assyrian, and Babylonian societies were founded here. The plow, wheel, and writing are believed to have originated in

this eastern half of the Fertile Crescent. The Tigris River runs through the capital, Baghdad, once the heart of the Arab Empire and a 9th century center of learning. The city was later ravaged by Mongol invaders and never regained its stature. Border disputes precipitated Iraq's invasion of Iran in 1980 and Kuwait in 1990. The latter brought on a U.S.-led, U.N. response termed "Operation Desert Storm." Victory was swift, and Iraq was ousted from Kuwait. The U.N. placed an economic and military embargo on Iraq (intended to incite a rebellion against Saddam Hussein, who remained in power). His interference with the U.N.'s ability to inspect and destroy Iraq's war-making capacity—pursuant to terms of Iraq's surrender in 1990—created numerous crises in the 1990s.

ARABIAN PENINSULA ∴

BAHRAIN ᴄ

Area: 250 sq mi(663 km²). **Population:** 620,000. **Capital:** Manama, 150,000. **Government:** Sheikdom. **Language:** Arabic. **Religion:** Islam. **Exports:** Oil and oil products. **Climate:** Hot summers; warm winters. ☐ This tiny country consists of a 33-island archipelago in the Persian Gulf. Bahrain is the name of the largest island, which is connected to Saudi Arabia by a causeway. Bahrain derives its fortune from oil reserves and by refining Saudi Arabian oil in huge facilities. It has become a Middle Eastern banking and finance center and a headquarters for international corporations. The ruler of this sheikdom, called an emir, controls the cabinet that governs the country.

KUWAIT ᴅ

Area: 6,870 sq mi(17,793 km²). **Population:** 1,900,000. **Capital:** Kuwait City, 450,000. **Government:** Emirate. **Language:** Arabic. **Religion:** Islam. **Exports:** Oil and natural gas. **Climate:** Hot summers. ☐ Kuwait, wedged into the northwest corner of the Persian Gulf, may have as much as 10–15% of the world's oil reserves. The wealthy Kuwaiti citizens are outnumbered by a Muslim labor force from India, Iran, and Pakistan, along with 250,000 Palestinian refugees. In 1990, a dispute over oil-rich border lands and access to the Gulf led to an invasion by Iraq. An international military force led by the U.S. ousted Iraq.

OMAN ᴇ

Area: 82,025 sq mi(212,445 km²). **Population:** 1,600,000. **Capital:** Muscat, 360,000. **Government:** Sultanate. **Language:** Arabic. **Religion:** Islam. **Exports:** Oil, dates, fruit, tobacco. **Climate:** Extremely hot. ☐ The ancient seafaring nation of Oman (o mahn') occupies the southeastern corner of the Arabian Peninsula. A detached tip of Oman juts into the Strait of Hormuz, separating the Persian Gulf and the Gulf of Oman. Population and agriculture are concentrated on the northern and southern coasts, where there is fertile land and some rainfall. A network of ancient underground canals provides additional water. Oman is rapidly modernizing with the aid of its oil income.

QATAR ꜰ

Area: 4,250 sq mi(11,008 km²). **Population:** 700,000. **Capital:** Doha, 300,000. **Government:** Emirate. **Language:** Arabic. **Religion:** Islam. **Exports:** Oil. **Climate:** Very hot and dry. ☐ Qatar is a barren, oil-rich peninsula that juts into the Persian Gulf from its boundary with Saudi Arabia. Oil has transformed Qatar from a poor, sea-dependent country to a bustling, almost entirely urban society, even richer, per capita, than Kuwait. Less than a third of the rapidly expanding population is native-born. The ruling emir is a member of a family that has been in power for over 100 years.

SAUDI ARABIA ɢ

Area: 830,000 sq mi(2,149,700 km²). **Population:** 20,800,000. **Capital:** Riyadh, 3,000,000. **Government:** Monarchy. **Language:** Arabic. **Religion:** Islam. **Exports:** Oil. **Climate:** Hot except for highlands. ☐ Saudi (saw dee' or sow

dee') Arabia, which occupies most of the Arabian Peninsula, is the leading oil exporter (the U.S. and Russia are the largest producers). The nation has about a quarter of the world's reserves. The government has used its enormous income to invest widely abroad and to build large industrial complexes at home. Agriculture in this hot, riverless country is limited to highlands bordering the Red Sea and to fertile desert oases. Nearly half of the Saudis are nomadic Bedouins, but no one lives in the southern Rub al Khali Desert—the "Empty Quarter." The presence of Mecca and Medina, Islam's holiest shrines, gives Saudi Arabia preeminent status among nations in the Arab world. Muhammed (or Mohammed), the founder of Islam, was born in Mecca in 570 AD. The revelations he received from God (Allah), cover all aspects of daily life and are written in the Koran. Muhammed believed that he was the last in the Judaic and Christian tradition of great prophets. Islam preaches the existence of a single God but tolerates other faiths. Five times a day, nearly a billion Muslims kneel in the direction of Mecca to pray. Many visit the city to pray at the Great Mosque and its most sacred shrine, the Kaaba, a small black cube-shaped building believed to have come from God. Because the Koran requires Muslims to visit Mecca at least once in their lives, over a million pilgrims arrive each year. Providing accommodations for that many people is an industry in itself. Medina, the city Muhammed governed, and in which he died, is almost as holy as Mecca. Most Arab Muslims are members of the Sunni sect. Shiite Muslims make up half of the Iraqi and almost all of the Iranian populations. Shiites are often the poorer and persecuted minority in Arab countries. Militant Shiites in Iran have challenged Saudi Arabia's moral qualifications to be the spiritual leader of Islam because of its moderate stance toward the West. The conflict between Sunnis and Shiites is an ancient disagreement over who inherited the leadership of Islam after the death of Muhammed. The Islamic religion bans idolatry and the depiction of living things in art—representing humans, animals, and plants realistically is thought to usurp the role of Allah as the sole creator. These restrictions have fostered the creation of the decorative and colorful abstract motifs that characterize Islamic art and architecture. Outstanding examples are found in the mosque (house of worship), the minarets (tall, slender towers) that adorn it, and the familiar Persian rug.

UNITED ARAB EMIRATES ʜ

Area: 32,285 sq mi (83,618 km²). **Population:** 2,300,000. **Capital:** Abu Dhabi, 400,000. **Government:** Emirate federation. **Language:** Arabic. **Religion:** Islam. **Exports:** Oil and oil products. **Climate:** Very hot and dry. ☐ In 1971, when Britain withdrew as the protector and foreign advisor of the Trucial States, a group of seven independent emirates banded together to form a federation known as the United Arab Emirates, each with its own ruling emir. The largest emirate, and the wealthiest state in the world, is Abu Dhabi. The federation takes care of matters of defense, foreign policy, and economic planning.

YEMEN ᵢ

Area: 207,050 sq mi (528,489 km²). **Population:** 16,400,000. **Government:** Republic. **Capital:** Sana, 2,000,000. **Language:** Arabic. **Religion:** Islam. **Exports:** Coffee, cotton, cotton, dates, tobacco, fish, oil. **Climate:** Hot and humid. ☐ For over 300 years, Yemen (North Yemen) and South Yemen were ruled separately. In recent times, South Yemen (The People's Democratic Republic of Yemen) was the only communist country in the Arab world. Relations between the two Yemens were hostile. But when democratic change, prompted by events in eastern Europe, came to merge Europe, the nations agreed to merge in 1990. Sana is the political capital, and Aden, with its oil refineries and shipbuilding facilities, is the economic capital. Northern Yemen has tall mountains and fertile fields—unusual geography for the Arabian Peninsula. Its importance in trade between Africa and Asia dates back to Biblical times. The best-known export is the coffee named after the port of Al Mukha (formerly Mocha). In the north, rival Muslim factions have disagreed over the country's course of development. Shiites, who have political and religious power, have opposed the modernization policies of the business-minded Sunnis.

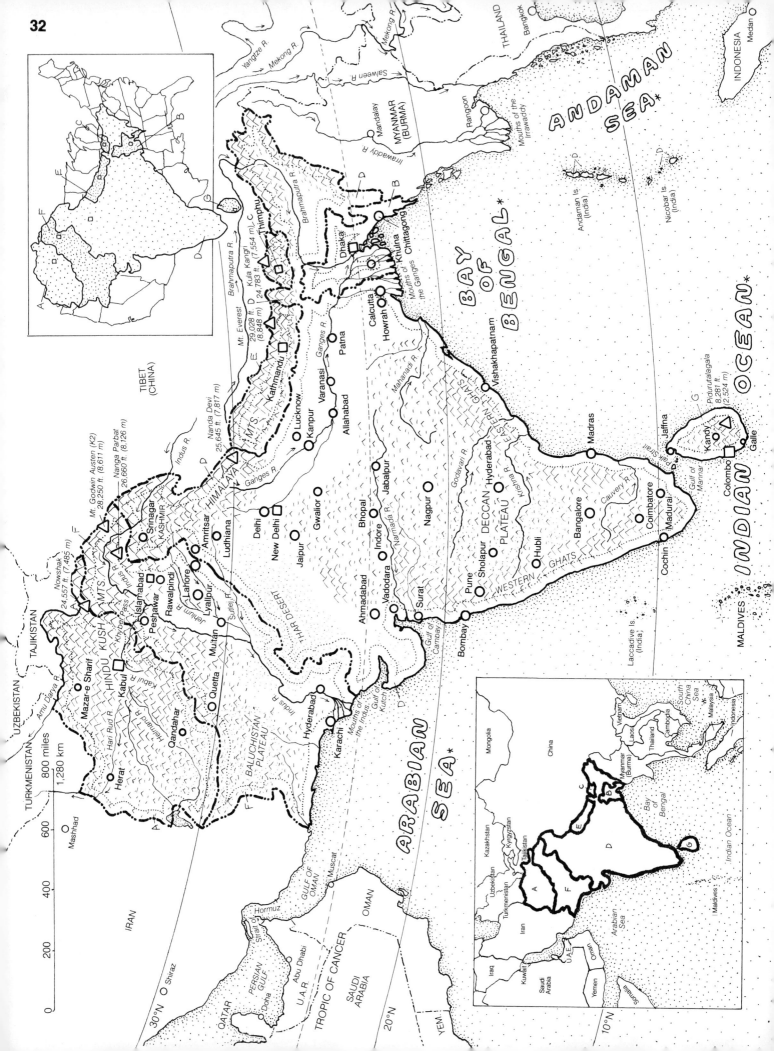

ASIA: SOUTHERN

The Himalaya Mountains have been a natural barrier between Mongoloids (speaking the Tibeto-Burman languages of Central Asia) and Caucasoids (speaking the Indo-Aryan languages of South Asia). Invasions and occupations have added many languages and dialects to South Asia's native tongues. The most influential, English, is the unifying language spoken in politics, business, education, and science. India, Bangladesh, and Pakistan made up most of the British-Indian Empire. In 1947, predominantly Hindu India was granted independence. Pakistan was created as a homeland (in two separate parts) for Muslim minorities. East Pakistan seceded in 1971 and became Bangladesh. Afghanistan and the Maldives also are Islamic nations. Nepal is Hindu, and Bhutan and Sri Lanka have Buddhist majorities. South Asia is the most densely populated region in the world. Huge populations are confined to coastal areas, river deltas, and river basins (the Ganges and Brahmaputra in India and Bangladesh, and the Indus in Pakistan). These countries have agricultural economies that depend upon the critical timing of summer monsoon rains. Local populations continue to grow rapidly because of improved health care and sanitation. There is significant loss of life from animal attacks. Tigers, leopards, elephants, rhinoceroses, and poisonous snakes are just a few of a vast variety of wild and dangerous animals that fill the jungles of the subcontinent. An "abominable snowman" is reputed to be alive in the Himalayas, but its existence has never been proved.

AFGHANISTAN A

Area: 253,000 sq.mi.(655,270 km^2). **Population:** 24,800,000. **Capital:** Kabul, 1,500,000. **Government:** One-party republic. **Language:** Pashto; Dari. **Religion:** Islam. **Exports:** Natural gas, hides, dried fruits, cotton. **Climate:** Dry, with continental extremes. ☐ Landlocked Afghanistan is a starkly beautiful mountainous nation. The only fertile land is found north of the Hindu Kush Mountains that cross the heart of the country. South of these towering peaks is a barren desert. For most of the 1980s, the nation was gripped by a bloody civil war in which the Soviet Union intervened on behalf of the communist government in power. The rebel faction was made up of many different Muslim groups fighting to preserve a feudal way of life. These "Mujahadeen" (holy warriors), armed by the U.S., forced a Soviet withdrawal. Religious-based resistance to centralized authority has been a common thread throughout Afghanistan's history. In the 1990s, rebel forces seeking to impose strict Islamic law were gaining control of the country. For centuries, Afghanistan was considered the gateway to the riches of the Indian subcontinent and was the customary route for invading armies. The famed Khyber Pass across the Hindu Kush is still the major road for Afghan exports traveling to the Pakistani port of Karachi.

BANGLADESH B

Area: 55,575 sq.mi.(143,940 km^2). **Population:** 127,700,000. Capital: Dhaka, 1,500,000. **Government:** Republic. **Language:** Bengali. **Religion:** 85% Islam; Hinduism. **Exports:** Jute, tea, fish products, hides. **Climate:** Tropical, with very heavy rainfall. ☐ Bangladesh is essentially a huge delta (the world's largest), formed by five rivers, including the Ganges and Brahmaputra. The boat-filled countryside is subject to annual flooding brought on by some of the heaviest rainfall on the planet. Cyclone-driven tidal waves are an added threat; one of them killed over a quarter of a million people in 1970. During the British occupation of India, Bangladesh was the Muslim, Bengali-speaking eastern half of the state of Bengal. It became the eastern portion of the newly formed Pakistan in 1947. Except for the Islamic religion, the two halves of the country, 1,000 mi.(1,600 km) apart, had nothing in common. After many years of exploitation by the government in the West, East Pakistan seceded in 1971. The West attacked, but with the aid of India, Bangladesh ("Bengal Nation") was created. The nation's food production is no match for the exploding population; this is one of the world's poorest and most crowded areas. Bangladesh is the world's top producer of jute, used to make rope and sack-cloth.

BHUTAN C

Area: 18,145 sq.mi.(46,980 km^2). **Population:** 1,900,000. **Capital:** Thimphu, 34,000. **Government:** Monarchy. **Language:** Dzongkha (Tibetan dialect), Nepali. **Religion:** Buddhism 70%; Hinduism. **Exports:** Timber, fruit, whiskey. **Climate:** Very wet; temperature varies according to altitude. ☐ This remote Himalayan kingdom wedged between India and Tibet closely resembles the fabled Shangri-la. The name Bhutan (boo tahn') means "Land of the Dragons," and the mythical animal graces the country's flag. Most of the people live in the foothills and river valleys. About two-thirds of them are Buddhists of Tibetan ancestry. Many monks live in hundreds of fortress-like monasteries. Until the 1960s, Bhutan was an isolated, almost totally illiterate nation. A program of modernization is now in progress. Road and air travel to India has been improved. India serves as Bhutan's protector in foreign affairs.

INDIA D

Area: 1,270,000 sq.mi.(3,289,300 km^2). **Population:** 985,000,000. **Capital:** New Delhi, 4,000,000. **Government:** Republic. **Language:** Hindi and English (official); 850 other languages and dialects. **Religion:** Hinduism 83%; Islam 11%. **Exports:** Iron ore, tea, cotton, hides, textiles, rubber. **Climate:** Tropical, with three seasons: cool, hot, and wet. ☐ India, the world's largest democracy, is a third the size of the U.S. but has more people than any country except China. Paradoxically, this poor, predominantly agricultural country is rapidly becoming a major industrial nation and is one of the leaders in producing scientists and skilled technicians. India is unified by the Hindu religion, but divided by over 800 languages and dialects. The Hindu religion, one of the world's oldest, supports a caste system with a rigid class structure that determines how the members of each caste shall live. One can never leave the caste of birth. There are four main castes, each with hundreds of subcastes. A person's standing in society depends upon his or her caste. Fifteen percent of Hindus are "untouchables"; their unfortunate position is at the bottom of the entire caste system. Modern laws prevent discrimination based on caste, but age-old traditions die slowly and class distinction remains a way of life. All Hindus believe in a reincarnation in which it is possible for a human to return as an animal in the next life. Consequently, most Hindus do not eat meat. Cows, whose milk is used, are considered sacred and are allowed to wander around business districts of major cities and to graze on valuable farmland. Other religions are represented in India: Muslims (11%) live mainly in the north; Christians (3%) live in the northeast; bearded, turban-wearing Sikhs (2%) have violently demanded greater autonomy in the northern state of Punjab; Buddhists (1%) were once the majority; and Jains (1%) extend reverence for life to all living creatures. Southern Indians are dark-skinned descendants of the Dravidians, the earliest known inhabitants of India. They were driven south by the Aryans, the ancestors of the light-skinned northerners. Indians of the north and south represent completely different cultures.

India is bordered on the north by the Himalayas. The fertile region just to the south is the world's largest alluvial plain. This densely populated region contains three river basins: the Indus, the Brahmaputra, and the temple-lined Ganges. Pilgrims come to bathe and spiritually cleanse themselves in the sacred waters of the Ganges. The triangular Indian Peninsula is a tropical plateau (the Deccan Plateau) rimmed by mountain ranges called the Western and Eastern Ghats. Ten cities in India have over one million residents. Many have modern sections, built by the British, that are currently occupied by wealthy or politically influential Indians. Thousands of homeless people bed down on the streets of the major cities. At Agra, in northern India, stands the Taj Mahal, one of the world's most beautiful structures. The white marble building was built as an Islamic tomb by an Indian prince for his favorite wife. The Indian non-violent movement for independence from Great Britain, led by Mohandas Gandhi, set an example for American civil rights activists of the 1960s, but India itself has not been non-violent in boundary disputes with China and Pakistan. After many battles between India and Pakistan over the beautiful state of Kashmir, a United Nations-mediated treaty established the current boundary lines. In 1990, tensions flared again as India accused Pakistan of aiding the Muslim separatist movement in Kashmir.

MALDIVES. This Islamic nation, consisting of 2,000 coral atolls, lies about 300 mi.(480 km) southwest of the tip of India. Most of the 200,000 people, who inhabit 200 of the islands, are descendants of Sri Lankans. Male (35,000) is the capital. Fish, coconuts, and tourism are the main industries of the islands.

NEPAL E

Area: 54,588 sq.mi.(141,383 km^2). **Population:** 22,500,000. **Capital:** Kathmandu, 500,000. **Government:** Democracy. **Language:** Nepali 50%; many others. **Religion:** Hinduism 90%; Buddhism 10%. **Exports:** Food products, wood, hides. **Climate:** Varies from alpine to tropical. ☐ The Himalayas, which include eight of the world's ten tallest peaks, occupy 90% of Nepal (nuh pawl'). The terrain consists mostly of mountain slopes. The country is less than 100 mi.(160 km) wide. It plummets from snowy Himalayan peaks to a swampy tropical plain on the southern border. Nepal is the birthplace of Gautama Buddha (560 BC), the founder of Buddhism. Nepal's famous Sherpa guides accompany many mountain-climbing expeditions originating in the country. The renowned Gurkha soldiers have distinguished themselves in the British and Indian armies; they were instrumental in making Nepal the only state in South Asia to have been able to resist British occupation. There is hope that the arrival of democracy will bring prosperity to this very poor nation.

PAKISTAN F

Area: 310;400 sq.mi.(803,936 km^2). **Population:** 135,200,000. **Capital:** Islamabad, 230,000. **Government:** Republic. **Language:** Urdu. **Religion:** Islam. **Exports:** Natural gas, cotton products, textiles, carpets, rice. **Climate:** Very dry and continental. ☐ In a land where monsoons blow hot or cold, but almost never wet, Pakistan is completely dependent on the mighty Indus River and its six major tributaries. The rivers provide water for the world's largest irrigation system. The Indus Valley was the site of many advanced, ancient South Asian civilizations. The largest industry is cotton and cotton goods. Most exports pass through the former capital, Karachi (5,350,000), a port on the Arabian Sea. Pakistan is a nation of many ethnic groups, each with its own language; fewer than 10% of the people speak Urdu, the official tongue. The nation was created as a homeland for Muslim minorities living in India. Although a conservative Islamic nation, where women are severely restricted, Pakistan elected Benazir Bhutto, the daughter of a former leader, as Prime Minister in 1988. She was removed from office in 1990 amid charges of corruption. Three years later she was reelected but she was removed again in 1996 for similar reasons.

SRI LANKA G

Area: 25,330 sq.mi.(65,605 km^2). **Population:** 19,000,000. **Capital:** Colombo, 2,000,000. **Government:** Republic. **Language:** Sinhala; Tamil. **Religion:** Buddhism 75%; Hinduism 18%. **Exports:** Tea, rubber, coconuts, graphite. **Climate:** Tropical. ☐ Sri Lanka (sree lahn'kuh), formerly Ceylon, is a beautiful, tropical island linked to the southeastern tip of the Indian mainland by a 20 mi.(32 km) chain of sandy islands called "Adam's Bridge." The Buddhist Sinhalese majority (75%) and the Hindu Tamil minority (18%) came originally from India. Tamil groups have been waging a guerrilla war for an independent northern state. A factional war has destroyed Sri Lanka's earlier promise of becoming an economically prosperous nation. Most people live in the wet and hilly southwest region, which is ideal for growing tea; Sri Lanka is the world's number-two producer of tea and the leading producer of high-quality graphite.

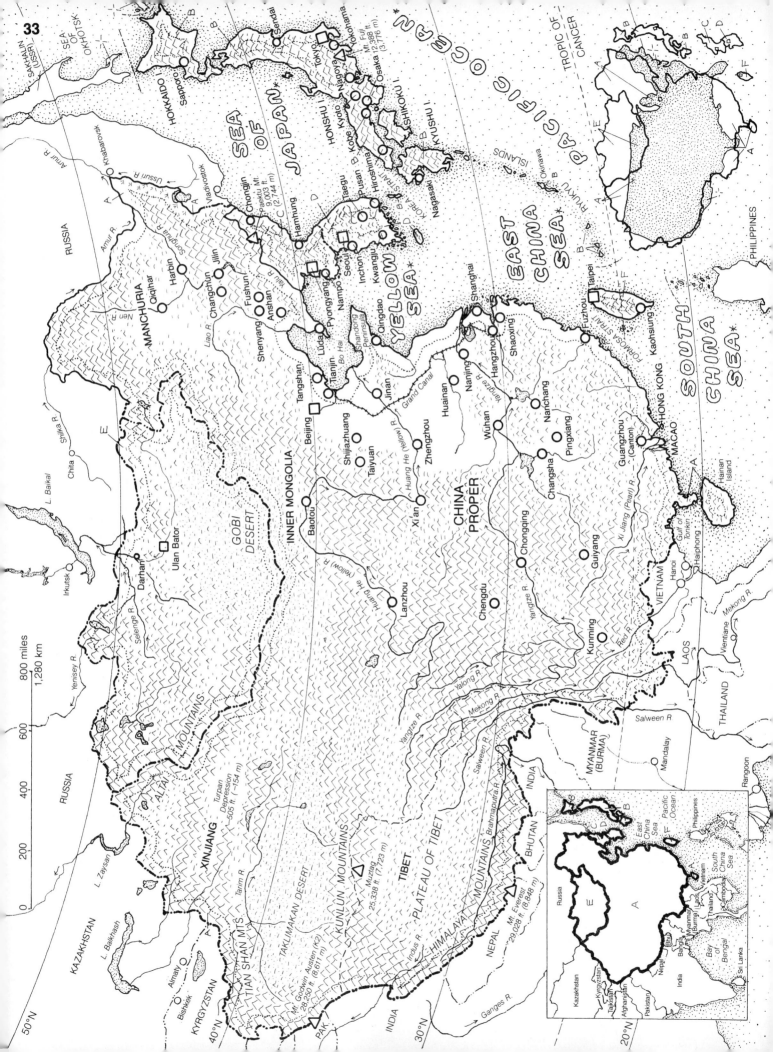

33

ASIA: EASTERN

Bordering the densely populated nations of the Far East is the world's most deserted country, Mongolia. The Mongolians were the only Asian communists to be influenced by the European reform movements of the 1980s. Although China maintains a politically hard-line, communist system, it has allowed its economy to follow the model of Japan and the "Little Dragons" (Hong Kong, South Korea, Singapore, and Taiwan). Until their financial collapse, these nations led the world in economic growth. But their success was not based on sound business practices; the inevitable burst of the bubble came in 1997.

For centuries, ancient China dominated this region. Surrounding nations bear the influence of China on their language, religion, and culture. The early 20th century marked the beginning of Japanese expansionism. By the beginning of World War II, Japanese aggression created an empire that reached the Indian border and covered Manchuria (northeastern China, which Japan called Manchukuo), eastern China, Korea (Chosen), Taiwan (Formosa), the Philippines, countries of southeast Asia, Indonesia, the islands of the western Pacific, and Alaska's Aleutian Islands—Japan's North American conquest.

CHINA A

Area: 3,685,000 sq.mi.(9,544,150 km²). **Population:** 1,237,000,000. **Capital:** Beijing, 6,000,000. **Government:** Republic (communist). **Language:** Mandarin (Northern Chinese); Cantonese. **Religion:** Confucianism 20%; Buddhism 7%; Taoism 2%. **Exports:** Manufactured goods, cotton, silk, textiles, tea, oil, asbestos. **Climate:** North and west are dry and continental; the east is moderate; the southeast is subtropical, with heavy summer rain. ☐ China, the world's oldest civilization, is the third-largest nation (after Russia and Canada) and has the largest population. A billion people are packed into the eastern part of China—almost one out of every five human beings on the planet. Although 80% of the population live a rural life, more than 30 cities have over one million residents. China is struggling to curb its population growth by severely penalizing families who have more than one child. There are numerous minorities in China, but their total number, though in excess of 50 million, represents less than 6% of the primarily Han Chinese population. Over two-thirds speak the northern dialect foreigners call Mandarin. All Chinese write with the same pictorial characters. A western alphabet that spells Chinese words phonetically is now used for foreign communications: Beijing (not Peking) is the capital, and Mao Zedong (not Mao Tse-tung) was the leader of the communist revolution. China is composed ot 22 provinces and 5 autonomous regions. Tibetan refugees, who witnessed the destruction of their culture by the Chinese government, would not agree that their homeland on the southwestern border is "autonomous." Manchuria, in the northeast, is a major industrial and agricultural region. Inner Mongolia, in the northcentral uplands, includes part of the Gobi Desert. Xinjiang, the sparsely populated northwest region, includes the Tian Shan Mountains, the Taklimakan Desert, and the Turpan Depression. China proper consists of the eastern lowlands and the central and southeastern uplands. The humid subtropical climate of south China permits multiple rice harvests. The fertile valleys of eastern China have been created by the Huang He and Yangtze Rivers, which originate in Tibet. The Huang He is called the Yellow River because its waters are tinted by the yellow soil it transports. Accumulations of silt can raise the riverbed, causing devastating floods and even a change of course. The Yangtze flows through the heart of China's commercial region and reaches the sea at one of the world's busiest ports, Shanghai. Virtually every village and city in southeast China is located on one of the hundreds of rivers and canals that serve as local highways. For centuries, junks with quilted sails and sampans looking like quonset huts have been used for fishing, transportation, and shelter. Close to the east coast is the longest artificial waterway in the world, the Grand Canal (1,105 mi.,1,768 km), which links Beijing with Hangzhou. The world's longest structure is the Great Wall, which snakes across the northern mountains and valleys for 1,500 mi.(2,400 km). It was designed to keep out marauding nomads and was built over a period of 2,000 years. The world's largest structure will be completed in 2009 when the Three Gorges Dam, a hydroelectric project on the Yangtze, will become operational. The communist revolution of 1949 eliminated disease and starvation but failed to raise the standard of living. In the 1980s, China's leaders allowed the introduction of economic reforms and free-market incentives but have brutally suppressed all moves for political freedom. The most poignant attempt was the student uprising in Beijing's Tiananmen Square, watched on television by millions around the world.

HONG KONG. Six million people (98% Chinese) are packed into this tiny peninsula and island group (410 sq.mi.,1,062 km²) off the south China coast. In 1997, after 150 years of British rule, ownership of Hong Kong reverted to China. It is not expected that there will be significant interference with the "Little Dragon's" capacity for doing business—the Asian financial crisis, in the same year, has had more of a dampening effect. In recent years, Hong Kong has transferred most of its production to the Chinese mainland in search of cheap labor. Hong Kong still functions as one of the Far East's centers for trade, finance, and tourism. Tall buildings in the capital, Victoria (1,000,000), overlook the harbor and Kowloon, its companion city on the peninsula.

MACAO. On the south China coast, 40 mi.(64 km) west of Hong Kong, is this Portuguese overseas province that, in 1999, will revert to China after 450 years of Portuguese rule. The Portuguese influence is evident in the city's Mediterranean appearance. Because gambling is banned in Hong Kong, tourists jetfoil to Macao's casinos. Back when it was a center for drug trading and foreign intrigue, Macao was called the "Casablanca of the Far East."

JAPAN B

Area: 145,745 sq.mi.(377,480 km²). **Population:** 126,000,000. **Capital:** Tokyo, 8,500,000. **Government:** Constitutional monarchy. **Language:** Japanese. **Religion:** Shintoism; Buddhism. **Exports:** Cars, trucks, ships, consumer electronics, chemicals, fish, textiles. **Climate:** Temperate, with heavy seasonal rainfall; north is cold. ☐ The "Land of the Rising Sun" has risen from the ruins of World War II to become the world's second-largest industrial economy. The military-dominated monarchy that directed Japan's war machine was modified by U.S.-imposed democratic institutions along with aid to rebuild its factories. Despite the cost of importing all of its raw materials, revenues from sales of its very fine products briefly made Japan the world's number one financial power. This was prior to a severe economic downturn that preceded the East Asian collapse by several years. Three-quarters of the population live on Honshu Island; the metropolitan area of Tokyo and Yokohama has over 25 million people. Every square inch of arable land in Japan is carefully cultivated. The northernmost island, Hokkaido, is a winter recreation area and the home of Japan's only minority group, the Ainu tribe, 15,000 Caucasians who are descendants of the country's original inhabitants. Hokkaido is cooled by the offshore Oyashio Current; the southern islands are warmed by the Kuroshio (Japan) Current (p. 57). The Shinto religion reveres many gods in nature. This reverence for natural forms is seen in Japanese architecture and landscaping. Most people practice a blend of religions and observe the holidays of both Buddhism and Shintoism. There is also a strong Confucian influence from China which advocates respect for family and authority. Today's Japanese culture is a fascinating combination of the traditional and the modern. This can be seen in the summer ritual of climbing Mt. Fuji, the sacred extinct volcano. What was formerly the practice of religious pilgrims is now a recreational pursuit of tens of thousands who seek the exercise and the magnificent view. The cone-shaped, snow-capped Fuji is the most familiar graphic symbol of Japan.

NORTH KOREA C

Area: 46,700 sq.mi.(120,953 km²). **Population:** 22,200,000. **Capital:** Pyongyang, 2,750,000. **Government:** Republic (communist). **Language: Korean. Religion:** Officially discouraged. **Exports:** Fish, graphite, iron ore, copper, lead, zinc. **Climate:** Very cold winters. ☐ When Japan was forced to withdraw from "Chosen" (Korea) after its defeat in World War II, the North was occupied by the Soviets and the South by the U.S. When these nations withdrew, the peninsula remained permanently divided. In 1950, communist North Korea invaded South Korea; U.S. and U.N. troops came to the aid of the South. After three years of fighting, the armies were stalemated along the 38th parallel, the original dividing line. For nearly 50 years, U.S. troops have protected South Korea. North and the South are about the same size, but the North has mineral wealth, and South Korea has arable land and twice the population. The Korean language shows the influence of earlier Chinese and Japanese occupations. Under Kim Il Sung's 45-year reign, North Korea was the incarnation of George Orwell's 1984. Sung's picture was everywhere, and there were daily indoctrinations. Following his death in 1991, the harsh conditions continued under his son's rule. Food shortages have added to the misery.

SOUTH KOREA D

Area: 38,025 sq.mi.(98,485 km). **Population:** 46,500,000. **Capital:** Seoul, 8,400,000. **Government:** Republic. **Language:** Korean. **Religion:** Buddhism; Confucianism; Christianity. **Exports:** Consumer electronics, cars, textiles, chemicals. **Climate:** Moderate, with hot, damp summers. ☐ In two decades, South Korea had changed from a poor, agricultural nation to one of the "Little Dragons." The accumulation of great wealth was being flaunted, contrary to Confucian tradition (which relegates the merchant to the lowest class), and it fueled tensions between rich and poor. But the crash of 1997 suddenly reduced the ranks of the moneyed class. The Christian minority (25%) is the largest on the Asian mainland. The Unification Church, which sponsors the "Moonies" cult, was founded in South Korea by Sun Myung Moon.

MONGOLIA E

Area: 604,248 sq.mi.(1,565,002 km²). **Population:** 2,600,000. **Capital:** Ulan Bator, 625,000. **Government:** Republic. **Language:** Mongolian. **Religion:** Buddhism. **Exports:** Livestock, animal products, fluorspar, tungsten. **Climate:** Dry and extremely continental. ☐ Trees are even scarcer than people in the world's largest landlocked nation. The Gobi, the most northern desert, occupies the southern third of the land. In the 13th century, a Mongolian, Ghengis Khan, reigned over the largest land empire in history, from eastern Europe to the coast of China (p. 48). The barbarian empire dissolved, and the region became "Outer Mongolia" under a long period of Chinese control. Mongolia was a staunch ally of the Soviet Union, which helped liberate it from China in 1924. The once-nomadic people formerly herded livestock and lived in large, round, felt-covered tents called "yurts"; most now live in villages constructed by the former communist government (which still wields influence).

TAIWAN F

Area: 13,910 sq.mi.(36,027 km²). **Population:** 22,000,000. **Capital: Taipei, 2,700,000. Government: Republic. Language:** Mandarin Chinese; dialects. **Religion:** Buddhism; Confucianism; Taoism. **Exports:** Clothing, consumer electronics, plastics. **Climate:** Subtropical. ☐ In 1949, Taiwan became "Nationalist China" when Chiang Kaishek, his army, and followers left China after the communist victory. For many years the Nationalists vowed to return to the mainland, 100 mi.(160 km) away. But in 1971, the improbable dream became less possible when communist China replaced Taiwan at the United Nations. Taiwan has gone on to become one of the Far East's prosperous "Little Dragons." Most residents live and work in the coastal areas. Dense mountain forests provide over half the world's supply of camphor, a substance used in chemicals, pharmaceuticals, and cosmetics.

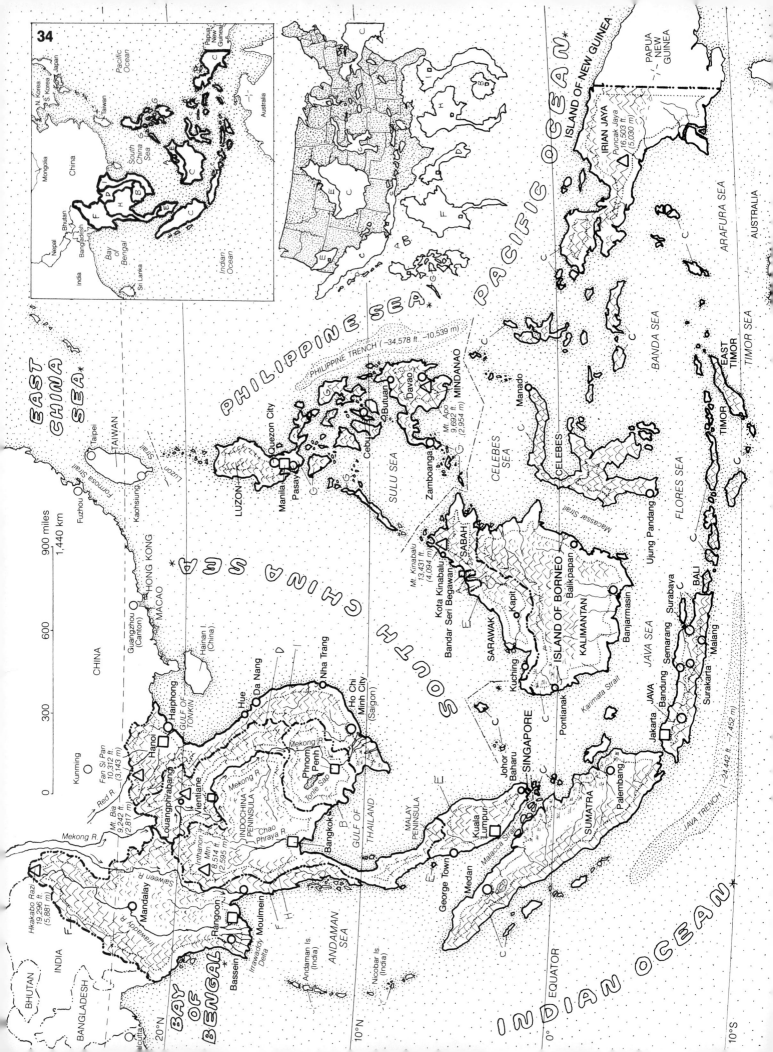

34

BHUTAN
BANGLADESH
INDIA
Calcutta

Hkakabo Razi
19,296 ft
(5,881 m)

Mekong R.
Kunming

CHINA

Fan Si Pan
10,312 ft
(3,143 m)
Red R.

Mt. Bia
9,242 ft
(2,817 m)

Hanoi
Haiphong
GULF OF TONKIN

Hainan I. (China)

Guangzhou (Canton)
MACAO
HONG KONG

Fuzhou

900 miles
1,440 km
600
300
0

Kunming

Mandalay
Irrawaddy R.
Salween R.

Louangphrabang
Vientiane

Inthanon Mtn.
8,514 ft
(2,595 m)

Mekong R.

INDOCHINA PENINSULA
Chao Phraya R.

Phnom Penh
Tonle Sap

Bangkok

GULF OF THAILAND

Da Nang
Hue
Nha Trang
Ho Chi Minh City (Saigon)

Rangoon
Moulmein
Bassein
Irrawaddy Delta

Andaman Is. (India)

ANDAMAN SEA

Nicobar Is. (India)

MALAY PENINSULA

George Town
Medan
Malacca Strait

Kuala Lumpur

Johor Baharu
SINGAPORE

SUMATRA

Palembang

EAST CHINA SEA*

Taipei
TAIWAN
Formosa Strait
Kaohsiung
Luzon Strait

SOUTH CHINA SEA*

PHILIPPINE SEA*

PHILIPPINE TRENCH (−34,578 ft, −10,539 m)

LUZON
Quezon City
Manila
Pasay

Cebu

Butuan
Davao
MINDANAO
Mt. Apo
9,692 ft
(2,954 m)
Zamboanga

SULU SEA

Mt. Kinabalu
13,431 ft
(4,094 m)
Kota Kinabalu
SABAH
Bandar Seri Begawan
Kapit
SARAWAK
Kuching

ISLAND OF BORNEO
KALIMANTAN
Balikpapan
Banjarmasin
Pontianak
Karimata Strait

CELEBES SEA
Manado
CELEBES
Ujung Pandang
Macassar Strait

PACIFIC OCEAN*

ISLAND OF NEW GUINEA

IRIAN JAYA
Puncak Jaya
16,503 ft
(5,030 m)

PAPUA NEW GUINEA

ARAFURA SEA

AUSTRALIA

BANDA SEA

TIMOR
EAST TIMOR
TIMOR SEA

FLORES SEA

BALI
Surabaya
Surakarta
Malang
Semarang
JAVA
Bandung
Jakarta

JAVA SEA

JAVA TRENCH (−24,442 ft, −7,452 m)

INDIAN OCEAN*

EQUATOR
0°
10°S
10°N
20°N

BAY OF BENGAL

Inset map labels: N Korea, S Korea, Japan, Mongolia, China, Nepal, Bhutan, Bangladesh, India, Sri Lanka, Taiwan, Pacific Ocean, South China Sea, Bay of Bengal, Indian Ocean, Australia, Papua New Guinea

ASIA: SOUTHEASTERN -;-

The nations of southeastern Asia are located on the Indochina and Malay Peninsulas, in the Malay Archipelago, and in the Philippines. The region is mountainous, forested, very warm, extremely humid, and seasonally drenched. Rivers and canals serve as roads in many areas. Native homes are perched on stilts for protection against flooding and wild animals. The watery landscape sustains the basic diet of fish and rice. Malaysia, Indonesia, and Thailand are the world's leading tin and rubber producers. Rubber trees were imported from South America in the 19th century. Tropical hardwood forests of teak, ebony, mahogany, and rosewood take 200 years to grow but are being cut down at an ever-increasing rate. Elephants are used as trucks by the timber industry, and water buffalos serve as local farm tractors. Brunei, Malaysia, and Indonesia are Islamic nations. The countries on the Indochina peninsula, including Cambodia and communist Laos and Vietnam (all part of French Indochina), are Buddhist; temples and monasteries are everywhere. The British once occupied Brunei, Burma, Malaysia, and Singapore. The Dutch governed Indonesia (the Dutch East Indies) for 300 years. Spain and the U.S. ruled the Philippines, the only Christian nation in eastern Asia.

BRUNEI A

Area: 2,226 sq.mi.(5,765 km²). **Population:** 315,000. **Capital:** Bandar Seri Begawan, 23,000. **Government:** Sultanate. **Language:** Malay; English. **Religion:** Islam. **Exports:** Oil, natural gas, rubber. **Climate:** Hot and wet. ☐ Brunei is similar to an oil-rich Arab state: the ruling sultan is the world's richest man; wealthy residents are mostly Muslims; there are no taxes; and social services are free. The Chinese minority (20%) operate most of the businesses. Brunei is split in two by a strip of land belonging to Malaysia.

CAMBODIA B

Area: 69,890 sq.mi.(181,015 km²). **Population:** 11,500,000. **Capital:** Phnom Penh, 1,000,000. **Government:** Constitutional monarchy. **Language:** Khmer. **Religion:** Buddhism. **Exports:** Rubber, rice, timber. **Climate:** Tropical. ☐ Cambodia is saucer-shaped; its large central plain region surrounds the Tonle Sap (the "Great Lake"). The lake provides fish protein as well as widespread flooding for rice farming. Most Cambodians are Khmers, the region's oldest ethnic group. From 700 to 1200 AD they dominated Indochina. In 1975, radical communists, called the Khmer Rouge, unleashed a reign of terror that killed over a million Cambodians. The "Killing Fields" were filled with bodies of potential reactionaries: intellectuals, priests, reporters, professionals, librarians, teachers, merchants and their families—anyone with eyeglasses. Cities were emptied and citizens were sent to work in the country. In 1979, Vietnam drove the Khmer Rouge out of power. After Vietnam's withdrawal in 1989, the country moved toward democracy. A constitutional monarchy is now in place.

INDONESIA C

Area: 741,100 sq.mi.(1,919,449 km²). **Population:** 213,000,000. **Capital:** Jakarta, 8,000,000. **Government:** Military. **Language:** Bahasa Indonesia. **Religion:** Islam. **Exports:** Oil, timber, rubber, tin, coffee, palm oil, sugar. **Climate:** Tropical. ☐ Indonesia occupies more than 13,000 islands straddling the Equator. The Malay Archipelago is the world's longest island chain (wider than the U.S.). It has the world's fifth-largest population, making it the largest Islamic nation. Nearly 120 million people live on Java, an island the size of Alabama. A Chinese minority (2%) run many of the nation's businesses. They bore the brunt of mass rioting in early 1998 over harsh conditions brought on by the nation's financial collapse. Corruption of the Suharto family-run regime has contributed to the economic problems. General Suharto is also accused of executing upwards of half a million suspected communists in the 1960s, as well as the murder and suppression of thousands of residents of East Timor, the former Portuguese colony that Indonesia invaded and annexed in the 1970s. Buddhist and Hindu temples are reminders of a pre-Islamic past. Indonesians still believe in spirits and tend to follow their own brand of Islam. Bali, an island off the coast of Java, is a Hindu enclave filled with magnificent temples. Tourists are drawn by Bali's great beauty and its highly advanced culture based upon music and dance. Kalimantan, the largest state, occupies most of the oil-rich island of Borneo. Indonesia has many earthquakes and the greatest number of active volcanoes. In 1883, the largest explosion in history destroyed most of the island of Krakatoa. The blast was heard for thousands of miles, and dust spewed into the atmosphere affected the earth's weather for several years.

LAOS D

Area: 91,430 sq.mi.(236,804 km²). **Population:** 5,300,000. Capital: Vientiane, 425,000, **Government:** Republic (communist). **Language:** Lao. **Religion:** Buddhism. **Exports:** Tin, coffee, timber. **Climate:** Tropical. ☐ Landlocked Laos (lah' os) is a poor nation wedged between two historically dominant neighbors: Thailand and Vietnam. The Mekong River, which defines most of the western border, is the main highway in a country of few roads and no railroads. Most of the diverse people live in the Mekong Valley. Laos was heavily bombed by the U.S. in the Vietnam War, when its "Ho Chi Minh Trail" was used by the North Vietnamese to supply their forces in South Vietnam. Thousands of refugees have left Laos since the communist takeover in 1975.

MALAYSIA E

Area: 127,318 sq.mi.(329,754 km²). **Population:** 17,000,000. **Capital:** Kuala Lumpur, 550,000. **Government:** Constitutional monarchy. **Language:** Bahasa Malaysia; Chinese. **Religion:** Islam 50%; Buddhism 35%. **Exports:** Oil, tin, rubber, palm oil, spices. **Climate:** Tropical. ☐ Malaysia (muh lay' zhuh) is a federation of independent sultanates. Two of them, oil-rich Sarawak and Sabah, occupy the northern part of Borneo. Eighty percent of the population lives on the southern end of the Malay Peninsula. Tensions have existed between the politically powerful Malayan Muslim majority and the economically strong Chinese Buddhists. The nation is the top producer of rubber, tin, and palm oil. Malaysia had a rapidly expanding industrial sector until the crash of 1997.

MYANMAR (BURMA) F

Area: 261,750 sq.mi.(677,933 km²). **Population:** 47,300,000. **Capital:** Rangoon, 2,500,000. **Government:** Military. **Language:** Burmese. **Religion:** Buddhism. **Exports:** Rice, teak, sugar, gems, rubber. **Climate:** Tropical. ☐ Myanmar (mee ahn' mahr), whose name was changed from "Burma" by a military dictatorship, is politically isolated from the international community. The Irrawaddy River, the nation's main highway, flows through the agricultural and population centers. Its delta is the world's leading rice-producing area. At one of the mouths of the river is the capital, Rangoon. The city grew up around Shwe Dagon, a 2,500-year-old pagoda that is strikingly decorated with tall, ornately molded, gold leaf-covered spires. In the surrounding mountains live over 100 ethnically different native tribes.

PHILIPPINES G

Area: 115,835 sq.mi.(300,013 km²). **Population:** 78,000,000. **Capital:** Manila, 1,700,000. **Government:** Republic. **Language:** Filipino; English. **Religion:** Roman Catholic. **Exports:** Timber, copra, abaca, sugar, copper. **Climate:** Tropical and humid. ☐ The Philippines Archipelago consists of 7,000 mountainous, earthquake-prone, volcanically active islands. Only ten percent are inhabited. The largest, Luzon and Mindanao, are home to two-thirds of the population. The people are mostly of Malay and some Polynesian ancestry. In many ways the Philippines resembles a Latin American country: it was a Spanish colony for over three centuries (Magellan discovered it in 1521); it is predominantly Catholic; the capital, Manila, is a contrast between modern architecture and frightful slums; a left-wing guerrilla movement has arisen in response to widespread poverty; there has not been any land reform; and corrupt presidents have become rich while in office. The U.S. acquired the Philippines from Spain after the Spanish-American War of 1898 and governed the colony for 48 years. An American-style educational system was introduced, and English is widely used. The Philippines is the leading producer of copra (coconut meat) and abaca (Manila hemp). Rice is the principal crop; some of the paddies, high in the mountains, were terraced over 2,000 years ago.

SINGAPORE. Singapore has been the dominant shipping port of Southeast Asia since 1819, when an agent of the British East India Company created it for trading purposes. The tiny (227 sq.mi., 588 km) island republic is connected to the tip of the Malay Peninsula by a causeway less than a mile long. It is the richest of Asia's "Little Dragons." Singapore is a center for finance, general manufacturing, consumer electronics, oil and rubber processing, shipbuilding, and shipping. Three-fourths of the 2,700,000 people are Chinese, and almost all live in the city of Singapore, known for rigidly enforced laws for public behavior.

THAILAND H

Area: 198,450 sq.mi.(513,986 km²). **Population:** 60,000,000. **Capital:** Bangkok, 5,600,000. **Government:** Constitutional monarchy. **Language:** Thai. **Religion:** Buddhism. **Exports:** Rubber, tin, teak, bamboo, handcrafts. **Climate:** Tropical. ☐ In Thailand (ty' land), formerly called Siam, almost everything happens on or close to water. The capital, Bangkok, at the mouth of Chao Phraya River, is reminiscent of Venice. The river basin, a fertile alluvial plain, is the national "rice bowl," and the heart of a prosperous economy. Rivers are often filled with logs from Thailand's sizeable teak forests. Every village has its Buddhist temple, and each Thai boy is expected to spend a few months living the life of a monk. Thailand means "free country" and Thais have managed to enjoy 700 years of independence through skillful diplomacy. They gave military assistance to Britain and France in World War I in exchange for a guarantee of Thai sovereignty. Thailand is now burdened by refugees from its communist neighbors—Cambodia (formerly communist), Laos, and Vietnam.

VIETNAM I

Area: 127,240 sq.mi.(329,551 km²). **Population:** 76,000,000. **Capital:** Hanoi, 2,500,000. **Government:** Republic (communist). **Language:** Vietnamese. **Religion:** Buddhism. **Exports:** Coal, timber, rubber, rice. **Climate:** Tropical. ☐ Vietnam is mostly mountainous; its huge population is concentrated in two delta regions connected by a long, narrow strip of low coastal land. Hanoi, the capital, is located on the silt-colored Red River delta in the north. The Mekong River delta in the south lies 3,200 mi.(5,120 km) from its headwaters in Tibet. The nation's largest city is on the delta—French-flavored Ho Chi Minh City (4,400,000), formerly Saigon. Vietnamese communists under Ho Chi Minh led the resistance against the Japanese occupation in World War II. The French tried to reclaim their colony after the war, but they were defeated by the communists in 1954 after eight years of fighting. The country was then temporarily divided; elections were to determine reunification. The non-communist South, backed by the U.S., refused to hold the elections, fearing defeat by the immensely popular Ho Chi Minh. The Viet Cong guerrilla movement was formed to overthrow the unpopular South Vietnamese government, run by wealthy, pro-French, Roman Catholics (Vietnam is 80% Buddhist). Vietnam harbors a traditional hatred of China because of 900 years of occupation. Ignoring Vietnamese history, the U.S. mistook the communist-led nationalist movement for Chinese expansionism, and intervened to prevent the fall of South Vietnam. The North responded with its own intervention, and the war escalated, culminating with a U.S. withdrawal in 1973. The South soon surrendered, and Vietnam became the third-largest communist country (it is now the second). Thousands of refugees, including the desperate "boat people," have left Vietnam since the war. Like China, Vietnam has eased restrictions on trade and commerce, but Vietnam and China remain enemies.

35

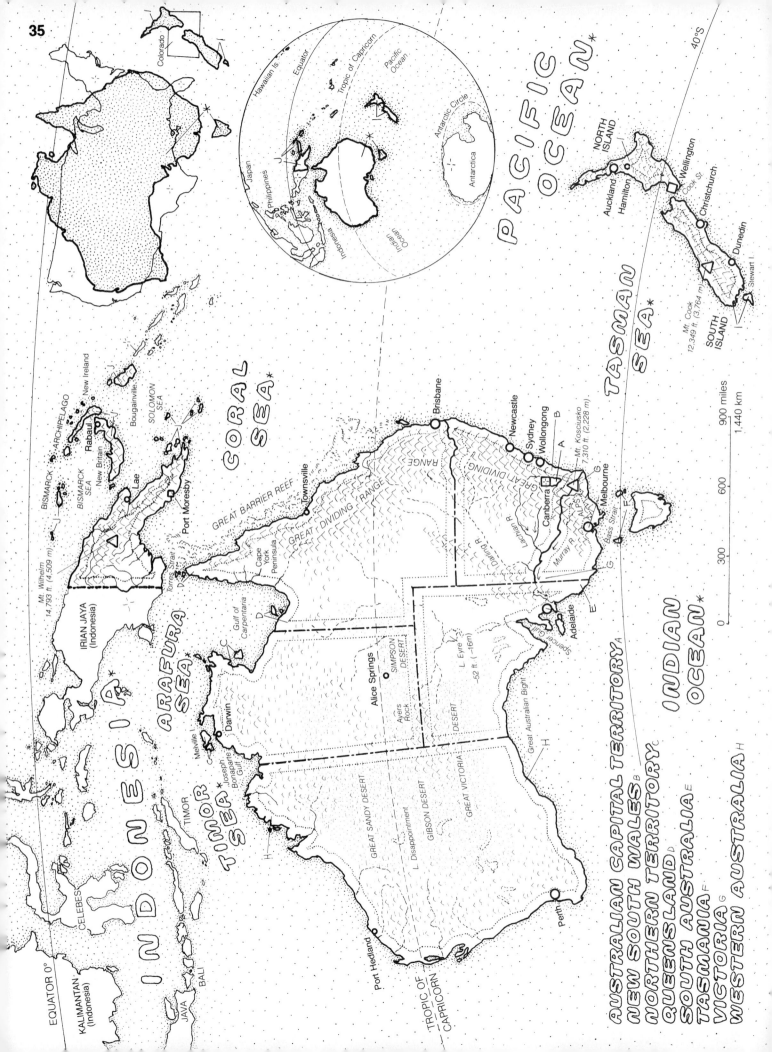

PACIFIC OCEAN*

TASMAN SEA*

NORTH ISLAND
Auckland
Hamilton
Wellington
Cook St.
Christchurch
Mt. Cook 12,349 ft. (3,764 m)
SOUTH ISLAND
Dunedin
Stewart I.

40°S

Hawaiian Is.
Equator
Tropic of Capricorn
Pacific Ocean
Japan
Philippines
Indonesia
Antarctic Circle
Antarctica
Indian Ocean

Colorado

CORAL SEA*

GREAT BARRIER REEF
Townsville
GREAT DIVIDING RANGE
Brisbane
Newcastle
Sydney
Wollongong
A
B
Mt. Kosciusko 7,310 ft. (2,228 m)
Canberra
ALPS
G
Melbourne
F
Bass Strait
G

GREAT DIVIDING RANGE
Cape York Peninsula

Lachlan R.
Darling R.
Murray R.
E

BISMARCK ARCHIPELAGO
New Ireland
Bougainville
SOLOMON SEA
Rabaul
New Britain
Lae
BISMARCK SEA
Port Moresby
Mt. Wilhelm 14,793 ft. (4,509 m)
IRIAN JAYA (Indonesia)

Torres Strait I.
Gulf of Carpentaria
D
C

ARAFURA SEA*

SIMPSON DESERT
Alice Springs
Ayers Rock
L. Eyre −52 ft. (−16m)
DESERT
Spencer Gulf
Adelaide
E

TIMOR SEA*
TIMOR
Joseph Bonaparte Gulf
Melville I.
Darwin
H

Great Australian Bight
H

INDONESIA*
CELEBES
KALIMANTAN (Indonesia)
JAVA
BALI

GREAT SANDY DESERT
L. Disappointment
GIBSON DESERT
GREAT VICTORIA DESERT
H
Port Hedland
Perth

EQUATOR 0°
TROPIC OF CAPRICORN

900 miles
1,440 km
600
300
0

INDIAN OCEAN*

AUSTRALIAN CAPITAL TERRITORY A
NEW SOUTH WALES B
NORTHERN TERRITORY C
QUEENSLAND D
SOUTH AUSTRALIA E
TASMANIA F
VICTORIA G
WESTERN AUSTRALIA H

OCEANIA I

Oceania is a broad expanse of water that surrounds a continent along with thousands of tiny, widely scattered islands. Oceania (also called Australasia) encompasses a region 8,000 mi. (12,800 km) wide in the Southern Hemisphere, from Australia to the central Pacific Ocean. Australia is by far the largest of the thousands of islands. It is large enough to be classified as a continent, but when compared to other continents, it is the smallest, driest, lowest, flattest, and oldest. Over time, the three-billion-year-old landscape has been worn down, making Australia the only continent without a tall mountain range. The Great Dividing Range (Australia's continental divide) is close to the east coast and has the continent's highest mountains. The Australian Alps, on its southern end, provide winter recreation and the summer runoff that irrigates the lowlands to the west. The farther south one travels in Australia and New Zealand, the cooler it gets. The two main islands of New Zealand, lying 1,200 mi.(1,920 km) to the southeast, are the opposite of flat, hot, and dry Australia. New Zealand is green and hilly, with a moist climate. Some of its southern mountain peaks have a permanent snow cover.

The western two-thirds of Australia is a low plateau with the most deserts of any continent. Between this plateau and the fertile east coast's Great Dividing Range are dry lowlands called artesian basins. Artesian water is underground water under pressure, which rises anywhere there is a crack in the Earth's surface. The underground water percolates from the eastern mountains. Because of a very high salt content, the water isn't drinkable and has only a limited crop application, but it does support the millions of sheep that dot the "outback." The westward-blowing Trade Winds (p. 57) are stripped of their moisture by the Great Dividing Range, causing rivers and lakes in central and western Australia to stay bone-dry most of the year (they are represented on the map by broken lines).

The Great Barrier Reef, the world's largest coral reef, runs 1,250 mi.(2,000 km) on the northeast coast of Australia. This multi-colored mecca for scuba divers is a collection of 2,500 individual reefs supporting thousands of species of marine life. The reefs are made from the skeletons of hundreds of species of sea coral. In order for reef building to take place, water temperature must exceed 65° F (18.3° C).

Australia and the other islands of Oceania are noted for their unusual wildlife. Because the lands have been separated from other continents for millions of years (p. 2), many plants and animals are unique to this part of the world. Most of the world's marsupials (mammals that raise their immature young in external pouches), including the well-known kangaroo, live only in Australia. Two unusual non-marsupials are the platypus and the echidna (the spiny anteater), the only mammals in the world that lay eggs. There is only one placental mammal found everywhere in Oceania: the bat. No other mammal had the ability to cross the ocean. The dingo, a wild dog that has roamed Australia for thousands of years, originally came from Asia with the Aborigines. The emu and cassowary, large flightless birds that resemble ostriches, also are confined to this continent. The koala, a leaf-eating, teddy bear-like marsupial, is a common resident of the tall eucalyptus trees. There are over 400 eucalyptus species and 600 species of the acacia tree. These trees and other varieties of Australian plant life have been transplanted to other parts of the world with similar climates, such as California.

AUSTRALIA *

Area: 2,970,000 sq.mi.(7,692,300 km²). **Population:** 18,700,000. **Capital:** Canberra, 285,000. **Government:** Constitutional monarchy. **Language:** English. **Religion:** Protestant. **Exports:** Wool, iron ore, coal, bauxite, beef, cereals, sugar. **Climate:** The tropical north gets heavy winter rains; the east coast is mild; Tasmania is cool and damp. The rest of the country is very dry and seasonally hot. ☐ Australia, the smallest continent but one of the largest countries (about the size of the U.S. without Alaska), has been called the "Land Down Under" because of its location south of the Equator. The name Australia comes from the Latin "australis," meaning "southern." Coastal parts of Australia were first explored by the Netherlands, Portugal, and Spain in the early 17th century. In 1770, Captain Cook arrived and claimed the continent for Great Britain. The earliest immigrants were inmates of Britain's new overseas penal colony. For many years the only European settlers were British and Irish immigrants; Australia was a huge, isolated, English outpost in Southeast Asia. The nation became completely independent in 1901, and after World War II, other ethnic Europeans were admitted. Immigration laws were later expanded to include Asian nationals, and today, Australia is a multi-racial and multi-cultural nation; close to a quarter of the population is foreign-born. When the first Europeans arrived, approximately 300,000 Aborigines were living on the continent. Their ancestors migrated from Asia 20,000 to 50,000 years ago. These dark-skinned nomads were nearly wiped out by European disease and mistreatment. Their numbers are on the rise (currently 140,000) as the government is making restitution for the wrongs of the past. They are now able to retain their tribal lands and culture, or have the choice of becoming thoroughly integrated into Australia's modern society. One well-known artifact, the boomerang, is an ancient Aborigine invention. Aboriginal artwork has been gaining international interest.

Even with growing industries, Australia's greatest wealth continues to come from mining and agriculture. The country is the top exporter of bauxite, and it has large reserves of iron, coal, lead, zinc, and other minerals. Australia is the world's leading producer of wool. Most of the sheep are the incredibly wooly merinos, which thrive in the hot, dry interior ("outback"). Not surprisingly, the country is also the leading exporter of lamb and mutton. Cattle do well on the relatively barren ranches (called "stations"); the weather is mild and there is plenty of land for each animal. Australia is so large, and its meager population so scattered, that in the outback, medical aid has to be rendered by airplane, and children are taught at home by radio and mail correspondence.

Most Australians live in the cities. The populated areas, farms, and industries are located along the southeast coast in the state of *New South Wales*. Sydney (3,800,000), the largest city, has one of the world's great natural harbors. Sydney's ultra-modern opera house is a startling sight; its many pointed roofs resemble the sails of passing boats. Australia is made up of six states and two territories. The tiny *Australian Capital Territory,* located within New South Wales, is the site of the nation's capital, Canberra, which is the only major city that isn't on the coast. *Victoria,* an important farming state, is on the southeast coast. Its principal city and major port is Melbourne (3,250,000). The smallest state, *Tasmania,* is an island off the southeastern tip of the mainland. The Great Dividing Range submerges beneath the Bass Strait and surfaces to form the island. This important apple-growing region is a popular resort area. *Queensland* occupies the northeast corner of the nation. The savanna-like land is ideal for cattle raising. Three states—the *Northern Territory, South Australia,* and *Western Australia*—make up two-thirds of the nation. Except for a narrow Mediterranean climate belt along the south coast, these states are mostly barren desert. The remarkable Ayers Rock, located near the center of Australia, may be the world's largest monolith. This enormous oval boulder is sacred to the Aborigines. It rises abruptly from the flat desert floor to a height of 1,140 ft.(348 m) and is nearly 2 mi.(3 km) long. It can be compared to an iceberg in that only about 5% of the rock is above ground.

NEW ZEALAND /

Area: 103,775 sq.mi.(268,777 km²). **Population:** 3,700,000. **Capital:** Wellington, 350,000. **Government:** Constitutional monarchy. **Language:** English; Maori. **Religion:** Protestant. **Exports:** Dairy products, lamb, wool, fruit, fish, paper products. **Climate:** Mild; temperatures vary with latitude and altitude. ☐ Mild weather, ample rainfall, and heavy fertilization are responsible for New Zealand's fine pasturelands. The nation is the world's largest exporter of dairy products and lamb, and the second-largest producer of wool. Because its orchards are harvested when the Northern Hemisphere is in winter, New Zealand's produce is in much demand. Almost all shipping passes through the nation's largest city, Auckland (975,000), on the North Island, home to three-fourths of the population.

New Zealand was discovered by Dutch explorer Abel Tasman in 1642. He was driven off by the Maoris, who had arrived from Polynesia around 700 years earlier. It wasn't until Captain Cook established good relations with the natives in 1769 that Britain was able to settle the islands. By the early 19th century, the Maoris thought that immigration had gotten out of hand and began attacking the Europeans. British troops arrived to establish order, and eventually a peace treaty was signed in 1840, giving the British sovereignty over the islands but assuring land ownership to the Maoris. Though the exact interpretation of the treaty is still being debated, the Maoris are once again thriving because of a government effort to respect their civil and property rights. The Maoris make up about 10% of the population, and intermarriage between these Polynesians and New Zealanders of European descent is common. New Zealand has always been a politically progressive nation and was one of the first to enact social welfare legislation. In 1893, New Zealand's women were the first in the world to receive full voting rights.

Because New Zealand is located at the point where the Pacific tectonic plate passes under the Indo-Australian plate (p. 2), the islands are geologically very active. In the center of North Island is a barren plateau with a hellish environment: active volcanoes, steaming fumaroles (vapor vents), powerful geysers, boiling hot springs, and bubbling mud pools. The world's first steam-powered geothermal plant was built here in 1961. Rushing rivers also provide hydro-electric power. The tallest mountains are on the rugged South Island. Its southwest coast is called Fiordland. The most remarkable fiord (the English spell it with an "j") is Milford Sound, whose mile-high sea cliffs are the world's tallest. An unusual combination of low light, saltwater, freshwater, and warm water temperature create a unique sea ecology that contains the world's largest formation of black coral. New Zealand does not have any snakes, but a native reptile called a tuatara predates the dinosaurs. Moas were enormous birds (up to 12 ft., 3.7 m tall), hunted into extinction by the early Maori tribes. The kiwi is not just a popular fruit; it is also a wingless bird that is the national symbol and the name by which New Zealanders refer to themselves.

PAPUA NEW GUINEA J

Area: 178,800 sq.mi.(463,092 km²). **Population:** 4,600,000. **Capital:** Port Moresby, 255,000. **Government:** Constitutional monarchy. **Language:** English; Pidgin; nearly 700 dialects. **Exports:** Copper, coffee, cocoa, copra, timber. **Climate:** Tropical and damp. ☐ Papua New Guinea includes the eastern half of the island of New Guinea, the Bismarck Archipelago, Bougainville and Buka in the Solomon Islands, and about 600 smaller islands. Some of the world's most primitive peoples live in this country, parts of which remain unexplored. The dark-skinned Melanesian inhabitants represent a great number of cultures and speak over 700 dialects. Pidgin English, which is English heavily flavored by native dialects, is the closest thing to a common tongue. In the late 19th century, the northern part of the island was controlled by Germany, which gave it to Australia following World War I. Papua New Guinea received its independence in 1975, but continues to rely on Australian aid.

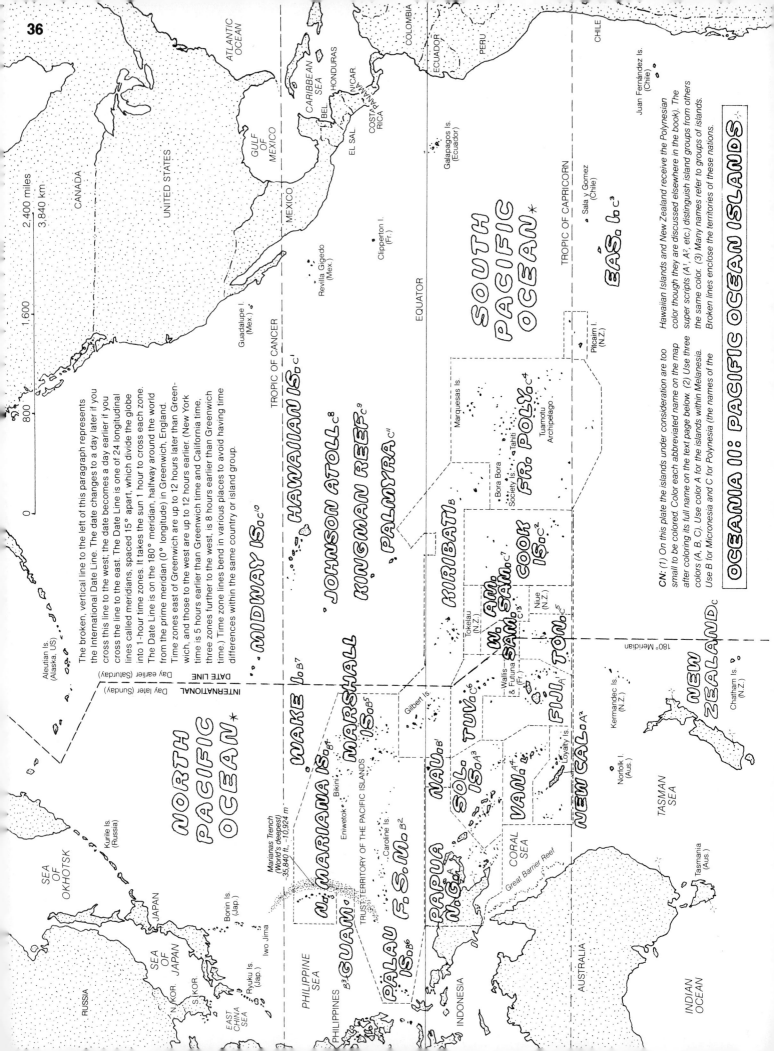

OCEANIA II: PACIFIC OCEAN ISLANDS

CN: (1) On this plate the islands under consideration are too small to be colored. Color each abbreviated name on the map after coloring its full name on the text page below. (2) Use three colors (A, B, C). Use color A for the islands within Melanesia. Use B for Micronesia and C for Polynesia (the names of the

Hawaiian Islands and New Zealand receive the Polynesian color though they are discussed elsewhere in the book). The super scripts (A[1], A[2], etc.) distinguish island groups from others the same color. (3) Many names refer to groups of islands. Broken lines enclose the territories of these nations.

The broken, vertical line to the left of this paragraph represents the International Date Line. The date changes to a day later if you cross this line to the west; the date becomes a day earlier if you cross the line to the east. The Date Line is one of 24 longitudinal lines called meridians, spaced 15° apart, which divide the globe into 1-hour time zones. It takes the sun 1 hour to cross each zone. The Date Line is on the 180° meridian, halfway around the world from the prime meridian (0° longitude) in Greenwich, England. Time zones east of Greenwich are up to 12 hours later than Greenwich, and those to the west are up to 12 hours earlier. (New York time is 5 hours earlier than Greenwich time and California time, three zones further to the west, is 8 hours earlier than Greenwich time.) Time zone lines bend in various places to avoid having time differences within the same country or island group.

INTERNATIONAL DATE LINE

Day later (Sunday)
Day earlier (Saturday)

NORTH PACIFIC OCEAN *

SOUTH PACIFIC OCEAN *

Map labels:
- SEA OF OKHOTSK
- RUSSIA
- Kurile Is. (Russia)
- Aleutian Is. (Alaska, US)
- N. KOR.
- S. KOR.
- EAST CHINA SEA
- JAPAN
- SEA OF JAPAN
- Ryukyu Is. (Jap.)
- Bonin Is. (Jap.)
- Iwo Jima
- PHILIPPINES
- PHILIPPINE SEA
- INDONESIA
- AUSTRALIA
- TASMAN SEA
- Tasmania (Aus.)
- CORAL SEA
- Great Barrier Reef
- INDIAN OCEAN
- Norfolk I. (Aus.)
- Kermandec Is. (N.Z.)
- Chatham Is. (N.Z.)
- NEW ZEALAND$_C$
- Marianas Trench (World's deepest) -35,840 ft., -10,924 m
- Bikini
- Eniwetok
- Caroline Is.
- TRUST TERRITORY OF THE PACIFIC ISLANDS
- Gilbert Is.
- Tokelau (N.Z.)
- Wallis & Futuna (Fr.)
- Niue (N.Z.)
- Loyalty Is.
- PALAU IS.$_{D}$B[6]
- GUAM B[3]
- N. MARIANA IS.$_{D}$B[4]
- F.S.M.$_{D}$B[2]
- MARSHALL IS.$_{D}$B[5]
- WAKE I.$_{D}$B[7]
- PAPUA N.G.$_A$
- NAU.$_{D}$B[1]
- SOL. IS.$_{D}$A[3]
- VAN.$_A$A[4]
- NEW CAL.$_{D}$A[2]
- FIJI.$_A$
- TUV.$_{D}$C[6]
- KIRIBATI$_B$
- W. SAM.$_{D}$C[7]
- AM. SAM.$_{D}$C[3]
- COOK IS.$_{D}$C[2]
- TON.$_{D}$C[5]
- MIDWAY IS.$_{D}$C[10]
- HAWAIIAN IS.$_{D}$C[1]
- JOHNSON ATOLL$_C$[8]
- KINGMAN REEF$_C$[9]
- PALMYRA$_C$[11]
- FR. POLY.$_{D}$C[4]
- EAS.$_{D}$C[3]
- Marquesas Is.
- Bora Bora
- Society Is.
- Tahiti
- Tuamotu Archipelago
- Pitcairn I. (N.Z.)
- Sala y Gomez (Chile)
- Juan Fernández Is. (Chile)
- Galapagos Is. (Ecuador)
- Clipperton I. (Fr.)
- Revilla Gigedo (Mex.)
- Guadalupe I. (Mex.)
- MEXICO
- UNITED STATES
- CANADA
- GULF OF MEXICO
- ATLANTIC OCEAN
- CARIBBEAN SEA
- HONDURAS
- BEL.
- GUAT.
- EL SAL.
- NICAR.
- COSTA RICA
- PAN.
- COLOMBIA
- ECUADOR
- PERU
- CHILE
- EQUATOR
- TROPIC OF CANCER
- TROPIC OF CAPRICORN
- DATE LINE
- 180° Meridian

Scale: 0 — 800 — 1,600 — 2,400 miles / 3,840 km

The islands of Oceania, most of which are located in the southwest Pacific, are grouped according to regions: *Melanesia*, *Micronesia*, and *Polynesia*. *Melanesia* ("black islands") refers to the skin color of the inhabitants of New Guinea and some of the islands that lie to the southeast. *Micronesia* ("small islands") describes the size of the islands that lie to the north of Melanesia. *Polynesia* ("many islands") includes a huge triangular-shaped area of the Pacific.

The islands of Oceania are coral reefs or the peaks of submerged volcanoes. The latter are prone to earthquakes and eruptions, and the moisture-capturing peaks make these islands wetter than the low-lying reefs. The soil of the older volcanic islands, such as Hawaii, is enriched by the gradual breakdown of volcanic rock. Coral reefs, formed by the skeletons of sea coral, have thin soil that is marginally fertile. Most reefs are in the form of an atoll—a ring of land surrounding a lagoon. The atoll is created when coral forms a fringe around the base of a volcanic peak, which eventually sinks beneath the sea, leaving in its place a lagoon surrounded by the coral.

The Pacific, the world's largest and deepest ocean, measures half the distance around the globe. The ocean isn't always as benign as its name implies. Typhoons (cyclones) of the western Pacific can inflict terrible destruction. These storms are generally larger and more powerful than the hurricanes of the eastern Pacific and the Atlantic Ocean. Oceania's climate is usually comfortable, since the normal Equatorial heat and humidity are tempered by the trade winds.

The islands of Oceania have been part of a romantic Western myth that pictures the "South Seas" as an island paradise with gleaming white beaches, swaying palms, crystal-clear waters, and attractive natives pursuing a carefree existence filled with food, song, and dance. This vision may have had some truth prior to the arrival of the Europeans in the 16th century, but in the years that followed, most native societies were forever changed by colonization, disease, Christian missionaries, commercial exploitation, and military battles. European and Asian immigrants, who on some islands are in the majority, have transformed the traditional village culture that easily lived within its environment into dependent modern societies.

MELANESIA [i-]

The residents of the "black islands," along with the Australian Aborigines, are the darkest-skinned inhabitants of Oceania. These short, powerfully built, curly-haired people are probably descended from the earliest migrants from Asia. Some of the tribes still live in the Stone Age. Social standing within a tribe is acquired by the accumulation or exchange of material goods.

FIJI [A1]

Area: 7,075 sq.mi.(18,324 km²). **Population:** 800,000. **Capital:** Suva, 69,000. **Government:** Republic. **Language:** English; Fijian; Hindi. **Religion:** Christianity 50%, Hinduism 40%. **Exports:** Sugar, copra, gold, oil. □ Two large volcanic islands make up 87% of the land area of Fiji (fee' jee). The 800 remaining islands are mostly low-lying coral atolls. Prosperous descendants of 19th century workers from India outnumber the landowning Melanesians.

NEW CALEDONIA [A2]

Area: 7,380 sq.mi.(19,114 km²). **Population:** 178,000. **Capital:** Noumea, 66,100. **Government:** French Overseas Territory. **Language:** French; native dialects. **Religion:** Roman Catholic 60%; Protestant 30%. **Exports:** Nickel, chrome, iron, cattle, coffee. □ New Caledonia, which includes the Loyalty Islands, was once the world's largest producer of nickel. European settlers have resisted the Melanesian minority's demand for independence.

SOLOMON ISLANDS [A3]

Area: 11,504 sq.mi.(29,795 km²). **Population:** 441,000. **Capital:** Honiara, 36,500. **Government:** Constitutional monarchy. **Language:** Pidgin; English; native dialects. **Religion:** Protestant 55%, Roman Catholic 20%. **Exports:** Fish, bananas, timber, copra. □ In the Solomons live some of the most primitive tribes in Oceania. Major battles were fought here in World War II.

VANUATU [A4]

Area: 5,700 sq.mi.(14,763 km²). **Population:** 185,000. **Capital:** Port Vila, 26,000. **Government:** Republic. **Language:** Bislama; English; French. **Religion:** Christianity. **Exports:** Copra. □ The 80 mountainous islands of Vanuatu (va noo' a too) were called the New Hebrides under English and French joint rule.

MICRONESIA [i-]

The "small islands" of Micronesia lie to the north of Melanesia. Though generally taller and lighter-skinned than Melanesians, Micronesians do not have any distinguishing physical characteristics. They tend to resemble their closest neighbors—Asians, Melanesians, or Polynesians. Leadership in the societies of Micronesia (and Polynesia) is based on royalty and inherited rank.

KIRIBATI [B]

Area: 300 sq.mi.(777 km²). **Population:** 85,000. **Capital:** Tarawa, 25,000. **Government:** Republic. **Language:** Kiribati; English. **Religion:** Roman Catholic 45%, Protestant 40%. **Exports:** Copra, produce, phosphates. □ Kiribati (kir' ih bahs) is a nation of 33 small islands (including the Gilberts) in an area nearly the width of the US. The economy relies on aid from Britain.

NAURU [B1]

Area: 8 sq.mi.(21 km²). **Population:** 10,600. **Capital:** Yaren, 600. **Government:** Republic. **Language:** Nauruan; English. **Religion:** Protestant 65%, Roman Catholic 30%. **Exports:** Phosphates. □ Nauru (nah' roo) is the third-smallest country in the world (after Vatican City and Monaco) and one of the richest, per capita. The island is a plateau of practically pure phosphate. In anticipation of the eventual depletion of reserves, revenues are being wisely invested. In 1993, a cash settlement was won from Australia, New Zealand, and the U.K. for exploiting Nauru's resources when they administrated the island.

U.S. TERRITORIES [B2-B7]

In 1947, the United Nations placed more than 2,000 islands, formerly occupied by the Japanese, under the *Trust Territory of the Pacific Islands* to be administered by the U.S. These islands were: the *Federated States of Micronesia (FSM*, formerly the *Caroline Islands)*; the *Marshall Islands*; the *Northern Mariana Islands*; and the 200 islands of *Palau.* The latter group became a sovereign nation in 1994. In this arrangement, the U.S. continues to give economic aid to all former Trust Territories and handles their defense. The Northern Marianas have U.S. commonwealth status, comparable to that of Puerto Rico. Another commonwealth in Micronesia is *Guam*, the most populous (164,000) of the U.S. islands territories. It was acquired (along with the Philippines and Puerto Rico) from Spain in 1899. Guam and the Northern Mariana chain are popular vacation spots for the Japanese. *Wake Island* is administered by the U.S. Air Force. *Bikini* and *Eniwetok* atolls in the Marshall Islands were U.S. nuclear test sites. After 40 years, Bikini is still radioactive.

POLYNESIA [i-]

Most of the "many islands" of Polynesia lie east of the International Date Line in the central Pacific. Polynesians, believed to be of Malay ancestry, are the tallest and have the lightest skin color of the natives of Oceania. Residents of these far-flung Pacific islands were the last immigrants in the movement from Southeast Asia. The Polynesians were superb sailors, capable of accurately navigating long distances without the aid of instruments. New Zealand (p. 35) and Hawaii (p. 9) are the largest islands in Polynesia.

COOK ISLANDS [c12]

Area: 92 sq.mi.(238 km²). **Population:** 20,000. **Capital:** Avarua, 9,550. **Government:** Self-governing citizens of New Zealand. **Language:** English; Maori dialect. **Religion:** Protestant. **Exports:** Copra and fruit. □ These 15 islands are named after the British Captain James Cook, who explored and charted much of Oceania. The natives are related to the Maoris of New Zealand.

EASTER ISLAND [c3]

Area: 46 sq.mi.(119 km²). **Population:** 2,100. **Capital:** Hanga Roa. **Government:** Territory of Chile. **Language:** Spanish. **Religion:** Roman Catholic. **Exports:** Wool. □ The island was named on Easter Sunday. No one has unraveled the mystery of more than 600 huge heads, ranging in size from 12 to 40 ft.(3.7-12.2 m), carved out of volcanic rock by an ancient civilization.

FRENCH POLYNESIA [c4]

Area: 1,500 sq.mi.(3,885 km²). **Population:** 238,000. **Capital:** Papeete, 24,000. **Government:** French Overseas Territory. **Language:** French; Tahitian. **Religion:** Protestant 50%, Roman Catholic 35%. **Exports:** Copra, sugar, fruit, mother-of-pearl. □ The Society Islands and the Marquesas are among the 130 islands of French Polynesia. The handsome natives of Tahiti, the largest island, were immortalized by the French painter Paul Gauguin.

TONGA [c5]

Area: 300 sq.mi.(777 km²). **Population:** 110,000. **Capital:** Nukualofa, 30,000. **Government:** Constitutional monarchy. **Language:** Tongan; English. **Religion:** Christianity. **Exports:** Copra and fruit. □ The "Friendly Islands" were so named by Captain Cook for the warm welcome he received. Tonga's 150 islands make up Polynesia's last kingdom. The people are very large and heavy set—the present king weighs well over 400 lbs. The weak economy depends upon paychecks sent home by Tongans working in New Zealand.

TUVALU [c6]

Area: 10 sq.mi.(25 km²). **Population:** 10,500. **Capital:** Funafuti, 3,900. **Government:** Constitutional monarchy. **Language:** Tuvaluan; English. **Religion:** Christianity. **Exports:** Copra. □ Tuvalu (formerly the Ellice Islands) is the world's fourth-smallest nation in size, and second-smallest in population. There were three times as many people living on Tuvalu in the 19th century, but the population was decimated by "blackbirders," Europeans who invaded island to seize natives to work the plantations on other islands.

U.S. TERRITORIES [c7-c11]

American Samoa, Jarvis Island, Johnson Atoll, Kingman Reef, Midway Islands, and Palmyra. About 85,000 people (more than twice the present population) have moved from American Samoa to the U.S. mainland. Both of the islands of Midway are administered by the U.S. Navy, whose decisive victory here over a Japanese fleet was the turning point of World War II in the Pacific. These territories have taken on new importance with the recent enlargement of U.S. coastal waters. In 1983, the U.S. followed other coastal nations and proclaimed a 200 mi.(320 km) "exclusive economic zone." In 1988, the U.S. widened the zone of "total sovereignty" from 3 mi.(4.8 km) to 12 mi.(19 km). Foreign ships may not enter the 200-mile zone except in free navigation or to lay seabed cables, and they cannot operate within 12 mi.(19 km) of a U.S. shoreline for any reason. This policy means that any U.S.-owned island, no matter how small, can control all fishing and mineral rights within 200 miles. A "wet America" has been created equal in size to continental America. Without international agreements, exclusive economic zones encourage the unrestricted exploitation of the oceans.

WESTERN SAMOA [c12]

Area: 1,098 sq.mi.(2,844 km²). **Population:** 225,000. **Capital:** Apia, 33,000. **Government:** Constitutional monarchy. **Language:** Samoan; English. **Religion:** Christianity. **Exports:** Copra and bananas. □ The people of these four islands are the tallest and most powerfully built in Oceania. Western Samoa was a colony of Germany before World War I, but it would be difficult to find a Teutonic influence on the easygoing, music- and dance-filled way of life.

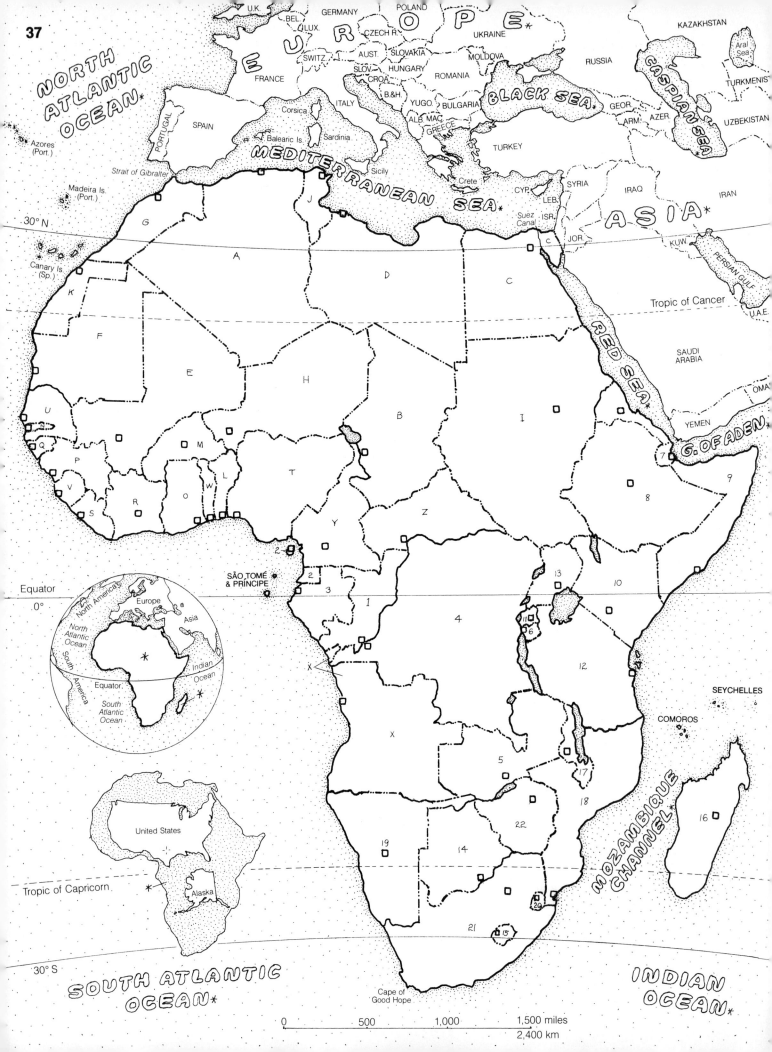

NORTHERN

ALGERIA(A) ALGIERS
CHAD(B) N'DJAMENA
EGYPT(C) CAIRO
LIBYA(D) TRIPOLI
MALI(E) BAMAKO
MAURITANIA(F) NOUAKCHOTT
MOROCCO(G) RABAT
NIGER(H) NIAMEY
SUDAN(I) KHARTOUM
TUNISIA(J) TUNIS
WESTERN SAHARA(K) AAIÚN

WESTERN

BENIN(L) PORTO NOVO
BURKINA FASO(M) OUAGADOUGOU
CÔTE D'IVOIRE (IVORY COAST) YAMOUSSOUKRO
GAMBIA(N) BANJUL
GHANA(O) ACCRA
GUINEA(P) CONAKRY
GUINEA-BISSAU(Q) BISSAU
LIBERIA(S) MONROVIA
NIGERIA(T) LAGOS
SENEGAL(U) DAKAR
SIERRA LEONE(V) FREETOWN
TOGO(W) LOMÉ

CENTRAL

ANGOLA(X) LUANDA
CAMEROON(Y) YAOUNDÉ
CENTRAL AFRICAN REP.(Z) BANGUI
CONGO(1) BRAZZAVILLE
DEM. REP. OF CONGO KINSHASA
EQUATORIAL GUINEA(2) MALABO
GABON(3) LIBREVILLE
ZAMBIA(5) LUSAKA

EASTERN

BURUNDI(6) BUJUMBURA
DJIBOUTI(7) DJIBOUTI
ERITREA ASMARA
ETHIOPIA(8) ADDIS ABABA
SOMALIA(9) MOGADISHU
KENYA(10) NAIROBI
RWANDA(11) KIGALI
TANZANIA(12) DAR ES SALAAM
UGANDA(13) KAMPALA

SOUTHERN

BOTSWANA(14) GABORONE
LESOTHO(15) MASERU
MADAGASCAR(16) ANTANANARIVO
MALAWI(17) BAMAKO
MOZAMBIQUE(18) MAPUTO
NAMIBIA(19) WINDHOEK
SWAZILAND(20) MBABANE
SOUTH AFRICA(21) PRETORIA
ZIMBABWE(22) HARARE

AFRICA: THE COUNTRIES

CN: (1) Color a country first, then its name. (2) Use light colors on the map of colonial Africa (24-30) so you can see the national boundaries. (3) The island nations of São Tomé & Príncipe, Comoros, Seychelles, Mauritius, and Cape Verde are too small to color; their names are not on the list.

Africa covers 11,700,000 sq.mi.(30,279,600 km^2); it is the second-largest continent (after Asia). It has 53 nations and 750 million people—the third-largest population (after Asia and Europe). The continent is divided into two racial and cultural zones by the Sahara Desert. Northern nations, bordering the Mediterranean, are populated by light-skinned, Arabic-speaking Muslims. Countries south of the Sahara are populated mostly by black Africans who speak hundreds of different languages (many are Bantu dialects). The most common form of Bantu is Swahili, which is the lingua franca of East Africa. Muslim missionaries and their Christian counterparts are making inroads among the native religions (mostly Animist) of black Africans. Islam is gaining because blacks regard it as an African religion, despite its Middle Eastern origins; however, Christianity (with origins similar to Islam) is considered European.

Although the Portuguese began establishing coastal colonies as early as the 15th century, it wasn't until the late 1800s that Europeans penetrated the African interior and began carving up the continent in earnest. At the outbreak of World War I (1914), only Liberia and most of Ethiopia remained free of foreign domination. The second half of the 20th century brought a great rush toward independence, and colonialism formally came to an end in 1990, with the free elections in Namibia. France, which had the largest African empire, maintains close relations with most of its former colonies, and many are dependent upon it for economic and military aid.

Independence did not mean freedom for most Africans; they generally were ruled by some form of dictatorship. The boundaries of the new countries were virtually the same as those drawn by the colonialists, who were either ignorant of or indifferent to traditional tribal divisions. Many nations suffer from arbitrary borders that separate related groups or confine traditional enemies within the same country. Tribal loyalties often take precedence over allegiance to the new nation.

The problems facing these young nations were enormous. Fertile land and rainfall are not generally plentiful in Africa. In the few places where agriculture is productive, gains have been erased by excessive population growth. The largest farms followed the colonial practice of growing cash crops for export instead of food for local consumption; when world commodity prices are depressed, cash crops do not provide enough income to buy food. Most African nations were dependent on declining foreign aid. Except for a few oil producers, the nations rich in natural resources were unable to mine and market them profitably. Potentially healthy economies were wrecked by communist mismanagement. Others have been victimized by brutal and corrupt leaders who squandered precious revenues on ill-conceived public works projects or monuments of self-aggrandizement. Still other countries have been torn apart by civil war. Hunger, poverty, disease (especially AIDS), and illiteracy are on the increase in many nations across the continent.

In the middle 1990s, a positive spirit began to sweep the continent. The conditions described above are giving way to new practices that augur economic growth and optimism. Governments are becoming more democratic, peace and security are replacing warfare, and free enterprise (accompanied by individual initiative) is producing signs of prosperity in many regions. Foreign aid has become more selective; help goes directly to businesspeople and ambitious villagers, and less and less to corrupt and inefficient governments. Africans are beginning to take charge of their own lives.

COLONIAL AFRICA

Except for Liberia and Ethiopia, the entire continent was under European control by the early 19th century.

BELGIAN(23)
BRITISH(24)
FRENCH(25)
GERMAN(26)
ITALIAN(27)
PORTUGUESE(28)
SPANISH(23)

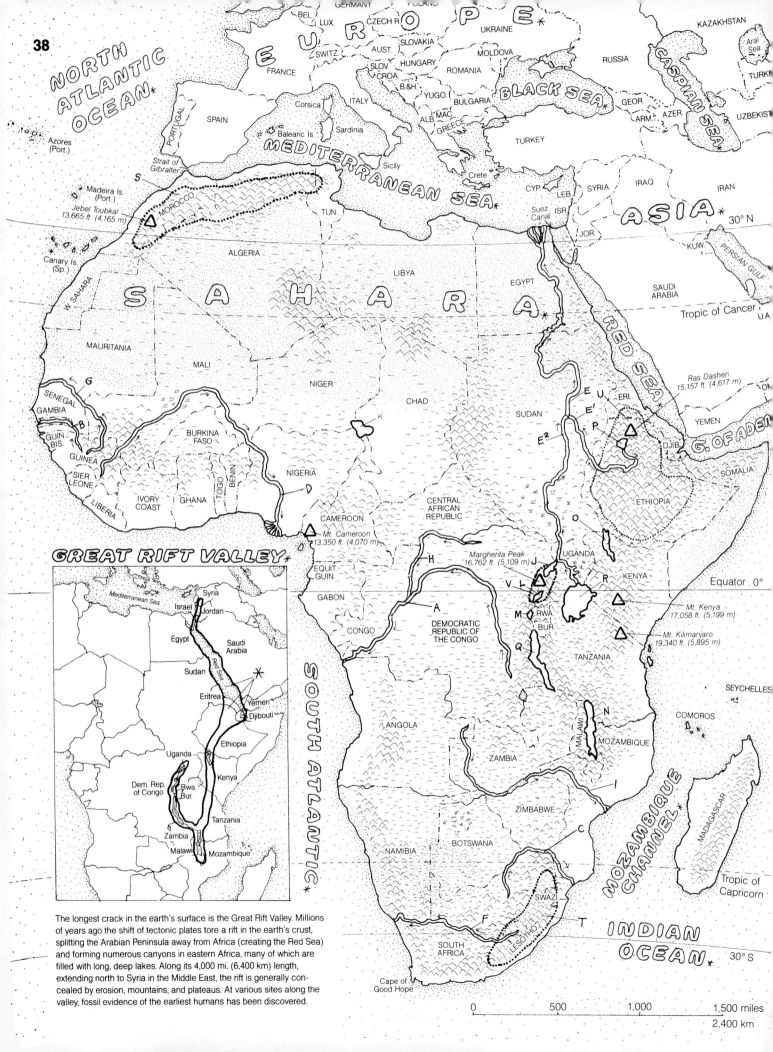

The longest crack in the earth's surface is the Great Rift Valley. Millions of years ago the shift of tectonic plates tore a rift in the earth's crust, splitting the Arabian Peninsula away from Africa (creating the Red Sea) and forming numerous canyons in eastern Africa, many of which are filled with long, deep lakes. Along its 4,000 mi. (6,400 km) length, extending north to Syria in the Middle East, the rift is generally concealed by erosion, mountains, and plateaus. At various sites along the valley, fossil evidence of the earliest humans has been discovered.

AFRICA: THE PHYSICAL LAND

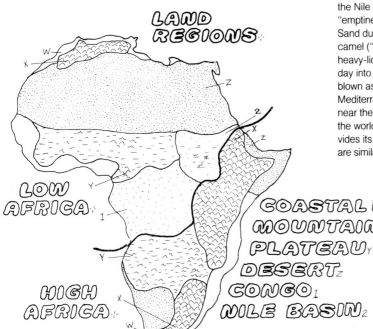

CN: (1) On the large map, use gray on the triangles representing important mountain peaks. (2) Use gray for the map of the Great Rift Valley on the far left. (3) Use light colors on the map of the land regions below.

Over 80% of Africa lies between the Tropics of Cancer and Capricorn; it has the largest tropical region of any continent. It has been said that the night hours are Africa's winter. Not all of Africa is warm; there are glacial areas in some eastern mountain ranges near the Equator. Curiously, Africa's Equatorial regions are not the warmest—it gets hotter the farther one goes from the Equator (except for Mediterranean and southernmost coasts, which have pleasant climates). Although most of the continent is very dry to semiarid, heavy rainfall occurs in the Equatorial regions, particularly in Central and Western Africa.

The African landscape has relatively little fertile territory; topsoil is generally thin, the deserts are huge, and most of the wetter regions are covered by a thick jungle. Tree-dotted, semi-arid grasslands (savannas) occupy wide areas of the continent and support an enormous population of large animals—elephants, giraffes, rhinoceroses, lions, and others. In the rainforests of central Africa, monkeys, chimpanzees, gorillas, reptiles, and birds live in the canopy of tall trees, high above the dark, dank jungle floor.

Africa is a plateau made of ancient rock. It is rimmed by narrow coastal lowlands. Most of the mountain ranges are in the eastern and southern portions (high Africa). Here the plateau reaches an altitude of 6,000 ft.(1,830 m) and slopes even higher to form the Drakensberg Mountains, which tower over the coast of Southern Africa. The most fascinating mountains are in the glacier-covered Ruwenzori Range between Lakes Edward and Albert, on the Dem. Rep. of Congo-Uganda border. There is an almost constant cloud cover, so the sight of glacier-covered peaks, nine of which reach over 16,000 ft.(4,878 m), is a rare and impressive experience. Ordinary plants have been known to grow to extraordinary sizes on the Ruwenzori slopes because of unusually favorable soil and weather conditions. To the east lies snowcapped Kilimanjaro, Africa's tallest mountain (19,340 ft., 5,895 m). It is one of a group of volcanic peaks formed by the Great Rift Valley.

One reason Africa was the last major continent to be explored and colonized by Europeans was that it presented formidable physical obstacles: an unusually smooth coastline with few peninsulas, islands, and natural harbors; a forbidding interior of deserts, jungles, and hot, arid plains; and a shortage of navigable rivers. Most African rivers, including the four major ones—Nile, Congo, Niger, and Zambezi—are interrupted by impassable rapids and waterfalls. The Nile (including the White Nile) is the world's longest river (4,150 mi., 6,640 km). Lake Victoria is credited as its source, but most of the White Nile (whose waters are pale green) is dissipated (through evaporation) in the swamps of southern Sudan. Nearly 90% of the water that flows along the main Nile through Egypt comes from Lake Tana, Ethiopia, via the shorter Blue Nile (whose waters are blue). The massive Congo is the world's second-largest river by volume (after the Amazon), and is 2,600 mi. (4,160 km) long. The Niger River is unusual in that it travels nearly that far to reach the sea, even though it originates only 150 mi.(240 km) from the coast. The major lakes are found in the Great Rift Valley. Lake Victoria, the world's second-largest (after Lake Superior), is actually situated on a plateau between two arms of the valley. Extremely deep Lake Tanganyika, on the Dem. Rep. of Congo-Tanzania border, is the world's longest lake (420 mi., 680 km).

The dominant geographical feature of Africa is the world's largest desert, the constantly expanding Sahara, currently the size of the continental United States. Any rainfall is scant and unpredictable. The only available water in the "Land of Thirst" is found in isolated oases and in the Nile River on its eastern edge. Yet as recently as 5,000 years ago the Sahara (Arabic for "emptiness") was a grassland. Today it is covered mostly by rock, gravel, and salt deposits. Sand dunes account for only one-fifth of the desert's surface. Because roads are so few, the camel ("ship of the desert") remains the most reliable form of transportation. The camel's heavy-lidded eyes and closeable nostrils enable it to withstand fierce sandstorms that can turn day into night while they cut a swath as wide as 300 mi.(480 km). Dust from the Sahara has blown as far north as the Swiss Alps. In the summer, desert winds bring intense heat to the Mediterranean region. The world's highest shade temperature, 136° F (58° C), was recorded near the Libyan coast. The Namib Desert, along the coast of Namibia in Southwest Africa, has the world's tallest sand dunes: some reach 1,000 ft. (305 m). Fog from the adjacent ocean provides its only moisture. The cold ocean currents that prevent rain from reaching the shore are similar to the conditions creating the deserts along the west coast of South America (p. 17).

PRINCIPAL RIVERS:
CONGO A
GAMBIA B
LIMPOPO C
NIGER D
NILE E
    BLUE NILE E'
    WHITE NILE E²
ORANGE F
SENEGAL G
UBANGI H
ZAMBEZI I

PRINCIPAL LAKES:
L. ALBERT J
L. CHAD K
L. EDWARD L
L. KIVU M
L. MALAWI N
L. TURKANA O
L. TANA P
L. TANGANYIKA Q
L. VICTORIA R

PRINCIPAL MOUNTAIN RANGES:
ATLAS MTS. S
DRAKENSBERG T
ETHIOPIAN HIGHLANDS U
RUWENZORI V

LAND REGIONS:

LOW AFRICA:
HIGH AFRICA:

COASTAL LOWLANDS W
MOUNTAINS X
PLATEAU Y
DESERT Z
CONGO 1
NILE BASIN 2

The dark line across this smaller map divides low Africa from the lands of the east and south, known as high Africa. The Congo Basin and the lowlands of west Africa are covered with rainforests. Surrounding these jungle areas are broad semi-arid plateaus, mostly covered by savannas (grasslands). The northern third of the continent is virtually all desert.

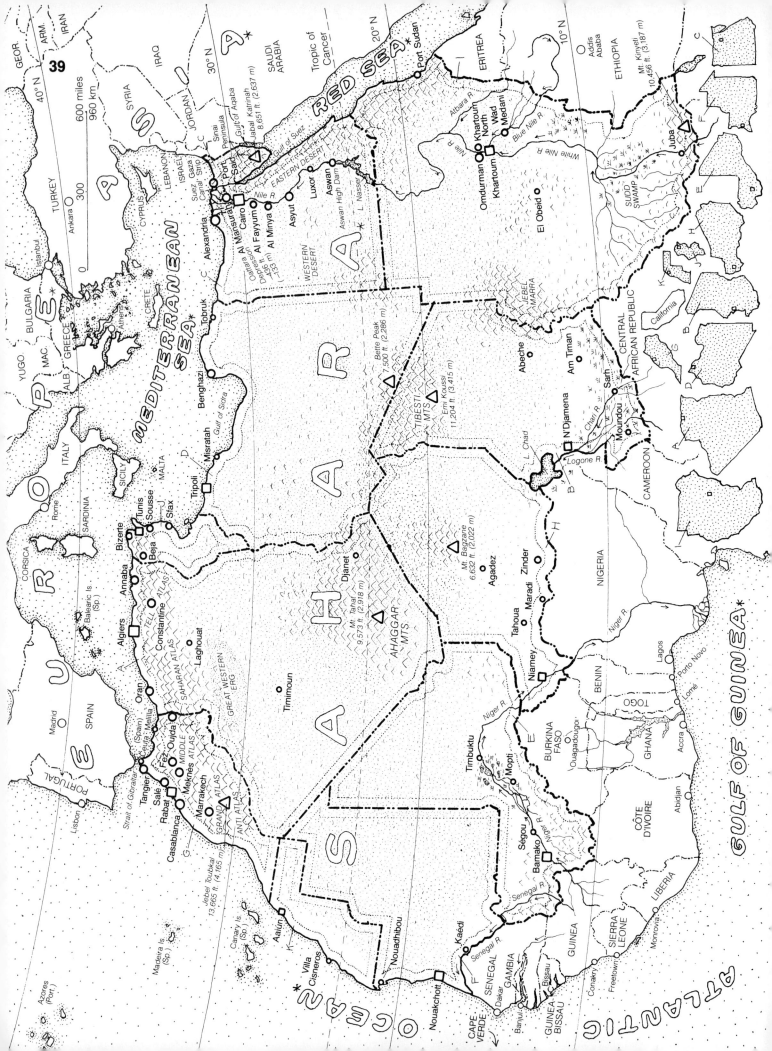

AFRICA: NORTHERN

Northern Africa is set apart from the rest of the continent by the Sahara Desert and by the Islamic culture. The people of this region are light-skinned Arabs, Berbers, Moors (a mixture of the two), and Egyptians. Their homelands are almost entirely desert, except for the northern coast, where most of the population lives. Mediterranean crops, including grapes, olives, citrus, and dates, are grown in abundance. The Sahara Desert is drifting to the south and west, covering the northern portions of sub-Sarahan Mauritania, Mali, Niger, Chad, and the Sudan. Nomadic residents of the desert regions are generally light-skinned Muslims who have little in common with the black populations in the southern parts of those nations. These sub-Saharan countries, many under military rule, are Africa's poorest. The three in the middle (Mali, Niger, and Chad) are landlocked. They have few natural resources and must depend upon foreign aid and a limited agriculture in small but fertile regions along their southern borders. Frequent droughts have added to their misery. In the early 20th century, French West Africa ruled all of northern Africa except Libya (Italy), Western Sahara and part of Morocco (Spain), Egypt (Britain), and Sudan (Britain and Egypt). France still plays an influential role.

ALGERIA A

Area: 952,580 sq.mi.(2,467,182 km²) **Population:** 30,500,000. **Capital:** Algiers, 1,600,000. **Government:** One-party republic. **Language:** Arabic; Berber; French. **Religion:** Islam. **Exports:** Oil and natural gas, citrus, wine, dates. **Climate:** Mediterranean on the coast; hot and dry inland. □ After 130 years of colonial rule, France regarded Algeria as its southern province. But by 1962, after a bloody war of liberation, France was forced to leave, along with one million French settlers. In the 1990s, Islamic fundamentalists, whose victories at the polls were nullified by the military, have begun to terrorize the population. Most of Africa's second-largest nation (after Sudan) is nearly deserted. Only nomads (a half million Tuaregs) brave the Sahara (which covers 80% of Algeria). In Tuareg society, the men wear the veils. The Sahara contains valuable oil and natural gas deposits. In the Great Western Erg, crescent-shaped sand dunes 400 ft.(122 m) high drift 100 ft. (30 m) each year.

CHAD B

Area: 495,800 sq.mi.(1,279,164 km²). **Population:** 7,400,000. **Capital:** N'Djamena, 550,000. **Government:** Military republic. **Language:** Arabic; many African dialects. **Religion:** Islam 45%, Christianity 30%. **Exports:** Cotton and peanuts. **Climate:** Hot and dry. □ Landlocked and sparsely populated Chad is virtually two separate countries. The nomadic Muslim minority in the north has fought an intermittent 25-year war with black African farmers in the south. In the late 1980s, French troops expelled Libyan forces aiding the Muslims. Lake Chad (also bordered by Niger, Nigeria, and Cameroon) is all that's left of a prehistoric inland sea. It contains fresh water, which is unusual for a desert lake. Although its depth and dimensions vary according to season, the lake continues to shrink due to the prolonged drought.

EGYPT C

Area: 386,660 sq.mi.(997,583 km²). **Population:** 66,000,000. **Capital:** Cairo, 7,000,000. **Government:** Republic. **Language:** Arabic; English. **Religion:** Islam. **Exports:** Cotton and cotton products, oranges, dates, oil, rice. **Climate:** Hot and dry, with mild winters. □ Over 50 million people live in the narrow Nile valley, the world's largest oasis. Occupying only 4% of Egypt, this strip of land is 750 miles (1,200 km) long and little more than 2–10 miles (3.2–16 km) wide. The Nile delta fans out to include 155 mi.(250 km) of Mediterranean coastline. Cairo, the overcrowded desert capital, is Africa's largest city. Egypt is the leading producer of high-quality cotton, and after South Africa, it is the most industrialized country in Africa. But the economy cannot match the growth of the population, and Egypt slips further into poverty. Five thousand years ago, Egypt was one of the great civilizations. Each year, a million tourists come to see the pyramids, Africa's largest structures, built as tombs for the pharaohs. The ancient structures have been remarkably well preserved by the desert climate. For millennia, life in Egypt was sustained by the annual flooding of the Nile. When the Aswan High Dam was completed in 1968, greatly enlarging the capacity of an older dam, it became possible to control the flow of the river, store water for irrigation (doubling the agricultural output), and produce hydroelectric power. But without flooding, no silt is being deposited and farmers are forced to buy chemical fertilizers, and the delta region is shrinking. Before the Suez Canal was built in the 19th century, ships had to sail around the tip of Africa. There are plans to enlarge the canal because supertankers (loaded with Middle Eastern oil) cannot use it. Egypt is the political hub of the Arab world, and Al-Azhar University in Cairo is the center for Islamic learning. In 1979, after years of hostility, Egypt signed a peace treaty with Israel, which returned the Sinai Peninsula but still controls Egypt's Gaza Strip (p. 30).

LIBYA D

Area: 679,358 sq.mi.(1,752,744 km²). **Population:** 5,700,000. **Capital:** Tripoli, 650,000. **Government:** One-party socialist republic. **Language:** Arabic. **Religion:** Islam. **Exports:** Oil and oil products. **Climate:** The coast is Mediterranean; the rest is hot and dry. □ In 1959, oil was discovered under the Sahara, and Libya, previously one of the world's poorest nations, rapidly became the richest (per capita) in Africa. Except for two separate strips of fertile coastline, the nation is almost entirely a gravel-covered desert. From 1911 to World War II, it was a colony of Italy. After the war, Britain and France occupied Libya until it became an independent monarchy in 1951. Eighteen years later, a coup brought Colonel Muammar Qaddafi to power. He transformed Libya into a modern industrial state with strong educational and social welfare programs. In 1986, the U.S. bombed Qaddafi's palace in Tripoli in retaliation for his alleged support of international terrorism. This was not the first time the U.S. attacked the Libyan capital. In 1804, in response to harassment of its shipping by pirates from the Barbary Coast (named after the local Berbers), the U.S. sent Marines [illegible] "Shores of Tripoli" to destroy their base.

[illegible] E

Area: 4[illegible]8,800 [illegible],221,042 km²). **Population:** 10,100,000. **Capital:** Bamako, 750,000. **Government:** Military republic. **Language:** French; native dialects. **Religion:** Islam 80%, indigenous religions 20%. **Exports:** Cotton, peanuts, dried fish. **Climate:** Desert region is hot; south is tropical. □ In the 14th century, the fabled city of Timbuktu ("from here to Timbuktu") was the center of Islamic learning and commerce in the vast North African kingdom of Mali (mah' lee). Today, it is a small desert city. In southern Mali, the Niger River forms an irrigation network of rivers and lakes (an "inland delta"). The largely black, Muslim population has faced drought, starvation, and disease.

MAURITANIA F

Area: 398,000 sq.mi.(1,030,820 km²). **Population:** 2,500,000. **Capital:** Nouakchott, 700,000. **Government:** Military republic. **Language:** Arabic; French. **Religion:** Islam. **Exports:** Fish, iron ore, copper, gum arabic. **Climate:** Hot and dry; south is moderate. □ An encroaching Sahara covers 80% of the land; desert sands are blowing out over the Atlantic. Refugees from the desert have expanded Nouakchott's population from 10,000 to 700,000 in less than 40 years. The Senegal River, which separates Mauritania from Senegal, is the heart of the nation's only fertile region. Atlantic fishing and iron mining are the chief industries. Most Mauritanians are Moors, many of whom are nomads.

MOROCCO G

Area: 172,415 sq.mi.(446 555 km²). **Population:** 29,200,000. **Capital:** Rabat, 1,250,000. **Government:** Constitutional monarchy. **Language:** Arabic; Berber; French. **Religion:** Islam. **Exports:** Phosphates, citrus, fish, produce, crafts. **Climate:** Mediterranean on the coast; mountain extremes; hot in the desert. □ Eight miles (13 km) from Spain, across the Strait of Gibraltar, lies Morocco, "Crossroads of Western and Islamic Culture." The old walled cities and native quarters within Casablanca and Marrakech are famous tourist attractions. Tangier, opposite Gibraltar, is an ancient trading port. The population is mostly mixed Arab and Berber. Arabic, French, and some Spanish are spoken in the cities; Berber is used by residents of the Atlas Mountains. Morocco is the world's leading exporter of phosphates and is a major supplier of winter fruits and vegetables to Europe. Morocco was colonized by both Spain and France prior to gaining independence in 1956. Spain retains control of Ceuta and Melilla, cities on Morocco's Mediterranean coast.

NIGER H

Area: 489,200 sq.mi.(1,267,028 km²). **Population:** 9,700,000. **Capital:** Niamey, 400,000. **Government:** Military republic. **Language:** French; native dialects. **Religion:** Islam. **Exports:** Uranium, peanuts, livestock, cotton, fish. **Climate:** Hot and dry, except in the South. □ Landlocked Niger (ni'jer) is named after a mighty river that flows only through the western edge of this drought-stricken nation. Lake Chad, in the southeast corner, is the other major source of water. Black Muslims make up the largest ethnic group.

SUDAN I

Area: 967,525 sq.mi.(2,505,890 km²). **Population:** 34,000,000. **Capital:** Khartoum, 550,000. **Government:** Military republic. **Language:** Arabic; English; native dialects. **Religion:** Islam 70%, indigenous 25%, Christianity 5%. **Exports:** Gum arabic, cotton, peanuts, sugar, textiles. **Climate:** Hot and dry in the north, humid in the south. □ The largest nation in Africa is composed of two warring societies: Arabic-speaking Muslims in the north and blacks in the south. The latter have violently resisted attempts by the government in the north to impose Islamic law throughout the nation. The southern part of the country is the heart of Sudan's farming region; the best land lies between the White and Blue Nile Rivers. The three largest cities (Omdurman, Khartoum, and Khartoum North) are located where the rivers join to form the Nile. In the far south is the Sudd, one of the world's largest swamps. Sudan's acacia trees yield 90% of the world's gum arabic, a sticky substance used in adhesives, inks, candy, cosmetics, perfumes, and medicines. Sudan is struggling with war, drought, and the presence of thousands of refugees from Chad and Ethiopia. The refugees strain an already depleted agricultural sector, raising the specter of famine in the former "Breadbasket of Africa."

TUNISIA J

Area: 63,200 sq.mi.(163,688 km²). **Population:** 9,400,000. **Capital:** Tunis, 800,000. **Government:** Socialist republic. **Language:** Arabic; French. **Religion:** Islam. **Exports:** Oil, phosphates, olive oil, grapes, dates. **Climate:** Mediterranean. □ Tunis, the capital, is located near the site of Carthage, a Phoenician city that dominated the Mediterranean for 1,000 years before being destroyed by the Romans in 146 B.C.; ruins of the later Roman colony attract many tourists. After independence from France in 1956, Tunisia pursued a policy of moderation in foreign affairs under the skillful leadership of Habib Bourguiba. It was the most European of the North African nations, but after 30 years of Western secularism, there are signs of an Islamic revival.

WESTERN SAHARA K

Area: 102,690 sq.mi.(265,967 km²). **Population:** 234,000. Capital: Aaiún, 34,000. **Government:** Controlled by Morocco. **Language:** Arabic; Berber. **Religion:** Islam. Exports: Phosphates. **Climate:** Mild and dry. □ When Spain gave up Spanish Sahara in 1975, Morocco received the phosphate-rich northern portion and Mauritania took the southern part. The Polisario, a national liberation organization, aided by Algeria, resisted both countries. Mauritania gave up its portion in 1979, and Morocco claims all of Western Sahara as part of historic "Greater Morocco." The resistance continues.

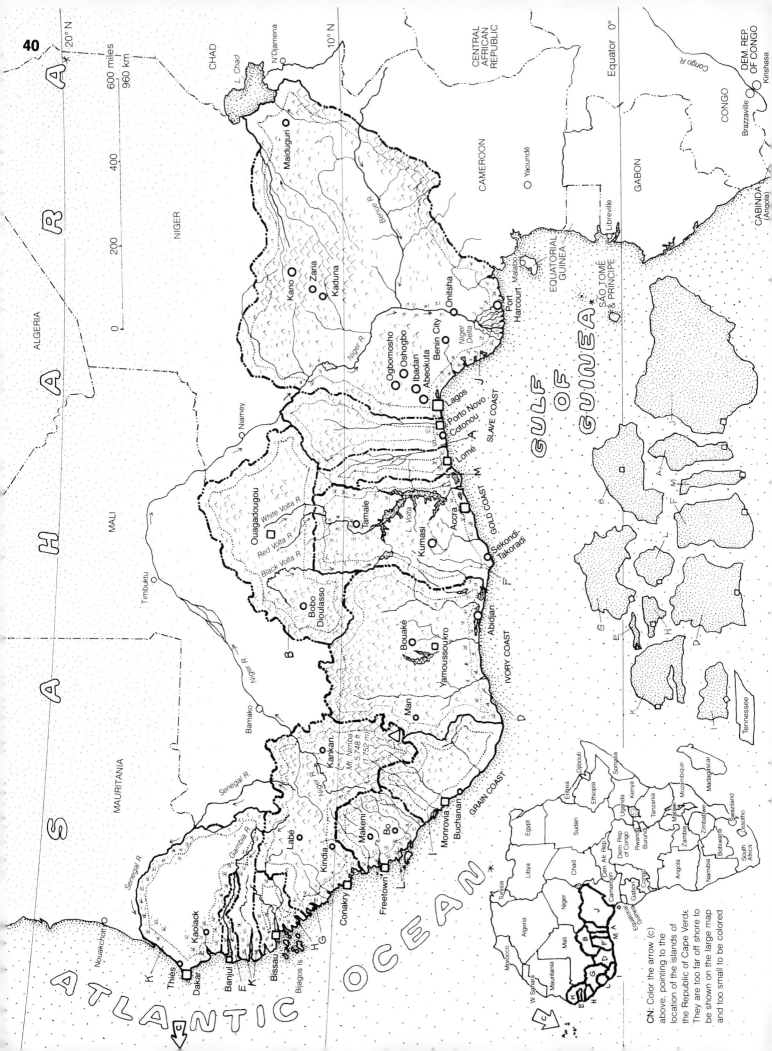

CN: Color the arrow (c) above, pointing to the location of the islands of the Republic of Cape Verde. They are too far off shore to be shown on the large map and too small to be colored

AFRICA: WESTERN

Many coastal nations on the western "bulge" of Africa share a similar landscape, climate, economy, religion, and history. Cash crops introduced by Europeans (cocoa, coffee, palm oil, and rubber) are the chief products of the coastal rain forests. Most of the world's cocoa and chocolate come from the seeds of local cacao plants. Peanuts (locally called groundnuts) and cotton are grown inland, on the higher and drier Sahel (the savanna bordering the Sahara). Harbors had to be built along the coast because the shallow waters and treacherous surf prevented large ships from reaching shore. During colonial days, sections of the coast were named for their principal trade activities. Even though the commerce has changed, the descriptive labels remain: Grain Coast (Liberia); Ivory Coast (Côte d'Ivoire); Gold Coast (Ghana); and Slave Coast (Togo, Benin, Nigeria). From the Slave Coast and other ports on the continent, an estimated 10 million African slaves were sent to the New World between the 16th and the 18th centuries. Many did not survive the voyage. Slave ships transported three different cargoes in a triangular route: they carried finished goods from Europe to Africa to be exchanged for slaves, who were shipped to the Americas, where the ships took on raw materials bound for Europe.

Christianity was introduced by European missionaries. Islam came from the sub-Saharan nations on the northern borders. But most West Africans are Animists, people who worship the dead souls and spirits believed to be part of the natural environment.

When these nations gained independence, they found it necessary to retain French, English, or Portuguese as the official language.

BENIN A

Area: 43,480 sq.mi.(112,613 km²). **Population:** 6,100,000. **Capital:** Porto Novo, 170,000. **Government:** Military republic. **Language:** French; African dialects. **Religion:** Animism 70%, Christianity 15%, Islam 15%. **Exports:** Palm oil, cotton, cocoa, peanuts. **Climate:** Hot, with coastal rainfall. ☐ Benin (beh neen'), formerly Dahomey, has had more coups than any of the other newly independent African nations. Most of its people practice Animism (Benin is considered the birthplace of West Indian voodoo and black magic).

BURKINA FASO B

Area: 105,875 sq.mi.(274,216 km²). **Population:** 11,300,000. **Capital:** Ouagadougou, 500,000. **Government:** Military republic. **Language:** French; African dialects. **Religion:** Animism 65%, Islam 25%, Christianity 10%. **Exports:** Livestock, peanuts, and cotton. **Climate:** Warm and generally dry. ☐ Burkina Faso, formerly Upper Volta, is an extremely poor, landlocked country with few natural resources. Although the Black, Red, and White Volta Rivers flow through the country, it is vulnerable to droughts. The thin, non-porous soil is unable to absorb the summer monsoon rains that sweep up from the Gulf of Guinea. In addition to aid from France, the economy depends upon paychecks sent home by citizens working on the Guinea coast.

CAPE VERDE C

Area: 1,557 sq.mi.(4,033 km²). **Population:** 400,000. **Capital:** Praia, 63,000. **Government:** Republic. **Language:** Portuguese; Crioulo. **Religion:** Roman Catholic. **Exports:** Coffee, bananas, salt, coal, fish. **Climate:** Hot and very dry. ☐ About 300 mi.(480 km) off the coast of Senegal lie the islands of Cape Verde (verd). They were uninhabited when the Portuguese arrived in the 15th century. Plantations were built, and slaves were brought from the mainland. Slaves bound for the Americas were often trained here. The prevalence of droughts has forced the emigration of half the population.

CÔTE D'IVOIRE (IVORY COAST) D

Area: 124,500 sq.mi.(322,455 km). **Population:** 15,500,000. **Capital:** Yamoussoukro, 110,000. **Government:** Republic. **Language:** French; African dialects. **Religion:** Animism 65%, Islam 23%, Christianity 12%. **Exports:** Coffee, cocoa, hardwoods, bananas, rubber, fish. **Climate:** Tropical, with coastal rainfall. ☐ Côte d'Ivoire, formerly Ivory Coast, was so named when there was an active trade in elephant tusks. It might better be called the "cocoa coast"—it is the world's top cocoa producer. It is also one of the leaders in hardwood and coffee production. Félix Houphouet-Boigny, the only president from 1960 until his death in 1993, had pursued a liberal economic policy that, until the commodity price decline of the 1980s, made the Ivory Coast the richest nation in black Africa. Prosperity returned in the late 90s, especially to the port of Abidjan (1,500,000), the "Paris of Africa." The center of commerce and manufacturing, it was the capital until Yamoussoukro was built inland. There the president financed the construction of the world's largest Christian church.

GAMBIA (THE) E

Area: 4,360 sq.mi.(11,292 km²). **Population:** 1,300,000. **Capital:** Banjul, 45,000. **Government:** Military. **Language:** English; African dialects. **Religion:** Islam. **Exports:** Peanuts. **Climate:** Tropical and humid. ☐ The smallest nation on continental Africa straddles the Gambia River. There are no bridges, so the river virtually creates two countries, each about 200 mi.(320 km) long and 10 mi. (16 km) wide. The Gambia resembles a crooked finger poking deep into Senegal. Although the people of the Gambia and Senegal share similar ethnic backgrounds and are allied in defense and foreign affairs, their dissimilar cultures (Gambia is English and Senegal is French) prevent unification. The Gambia is the country author Alex Haley went to to trace his history for the book "Roots."

GHANA F

Area: 91,910 sq.mi.(238,047 km²). **Population:** 18,500,000. **Capital:** Accra, 950,000. **Government:** Military republic. **Language:** English; African dialects. **Religion:** Animism 45%, Christianity 40%. **Exports:** Cocoa, hardwoods, bauxite, diamonds, gold. **Climate:** Tropical and wet. ☐ Ghana (gah' nuh) was named after a medieval African kingdom. Its former name, the Gold Coast, referred to a time when it was a gold-producing British colony. Ghana was once the world's leading exporter of cocoa, which is still its major crop. Lake Volta, behind a dam on the Volta River, is one of the world's largest artificial lakes. The dam supplies power for a growing aluminum industry.

GUINEA G

Area: 94,957 sq.mi.(245,939 km). **Population:** 7,500,000. **Capital:** Conakry, 1,250,000. **Government:** Military republic. **Language:** French; African dialects. **Religion:** Islam 70%, Animism 25%, Christianity 5%. **Exports:** Bauxite, iron ore, bananas, coffee, pineapples. **Climate:** Tropical, with heavy coastal rainfall. ☐ The word guinea (gihn' ee) is believed to be Berber, meaning "land of the blacks." Guinea, formerly French Guinea, was the first French colony to become independent (1958). Despite huge mineral reserves, Guinea's economy stagnated under years of communist rule. Guinea is the world's number-two producer of bauxite (after Australia). The English "guinea" coin was minted from gold once mined here. Three major rivers (the Niger, Senegal, and Gambia) originate in the plateau region.

GUINEA-BISSAU H

Area: 13,948 sq.mi.(36,125 km). **Population:** 1,250,000. Capital: Bissau, 210,000. **Government:** Republic. **Language:** Portuguese; African dialects. **Religion:** Animism 65%, Islam 30%, Christianity 5%. **Exports:** Peanuts, coconuts, palm oil, fish. **Climate:** Tropical and wet. ☐ Guinea-Bissau (bih sow'), formerly Portuguese Guinea, has the unhappy distinction of having lived under the longest period of colonial rule in history—over 500 years. It also endured 12 years of bitter fighting before it became Portugal's first African colony to achieve independence in 1974. Much of the country is a low-lying, swampy region. The offshore Bijago Archipelago, with a large network of waterways, is unusual—islands on the coast of Africa are quite rare.

LIBERIA I

Area: 42,950 sq.mi.(111,241 km²). **Population:** 2,800,000. **Capital:** Monrovia, 950,000. **Government:** Republic. **Language:** English; African dialects. **Religion:** Animism 75%, Islam 15%, Christianity 10%. **Exports:** Iron ore, rubber, cocoa, coffee, hardwoods. **Climate:** Tropical. ☐ Liberia became Africa's first independent black nation in 1847 (Haiti was the world's first). The land was purchased in 1822 by the American Colonization Society, an organization seeking to create an African homeland for freed slaves. Descendants of those original immigrants make up less than 5% of the population, but they have traditionally controlled Liberia. These English-speaking "Americo-Liberians" are well-educated and prosperous; they live an American lifestyle. In 1980, a military coup sought to share that power among the many ethnic groups. A corrupt government followed; the leader was killed in 1990 and a vicious civil war was waged among six tribes. They eventually formed an interim government, which held elections in the late 1990s. An interest in Liberia by the U.S. is the reason the nation was one of only two in Africa to escape colonization. Liberia is Africa's leading producer of iron ore and rubber and has the world's largest merchant navy; foreign ships use its registry because of low taxes.

NIGERIA J

Area: 356,700 sq.mi.(923,853 km²). **Population:** 105,000,000. **Capital:** Lagos, 1,750,000. **Government:** Military. **Language:** English; African dialects. **Religion:** Islam 45%, Christianity 35%, Animism 20%. **Exports:** Oil, cocoa, cotton, rubber, tin, palm oil, peanuts. **Climate:** Tropical, with a dry interior. ☐ The most populous nation in Africa has the potential to become the continent's superpower. The land is rich in natural resources and grows a wide variety of crops. The pyramids one sees in Nigeria are stacked sacks of peanuts awaiting shipment. Oil and natural gas produce 90% of the revenue that is financing industrial development. Nearly half of Nigerians are Muslims, most of whom live in the northern Sahel. The country tends to be divided along ethnic lines. The presence of over 250 different groups has led to numerous conflicts. In 1967, a bloody civil war was fought when the Ibos tried to create their own nation, Biafra, in eastern Nigeria.

SENEGAL K

Area: 76,120 sq.mi.(197,151 km²). **Population:** 9,800,000. **Capital:** Dakar, 850,000. **Government:** Republic. **Language:** French; African dialects. **Religion:** Islam. **Exports:** Peanuts, cotton, phosphates, fish. **Climate:** South coast is wet; interior is dry and hot. ☐ Until independence in 1960, Senegal (sen ih gawl') was France's favorite colony in French West Africa. Dakar, the "Gateway to Africa," was the colonial capital. Today it is a modern city; its natural harbor was the major supply port for Allied forces in Africa during World War II. Senegal's principal export is peanuts. The crop was introduced in Africa by Europeans to provide food for the slave trade.

SIERRA LEONE L

Area: 27,800 sq.mi.(72,002 km²). **Population:** 5,100,000. **Capital:** Freetown, 450,000. **Government:** Military. **Language:** English; native dialects. **Religion:** Animism 75%, Islam, Christianity. **Exports:** Diamonds, bauxite, cocoa, coffee. **Climate:** Tropical and wet. ☐ In 1787, Sierra Leone (lee own') was founded as the first colony for freed slaves by a British antislavery group. After the British abolished slavery in 1807, they liberated slave ships and sent the freed slaves to Sierra Leone. It remained a British colony for more than 150 years. The main industries are the mining of diamonds, iron ore, and bauxite. Democracy has had a very limited presence here in the 1990s.

TOGO M

Area: 21,650 sq.mi.(56,073 km²). **Population:** 5,000,000. Capital: **Lome,** 525,000. **Government:** Republic. **Language:** French; native dialects. **Religion:** Animism 60%, Christianity 25%, Islam 15%. **Exports:** Phosphates, cacao, coffee, palm oil. **Climate:** Tropical in the south. ☐ Tiny Togo is divided ethnically and physically. A mountain range separates northern Muslims (who are racially related to the people in Burkina Faso and Niger) from the southern Togolese (who share ethnic backgrounds with tribes in Ghana and Benin). Togo has been hurt by the drop in world phosphate prices.

41

NIGER

Neb.
Colo. | Kan. | Mo.
N.M. | Okl. | Ark.
Texas | La.

L. Chad

N'Djamena

CHAD

10° N

NIGERIA

Garoua

Bouar

Berbérati

Bangui

SUDAN

C

Niger R.

Benue R.

Logone R.

Chari R.

Ubangi R.

E

White Nile

B

Port
Harcourt

Mt. Cameroon
13,350 ft. (4,070 m)

Douala

Malabo

BIOKO

F

RÍO MUNI

H

Bata

PRÍNCIPE F

São Tomé

SÃO TOMÉ H

Port-Gentil

G

Yaoundé

Mbandaka

Congo R.

UGANDA

Margherita Peak
16,762 ft. (5,109 m)

L. Albert

Equator 0°

Kampala

Ogooué R.

Lambarene

Ubangi R.

Congo R.

Lualaba R.

L. Edward

L. Victoria

MITUMBA MOUNTAINS

Kigali

RWANDA

Bukavu

L. Kivu

Libreville

Stanley
Pool

Brazzaville

Kinshasa

Kasai R.

Kananga

Mbuji-Mayi

Bujumbura

BURUNDI

L. Tanganyika

TANZANIA

E

D

Pointe-Noire

CABINDA
(Angola)

A
E

A

Congo R.

Cuango R.

L. Mweru

ATLANTIC

Luanda

10° S

OCEAN

Lobito

Benguela

Mt. Môco
8,596 ft. (2,620 m)

Huambo

Lubumbashi

Kitwe

Ndola

L. Bangweulu

MUCHINGA MOUNTAINS

Lilongwe

MALAWI

L. Malawi

Cubango R.

Cuando R.

Zambezi R.

Lusaka

Victoria
Falls

Zambezi R.

L. Kariba

Harare

MOZAMBIQUE

Zambezi R.

ZIMBABWE

Cunene R.

A

NAMIBIA

0   200   400   600 miles

960 km

20° S

BOTSWANA

Morocco
Tunisia
Algeria
Libya
Egypt
W. Sahara
Mauritania
Mali
Niger
Chad
Sudan
Eritrea
Djibouti
Senegal
Burkina
Faso
Nigeria
Ethiopia
Gambia
Guinea-
Bissau
Guinea
Benin
Somalia
Sierra
Leone
Liberia
Côte
d'Ivoire
Ghana
Togo
Cameroon
Uganda
Kenya
B
C
F
G
D
E
A
Tanzania
Gabon
Congo
Malawi
Mozambique
Angola
Zambia
Zimbabwe
Namibia
Botswana
Madagascar
South
Africa
Lesotho
Swaziland

AFRICA: CENTRAL

Dense rain forests form a wide band across central Africa (also called Equatorial Africa). To Western explorers, the impenetrable jungle, with its opaque canopy of vegetation, presented a dank, forbidding, unknown world. They described Africa as "the Dark Continent." Few people live in the jungle; most live to the north or south, in the drier, tree-dotted savannas. Central Africa's sparse population isn't just the result of an inhospitable jungle. Diseases, particularly malaria and sleeping sickness, take thousands of lives. Sleeping sickness is spread by the tsetse fly, whose bite can be just as deadly to certain domestic animals. The fly makes it virtually impossible to raise cattle in this part of Africa (as well as many regions of western and eastern Africa). Because cattle droppings are the primary source of fertilizer for most Third World nations, the tsetse fly can also restrict crop production.

Most Central Africans speak variations of the Bantu language. The Bantus displaced the Pygmies some 2,500 years ago. The current Pygmy population of 200,000 live as hunter-gatherers in remote parts of the jungle. They average 4.5 ft.(1.4 m) in height. Pygmies are slowly giving up their life in the jungle because of the destruction of the rainforest.

The mighty Congo River and its hundreds of tributaries drain the world's second-largest river basin (after the Amazon). The waterways serve as national highways despite rapids and waterfalls. In some places, mini-railroad lines transport cargo around river obstacles. Brazzaville and Kinshasa, capital cities of the Republic of Congo and the Democratic Republic of the Congo (formerly Zaire), face each other from opposite banks of the river.

Portugal was the first European nation to explore and colonize lands that border Central Africa's Atlantic coast. But here, as in West Africa, the Portuguese lost most of their possessions to more powerful European nations.

ANGOLA A
Area: 481,360 sq.mi.(1,240,092 km²). **Population:** 10,900,000. **Capital:** Luanda, 1,500,000. **Government:** Republic. **Language:** Portuguese; Bantu dialects. **Religion:** Christianity 70%, indigenous 20%. **Exports:** Oil, coffee, diamonds, sisal. **Climate:** Generally mild. □ Any nation blessed with the rare combination of natural resources, accessible coastline, fertile land, and a favorable climate should be quite prosperous. But a 14-year war of independence, 14 more years of civil war, a communist-controlled economy, and the abrupt departure of the Portuguese in 1975 (Angola was the last Portuguese colony in Africa to gain independence) have left Angola in shambles. Only the flow of oil has been unaffected by the prolonged upheaval. The civil war, fueled by Cold War rivals (the U.S. and the Soviet Union), ended in 1989 and free elections were held in 1992. During the period of slavery, 2 million Angolans were sent to Brazil and other parts of Portugal's empire.

CAMEROON B
Area: 183,569 sq.mi.(475,444 km²). **Population:** 15,100,000. **Capital:** Yaoundé, 700,000. **Government:** One-party republic. **Language:** French; English; African dialects. **Religion:** Indigenous 50%, Christianity 35%, Islam 15%. **Exports:** Oil, natural gas, cocoa, coffee, timber, rubber, palm oil, cotton. **Climate:** Hot and extremely wet on the coast; drier in the north. □ For a nation with over 200 ethnic tribes, Cameroon has had an unsually stable government. Culturally, Cameroon reflects its multi-colonial past: first ruled by Germany in the late 19th century, and then divided between Britain and France after their victory in World War I. The British and French Cameroons united in 1960 (the northern part of British Cameroon joined Nigeria). Cameroon is the only country in Africa that uses both French and English as official languages. The climate and the mountainous landscape change dramatically from the arid north to the very wet and green coastal region. Each year, Mt. Cameroon (13,354 ft., 4,070 m), an active volcano and the tallest peak in this part of Africa, attracts 400 in.(1,016 cm) of rain, making it the wettest spot on the continent.

CENTRAL AFRICAN REPUBLIC C
Area: 240,530 sq.mi.(622,947 km²). **Population:** 3,400,000. **Capital:** Bangui, 600,000. **Government:** Republic. **Language:** French; African dialects. **Religion:** Animism 60%, Christianity 20%, Islam 10%. **Exports:** Diamonds, coffee, cotton. **Climate:** Tropical; dry in the northeast. □ When it was part of French Equatorial Africa, this poor, landlocked nation was called Ubangi-Shari. The name derives from two rivers: the Chari, which flows north to Lake Chad, and the Ubangi. The latter joins the Congo in the south and provides the only feasible route to the sea—1,000 mi.(1,600 km) of waterway with a short rail connection to the coast. For three terrible years, 1976-1979, the nation was called the Central African "Empire" and its leader, Jean-Bedel Bokassa, declared himself emperor. His inauguration was so opulent it received worldwide attention. Bokassa brutalized his people, and his personal trade in diamonds and ivory not only bankrupted the nation, it nearly wiped out the elephant population.

CONGO D
Area: 132,050 sq.mi.(342,010 km²). **Population:** 2,600,000. **Capital:** Brazzaville, 490,000. **Government:** Republic. **Language:** French; Bantu dialects. **Religion:** Indigenous 50%, Christianity 40%. **Exports:** Oil, timber, wood products, potash, uranium, palm oil, tobacco. **Climate:** Equatorial and wet. □ When the Congo (Republic of the Congo) gained independence in 1960, it became the first communist government in Africa. Communism was rejected in 1991, but free elections have not brought political stability.

Brazzaville was the former capital of French Equatonal Africa (Gabon, Chad, and the Central African Republic). Much of the Congo's export revenue has been squandered on poor planning and impractical projects. The nation is nearly covered by impenetrable jungle and non-arable soil; consequently, over half the population live in cities and towns—an unsually high percentage for Africa. This concentration of potential converts has aided the work of Christian missionaries.

DEM. REP. OF CONGO E
Area: 905,360 sq.mi.(2,335,828 km²). **Population:** 49,000,000. **Capital:** Kinshasa, 4,000,000. **Government:** Republic. **Language:** French; 200 Bantu dialects. **Religion:** Christianity 55%, indigenous 35%, Islam 10%. **Exports:** Cobalt, industrial diamonds, copper, oil, coffee, uranium, palm oil. **Climate:** Equatorial. □ After the overthrow in 1997 of the corrupt regime of Mobuto Sese Seko, a one-party president who held power since 1965, the nation's name was changed from Zaire to the Democratic Republic of the Congo (prior to independence it was the Belgian Congo). The widespread corruption and economic waste of the previous regime has left the nation in shambles. The country is huge, with enormous untapped mineral wealth and hydroelectric potential. Since independence in 1960, it has been beset by secessionist movements. The most serious challenge to national unity was the revolt of Katanga province (one of the world's richest mineral regions) from 1960 to 1963 and again in 1977. An obstacle to industrial development is the general lack of transportation facilities. Most of the D.R. Congo is an Equatorial jungle basin drained by the Congo and hundreds of tributaries. These waterways form the nation's highways. On a trip up the Congo, Joseph Conrad wrote his chilling short story "Heart of Darkness."

EQUATORIAL GUINEA F
Area: 10,830 sq.mi.(28,050 km²). **Population:** 454,000. **Capital:** Malabo, 32,000. **Government:** Republic. **Language:** Spanish; Bantu. **Religion:** Roman Catholic 80%, Animist 20%. **Exports:** Oil, cocoa, coffee, timber. **Climate:** Equatorial and very wet. □ Equatorial Guinea is a tiny nation of two parts: Bioko, a small island, and Río Muni, the larger mainland part. Malabo, the capital, is on Bioko, an island 100 mi.(160 km) northwest of Río Muni and closer to the coast of Cameroon. Under Spanish rule, the island's fertile volcanic soil (plus heavy rain) produced the world's finest cocoa. The colony had the highest per capita income in Africa. Equatorial Guinea is the only nation in Africa to use Spanish as its official language. Recovery from the brutal rule and economic corruption of the previous government has been accelerated by the recent exploitation of large oil and natural gas reserves.

GABON G
Area: 103,340 sq.mi.(267,651 km²). **Population:** 1,200,000. **Capital:** Libreville, 425,000. **Government:** Republic. **Language:** French; Bantu dialects. **Religion:** Animism 50%, Christianity 50%. **Exports:** Oil, manganese, iron ore, hardwoods, uranium. **Climate:** Equatorial and damp. □ At one time, Gabon's income came solely from the sale of hardwoods, but now oil and the world's largest production of manganese have made thinly populated Gabon (gah bone') the wealthiest black nation in Africa. But little of the prosperity has trickled down to the people. Tourists visit Lambarene to see the hospital built in 1913 by the young Albert Schweitzer. The great physician, musician, philosopher, and humanitarian spent 53 years ministering to the needs of the African people. Contrasted to that body of selfless work is the new $3 billion rail line that runs 400 mi.(640 km) from Libreville, the coastal capital, to Francoville, the birthplace of Gabon's president, Omar Bongo.

SÃO TOMÉ & PRÍNCIPE H
Area: 371 sq.mi.(957 km²). **Population:** 150,000. **Capital:** São Tomé, 43,500. **Government:** Republic. **Language:** Portuguese. **Religion:** Roman Catholic. **Exports:** Cocoa, coffee, bananas, palm oil, copra. **Climate:** Equatorial. □ The tiny nation of São Tomé & Príncipe (soun' tuh may' and prin' suh pay) consists of two mountainous islands and two islets 200 mi. (320 km) west of Gabon, in the Gulf of Guinea. São Tomé has 90% of the nation's land and a similar percentage of the population. The islands were uninhabited when the Portuguese discovered them in 1470. They were first used as a penal colony. Slaves were brought from the mainland to work the sugar plantations. Working conditions hardly improved after slavery was abolished; under Portuguese management, hundreds of workers were killed in labor protests. Much of the current workforce comes from the former Portuguese colonies: Cape Verde, Angola, and Mozambique. Cocoa is still the principal export, although production is less than it was when São Tomé, under Portuguese rule, was the world's leading producer.

ZAMBIA I
Area: 290,000 sq.mi.(748,200 km²). **Population:** 9,500,000. **Capital:** Lusaka, 900,000. **Government:** Republic. **Language:** English; many native dialects. **Religion:** Christianity 60%, Animism 40%. **Exports:** Copper, other metals. **Climate:** Tropical, but cooler at higher elevations. □ Mineral-rich but landlocked Zambia (formerly Northern Rhodesia) has struggled for years to get its exports to market, using the ports of neighboring Angola, Mozambique, and Tanzania. Railway lines through those nations have been the frequent target of guerrilla activity. In the 1980s the government developed a broad agricultural base to offset the critical reliance on copper production (Zambia is the world's third-largest copper producer). The landscape is a savanna-covered, relatively flat plateau with large swampy areas. The country was named after the Zambezi River, which forms the southern border with Zimbabwe. It was there that the English explorer David Livingstone discovered Victoria Falls, which he named in honor of his Queen. The falls present an awesome visual and auditory experience—the local name for it means "the smoke that thunders."

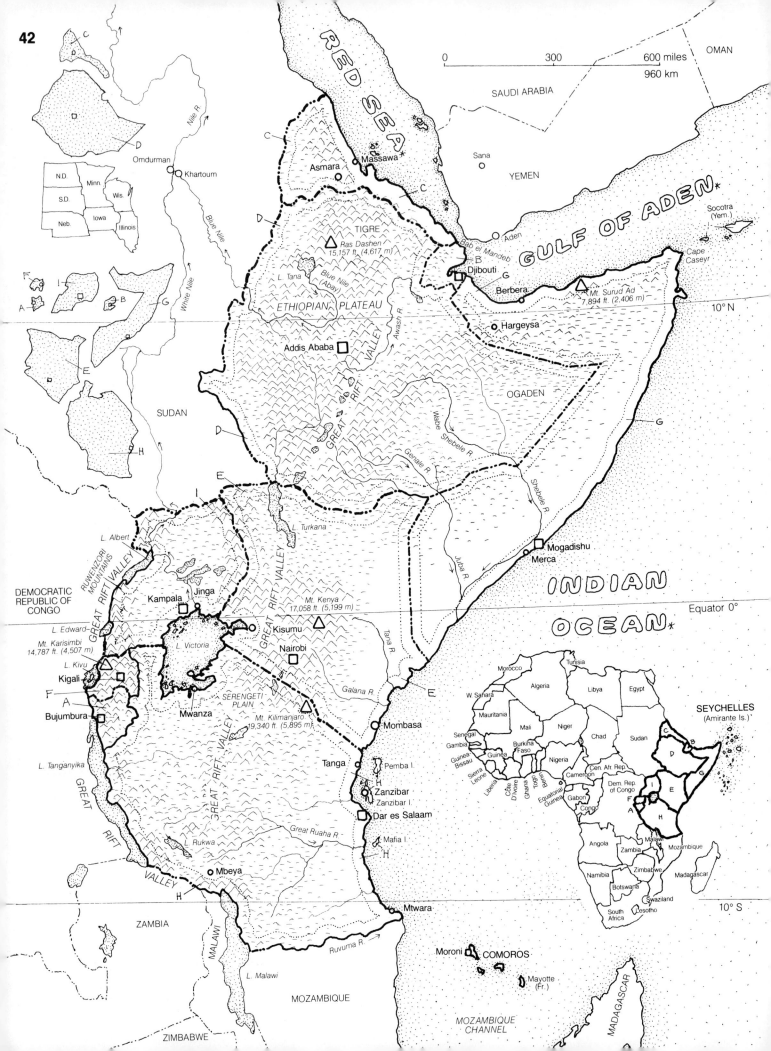

42

OMAN

SAUDI ARABIA

RED SEA

0   300   600 miles
           960 km

YEMEN

GULF OF ADEN

Massawa

Asmara

C

Sana

Socotra
(Yem.)

Cape
Caseyr

D

TIGRE

Ras Dashen
15,157 ft. (4,617 m)

Bab el Mandeb

Aden

B

Djibouti

G

10° N

L. Tana

Blue Nile
(Abay)

ETHIOPIAN PLATEAU

Berbera

Mt. Surud Ad
7,894 ft. (2,406 m)

Awash R.

Addis Ababa

Hargeysa

GREAT RIFT VALLEY

OGADEN

Wabe Shebele R.

Genale R.

Shebele R.

G

Nile R.

Omdurman

Khartoum

Blue Nile

N.D.   Minn.
S.D.   Wis.
Neb. Iowa   Illinois

White Nile

F   I

A   B   G

E

SUDAN

H

D

E

I

L. Turkana

GREAT RIFT VALLEY

Juba R.

Mogadishu

Merca

INDIAN

OCEAN

Equator 0°

L. Albert

RUWENZORI MOUNTAINS

GREAT RIFT VALLEY

Kampala   Jinga

Mt. Kenya
17,058 ft. (5,199 m)

DEMOCRATIC
REPUBLIC OF
CONGO

L. Edward

Mt. Karisimbi
14,787 ft. (4,507 m)

L. Kivu

Kigali

L. Victoria

Kisumu

GREAT RIFT VALLEY

Nairobi

Tana R.

E

Galana R.

F

A

Bujumbura

SERENGETI
PLAIN

Mwanza

Mt. Kilimanjaro
19,340 ft. (5,895 m)

Mombasa

L. Tanganyika

Tanga

Pemba I.

H

Zanzibar

Zanzibar I.

SEYCHELLES
(Amirante Is.)

GREAT RIFT VALLEY

Dar es Salaam

Morocco   Tunisia

Algeria   Libya   Egypt

W. Sahara

Mauritania   Mali   Niger   Chad   Sudan   C   B

Senegal   D

Gambia   Burkina   Nigeria   Cen. Afr. Rep.   I   G

Guinea   Faso

Guinea   Benin   Cameroon   E

Bissau

Sierra   Côte   Togo   Equatorial   Dem. Rep.   F

Leone   D'Ivoire   Ghana   Guinea   Gabon   of Congo   A   H

Liberia   Congo

Mafia I.

H

Great Ruaha R.

L. Rukwa

Angola   Zambia   Malawi   Mozambique

Namibia   Zimbabwe   Madagascar

Botswana

Swaziland

South   Lesotho

Africa

Mbeya

H

10° S

ZAMBIA

MALAWI

Mtwara

Ruvuma R.

Moroni   COMOROS

Mayotte
(Fr.)

MADAGASCAR

ZIMBABWE

L. Malawi

MOZAMBIQUE

MOZAMBIQUE
CHANNEL

AFRICA: EASTERN

Conditions that created the devastating famines of the 1980s in the Ethiopian region have improved—particularly the end of a prolonged civil war—but the specter of starvation in this part of Africa is always present. The dark-skinned and fine-featured people of this region are Caucasoids of Hamitic origin, related to the people of the Middle East. Christianity and Islam are the dominant religions. South of the Horn live mostly black Africans of the Swahili-speaking Bantu tribes. The nations of eastern Africa are considerably drier, higher, and cooler than other Equatorial countries of the continent. The Great Rift Valley, a giant ditch some 1,200 mi.(1,920 km) long, is the major geological feature that separates eastern Africa from the rest of the continent (p. 38).

BURUNDI A

Area: 10,747 sq.mi.(27,834 km²). **Population:** 5,540,000. **Capital:** Bujumbura, 350,000. **Government:** Republic. **Language:** Kirundi; French. **Religion:** Roman Catholic 70%, indigenous 25%. **Exports:** Coffee, cotton, tea. **Climate:** Mild. ☐ About 85% of the people of Burundi (buh run' dee) are Hutu farmers, but for three centuries they have been ruled by Tutsi (Watusi) cattle herders. The Tutsi are few in number but tall in physical stature (a height of 7 ft. is common). They own whatever wealth there is. Since an unsuccessful Hutu rebellion in 1972, ethnic strife has plagued this country.

COMOROS. Over 500,000 people of mixed African, Middle Eastern, and South Asian ancestry live on three main islands (694 sq.mi., 1,794 km²) in the Mozambique Channel. Most are Muslims who speak Swahili, Arabic, and French. Comoros, dependent on French aid, lacks raw materials and fertile soil. When Comoros became independent in 1975, Mayotte, one of the four main islands, chose to remain a French colony.

DJIBOUTI B

Area: 8,900 sq.mi.(23,051 km²). **Population:** 440,000. **Capital:** Djibouti, 255,000. **Government:** Republic. **Language:** Arabic; French; Somali; Afar. **Religion:** Islam. **Exports:** Livestock and hides. **Climate:** Hot and dry. ☐ Terribly hot Djibouti (jih boo' tee) is strategically located on the strait of Bab el Mandeb between the Gulf of Aden and the Red Sea. Only 20 mi.(32 km) away is the Arabian Peninsula. Djibouti, the capital city and chief port, serves as a shipping terminal for Addis Ababa, the capital of Ethiopia, 400 mi.(640 km) inland. This function as an entrepôt is the main source of Djibouti revenue.

ERITREA C

Area: 36,000 sq.mi.(93,600 km²). **Population:** 3,900,000. **Capital:** Asmara, 410,000. **Government:** Republic. **Language:** Tigrinya; Arabic; miscellaneous dialects. **Religion:** Eritrean (Coptic Christian) 45%, Islam 45%. **Exports:** Cotton, salt, coffee, copper. **Climate:** Hot and dry. ☐ For centuries, Eritrea was part of the Ethiopian Empire. It fell under Italian rule in the late 19th century (architecture in Asmara shows that influence). Britain took control in 1941 and made it part of Ethiopia in 1952. After a 30-year-long civil war, Eritrea broke away in 1991—taking with it Ethiopia's coastline on the Red Sea—and declared independence two years later. The nation is steadily rebuilding itself.

ETHIOPIA D

Area: 472,400 sq.mi.(1,223,516 km²). **Population:** 58,400,000. **Capital:** Addis Ababa, 1,800,000. **Government:** Republic. **Language:** Amharic; Galla; Arabic; Sidama. **Religion:** Coptic Christian 45%, Islam 45%. **Exports:** Coffee, hides, oilseeds, cotton, sesame. **Climate:** Extremely hot on the coast; cooler inland. ☐ Ethiopia (formerly Abyssinia) is one of the world's oldest Christian nations. Until the 44-year reign of Emperor Haile Selassie was terminated by a Marxist military government (now gone) in 1974, there was an unbroken chain of kings and emperors dating back to Biblical times. Most Ethiopians are dark-skinned (Ethiopia is Greek for "land of sunburned faces"). The ethnically and linguistically diverse population is divided into two groups: the Semitic language-speaking Christian ruling class of the north and the Cushitic language-speaking Muslims of the south. The plateaus on which they live are similarly divided by the Rift Valley. In the higher northern plateau, the Blue Nile begins at Lake Tana and wends its way through the world's largest gorge (longer and wider than the Grand Canyon) en route to the Nile in Sudan. The Blue Nile provides almost 90% of the water that flows up the Nile. Addis Ababa, the modern capital city, sits on an 8,000 ft.(2,439 m) plateau in the center of the country. Ethiopian exports are transferred by rail to Djibouti. Ethiopia relied on Djibouti's shipping ports even before it lost its Red Sea coastline when the northern province of Eritrea seceded in the early 1990s. Coffee is an ancient export; some believe the word comes from "Kaffa," a local region.

KENYA E

Area: 219,790 sq.mi.(637,391 km²). **Population:** 28,400,000. **Capital:** Nairobi, 1,200,000. **Government:** Republic. **Language:** Swahili; English; native dialects. **Religion:** Christianity 70%, indigenous 25%, Islam 5%. **Exports:** Coffee, tea, pyrethrum, cashews, sisal, cotton. **Climate:** Hot and humid on coast; mild in the highlands. ☐ White beaches, mountain scenery, a pleasant climate, and wildlife parks and game preserves have made Kenya an outstanding attraction. Revenue from tourism exceeds the sale of cof-

fee, the principal export. For years, mile-high Nairobi was famous as the African safari (Arabic for "trip") capital. Kenya's coast was first settled by Arabs 2,000 years ago. Mombasa (500,000), the second-largest city and chief port, was an Arab colony. Many nations have controlled the coast, but the British colonized all of Kenya. Most Kenyans speak Swahili, a Bantu tongue containing many Arabic and some Portuguese words. Kenya's earliest human history dates back two million years. Fossil bones of remote ancestors were discovered in the Great Rift Valley. Except for the fertile cooler highlands, most of Kenya consists of hot, arid plains, home to a wide variety of wildlife.

RWANDA F

Area: 10,170 sq.mi.(26,340 km²). **Population:** 8,000,000. **Capital:** Kigali, 230,000. **Government:** One-party republic. **Language:** Kinyarwanda; French. **Religion:** Roman Catholic 65%, Animism 35%. **Exports:** Coffee, tea, tin, tungsten, pyrethrum. **Climate:** Mild because of altitude. ☐ Rwanda (roo wahn' da) is Africa's most densely populated country. In 1959, the Hutus, who make up 90% of the population, overcame six centuries of rule by the Tutsi tribe. In the early 1990s, the Hutu president died in a plane crash. Government forces, blaming the Tutsi, unleashed a bloodbath against the Tutsi population and Hutus who had been lenient toward Tutsi. A Tutsi military force finally stopped the slaughter and overthrew the Hutu government. Rwanda and Burundi, formerly Ruanda-Urundi, were part of Germany's East African Empire prior to World War I. The area became a Belgian Mandate and was split into two nations 1962.

SEYCHELLES. This 90-island archipelago (170 sq.ml., 440 km²) in the Indian Ocean, 1,000 mi.(1,600 km) from the African mainland, is home to 75,000 residents of mixed African and European heritage. Portugal found the islands in the 16th century, but France created a colony in 1814. The Seychelles (say' shells) were given to England, which granted independence in 1976. Farming is limited because of the granite and coral soil. But only in the Seychelles can one find trees producing double coconuts weighing as much as 50 lb.(22.7 kg).

SOMALIA G

Area: 246,200 sq.mi.(637,658 km²). **Population:** 6,850,000. **Capital:** Mogadishu, 600,000. **Government:** None. **Language:** Somali. **Religion:** Islam. **Export:** Livestock, hides, bananas, frankincense, myrrh. **Climate:** Extremely hot and dry on the Aden coast; more moisture to the south. ☐ Somalia (so mah' lee uh or so mahl' ya), on the tip of the Horn of Africa, is a poor, hot, dry nation of nomads. The only arable land, irrigated by two nonnavigable rivers, is in the south. By African standards, the population of Somalia is remarkably uniform. Somali-speaking Muslims can be distinguished from each other only by membership in one of four clans. Despite a 2,000-year oral tradition, the Somali language was unwritten until a system was devised in the 1970s. Prior to independence in 1960, the nation was divided into British Somaliland on the Gulf of Aden and Italian Somaliland on the Indian Ocean. Somalia angered Ethiopia, Djibouti, and Kenya by encouraging Somali-speaking Muslims in those countries to secede and join Somalia. Civil war broke out among the clans in 1991; the U.S. and U.N. intervened.

TANZANIA H

Area: 364,890 sq.mi.(945,065 km²). **Population:** 30,700,000. **Capital:** Dar es Salaam, 800,000. **Government:** Republic. **Language:** Swahili; English. **Religion:** Animism 40%, Christianity 30%, Islam 30%. **Exports:** Cloves, coffee, tobacco, sisal. **Climate:** Coast is tropical; interior is mild. ☐ Tanzania (tan za nee' a) was created in 1964 when newly independent Tanganyika and Zanzibar united. "Zanzibar" refers to the group of offshore islands, to the largest island itself, and to its capital city. It is the world's leading producer of cloves. In the early 19th century, the city of Zanzibar was an Arab sultanate and the major slave-trading center for East Africa. Tanganyika was first a German colony, then a British protectorate. The capital city, Dar es Salaam, handles commerce for landlocked nations to the west. For many years it has been Africa's most important Indian Ocean port. Tanzania's many natural wonders include snow-capped Kilimanjaro, Africa's tallest peak (19,340 ft., 5,895 m); Lake Tanganyika, one of the world's longest and deepest lakes (with the most species of fish); Lake Victoria, the world's second-largest freshwater lake; Olduvai Gorge, where fossils of some of the earliest human ancestors have been found; Selous, the world's largest game park; and Ngorongoro, the world's second-largest volcanic crater (12 mi., 19 km), whose grass-covered floor is home to 30,000 large animals.

UGANDA I

Area: 91,140 sq.mi.(236,053 km²). **Population:** 21,000,000. **Capital:** Kampala, 600,000. **Government:** Republic. **Language:** Swahili; English. **Religion:** Christianity 60%, Animism 25%, Islam 15%. **Exports:** Coffee, tea, cotton, copper. **Climate:** Mild. ☐ Landlocked, beautiful Uganda (yoo gan' da), the "Pearl of Africa," should be prosperous: it has fertile land, a good climate, ample rainfall, hydroelectric power, minerals, and a rail link to the port of Mombasa in Kenya. But the economy is slowly recovering from years of horror—the seven-year rule of General Idi Amin (300,000 Ugandans died), plus numerous coups, invasions, civil wars, and tribal conflicts. Over 15% of Uganda is covered by lakes and rivers. On Lake Victoria and close to the capital is Entebbe airport, where in a famous raid, Israeli commandos freed a planeload of hostages.

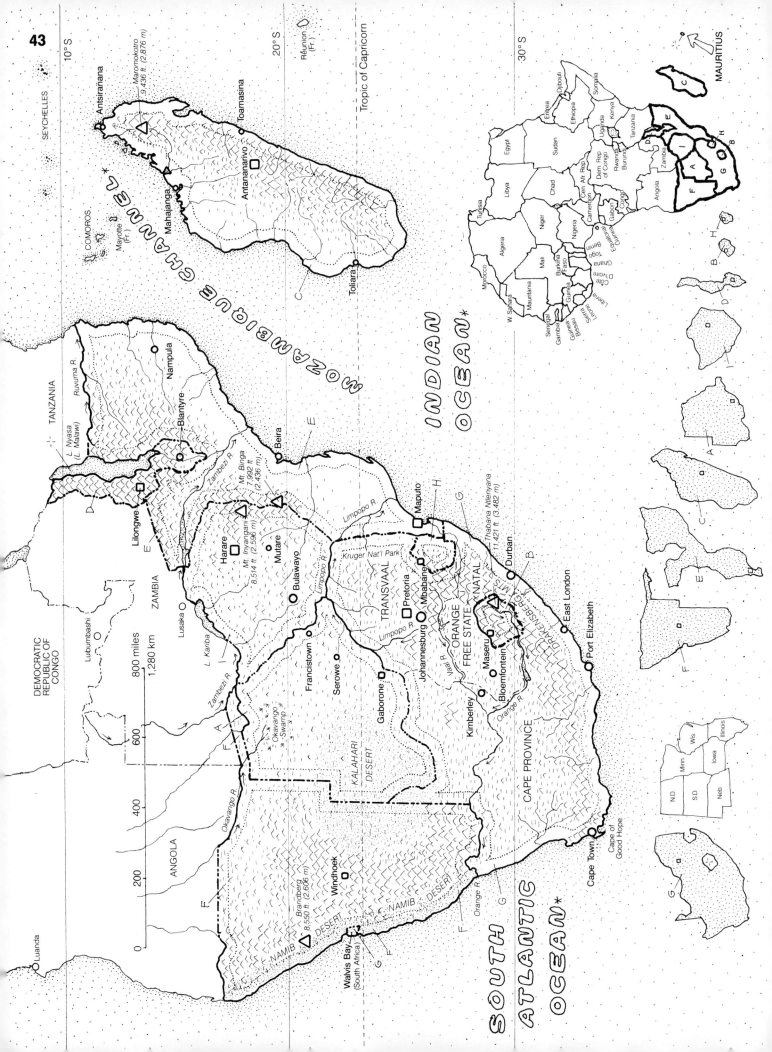

AFRICA: SOUTHERN J

Southern Africa is mineral-rich. Its great natural wealth and mild weather attracted the largest concentration of European immigrants in Africa. Consequently, this area has taken the longest for black majorities to regain control of their lands. Most of southern Africa is a high plateau, sloping eastward and upward to the Drakensberg range, looming over the coastline on the Indian Ocean. The western half is either all desert (the Namib) or semi-desert (the Kalahari).

BOTSWANA A

Area: 232,000 sq.mi.(603,200 km²). **Population:** 1,450,000. **Capital:** Gaborone, 130,000. **Government:** Republic. **Language:** English; Setswana. **Religion:** Indigenous 85%, Christianity 15%. **Climate:** Mild, with limited rainfall ☐ The discovery of huge diamond and other mineral deposits has brought prosperity to Botswana (bo tswa' na). Most of its limited population lives in a narrow, green region along the border with South Africa. Botswana is dependent on South Africa's business services, jobs, and rail links to the coast. Botswana was Bechuanaland, a British protectorate, until it gained independence in 1966. The government is one of Africa's rare non-racial, multi-party republics. The land is a flat plateau, much of which is covered by the Kalahari Desert, home to the Bushmen, one of Africa's oldest races. The ancestors of these small, yellowish-brown people retreated to the semi-desert centuries ago. Their language is distinctive for its use of clicking sounds.

LESOTHO B

Area: 11,720 sq.mi.(30,356 km²). **Population:** 2,100,000. **Capital:** Maseru, 150,000. **Government:** Constitutional monarchy. **Language:** Sesotho; English. **Religion:** Christianity 80%, indigenous 20%. **Exports:** Wool, meat, diamonds. **Climate:** Temperate. ☐ The principal income of Lesotho (luh so' to) consists of the paychecks sent home by the half of its male population working in South Africa. Until 1966, this small nation was the British protectorate Basutoland; that status kept it from being annexed by South Africa, which completely surrounds it. Wealth here is measured by ownership of livestock.

MADAGASCAR C

Area: 226,650 sq.mi.(587,024 km²). **Population:** 14,500,000. **Capital:** Antananarivo, 900,000. **Government:** Republic. **Language:** Malagasy; French. **Religion:** Indigenous 55%, Christianity 40%. **Exports:** Coffee, vanilla, spices, meat, fish. **Climate:** Tropical; cool highlands; wet east coast. ☐ Madagascar, 300 mi.(480 km) off the coast of Africa, is the world's fourth-largest island. Millions of years ago, it broke away from the mainland (p. 2). Madagascar was settled by Indonesians and Malayans from Southeast Asia nearly 2,000 years ago. It was colonized by France in 1896, and was granted independence in 1960. Most Madagascans are of African and Asian descent. An ethnic division exists between the races. While a black majority holds power, the Indonesian influence has moved Madagascar culturally closer to Asia. Almost all of the flora and fauna are unique to the island. The best-known animal is the monkey-like lemur (there are 22 species). Eggs of the prehistoric elephant bird (weighing as much as 20 lb., 9 kg) are still being found. The flightless birds were killed off by early settlers who began a process of extinction that is accelerating with the destruction of the eastern rainforest. The western region is a deciduous forest. An incredible variety of tree species is currently in danger of extinction.

MALAWI D

Area: 45,747 sq.mi.(118,485 km²). **Population:** 9,800,000. **Capital:** Lilongwe, 250,000. **Government:** Republic. **Language:** Chichewa; English. **Religion:** Christianity 80%, Islam 20%. **Exports:** Tea, tobacco, peanuts, cotton. **Climate:** Mild. ☐ The beautiful nation of Malawi (ma la' wee) was formerly Nyasaland, a British protectorate. "Malawi" was the name of an early Bantu kingdom. Lake Malawi (Nyasa) fills the southernmost trench of the Rift Valley. Malawi had only one leader from the time of independence in 1962 until a multi-party democracy was created in the 1990s and he lost in free elections. Dr. Hastings Kamuzu Banda, who had made himself president for life, was unusually conservative for a black African leader. Among his more restrictive policies was a personal dress code that forbade long hair on men and the wearing of shorts, pants, and short skirts by women.

MAURITIUS. This group of tropical islands in the Indian Ocean, 500 mi. (800 km) east of Madagascar (marked by an arrow to the right of the small map of Africa), has been colonized by the Dutch, French, and British. The main island, Mauritius (maw rish' us) (790 sq.mi., 2,046 km²) has one million multi-racial citizens; two-thirds are Hindus. Sugar plantations cover half of the land but tourism, fishing, and clothing manufacturing are the growing industries.

MOZAMBIQUE E

Area: 303,000 sq.mi.(784,770 km²). **Population:** 17,000,000. **Capital:** Maputo, 750,000. **Government:** Republic. **Language:** Portuguese; native dialects. **Religion:** Animism 65%, Christianity 20%, Islam 15%. **Exports:** Cashews, tea, textiles, copra, cotton. **Climate:** Tropical and humid along the coast. ☐ The end of a crippling civil war in 1992 marked the first time, since gaining independence in 1975, that Mozambique (mo zuhm' beek) was at peace. When 250,000 Portuguese fled the country in fear of retaliation for centuries of brutality (not counting 10 years of resisting the liberation movement), they abandoned skilled positions that Mozambique citizens were unprepared to handle. The 1980s then brought both drought and deluge. The communist government made progress in education, health, and women's rights, but failed to improve the economy (it has since renounced Marxism). A barbaric civil war was waged against right-wing rebels, organized and trained by white Rhodesians (before Rhodesia became Zambia and Zimbabwe) and South Africans, who for years had pursued a policy of destabilizing neighboring black governments. The countryside is finally free of the murderous guerrillas. Traffic now moves unimpeded on the nation's railroads and through the shipping ports, providing critical revenue from neighboring landlocked countries.

NAMIBIA F

Area: 318,260 sq.mi.(824,293 km²). **Population:** 1,700,000. **Capital:** Windhoek,165,000. **Government:** Republic. **Language:** English; Afrikaans; native dialects. **Religion:** Christianity 90%, indigenous 10%. **Exports:** Diamonds, copper, uranium, lead, fish. **Climate:** Temperate and very dry. ☐ Africa's last colony, Namibia (nuh mib' ee a), formerly South West Africa, became independent in 1990 after 100 years of German and South African rule. For 70 years, South Africa treated Namibia as its province, but international pressure and the militant actions of the Southwest Africa's People's Organization (SWAPO) forced South Africa to give it up. Namibia has less than two million people, including 70,000 whites from South Africa and about 30,000 Bushmen living in the Kalahari Desert. Most of Namibia is dry, particularly the Namib Desert along the Atlantic coast. The cold waters of Antarctica's Benguela Current (p. 57) are responsible for creating a coastal desert like those in Peru and Chile (p. 17). Namibia possibly has the continent's largest diamond and uranium deposits.

SOUTH AFRICA G

Area: 471,445 sq.mi.(1,225,757 km²). **Population:** 43,000,000. **Capital:** Pretoria, 450,000. **Government:** Republic. **Language:** Afrikaans; English; native dialects. **Religion:** Christianity 70%, indigenous 30%. **Exports:** Gold, coal, diamonds, food products, asbestos, metal ores. **Climate:** Temperate and dry; Mediterranean on the Cape coast. ☐ South Africa is the most industrial nation on the continent. It mines a wide array of minerals and half the world's supply of gold and diamonds. South Africa had been a strategic ally of the West during the Cold War, but in the 1980s, the U.S., Western Europe, and other nations imposed economic sanctions (restrictions on trade and investment) against it because of its racial policies. Since 1948, the practice of "apartheid" (uh part' tite), based on skin color, rigidly separated the population into four classes: blacks (who make up 70% of the population); whites (17%); colored (10%); and Asian (3%). Blacks were divided ethnically into ten separate "homelands," where they had to live if not employed elsewhere. These homelands, generally located in unproductive parts of the country, occupied less than 15% of the land. The whites in South Africa had segregated themselves as well: the white Afrikaners, the landowning majority, are concentrated in the interior regions of the Orange Free State and the Transvaal. These descendants of early Dutch, German, and French settlers speak Afrikaans, a dialect derived from Dutch and French. The Afrikaners controlled the military; their ruling Nationalist party created apartheid. The English minority (40% of the whites) lived mostly in the Cape and Natal Provinces. These descendants of English colonialists ran their own schools and many of the nation's business and industrial communities. The coloreds, who provided a skilled labor force in Cape Province, are mostly descendants of the earliest residents (Hottentots) and European settlers. The Asians living in Natal are descendants of contract laborers brought from India in the 19th century. The first Europeans to settle South Africa were the 17th century Dutch, called Boers ("farmers"). When the British arrived in the early 19th century, the fiercely independent Boers withdrew to the interior. When gold and diamonds were discovered in the interior, the British wanted their share, and this resulted in the Boer Wars of the turn of the century. The victorious British colonized all of South Africa, which became a self-governing nation in 1910. In the 1970s, the government began easing its racial restrictions. In 1990, it recognized the African National Congress (ANC), a militant black resistance organization, and released its leader, Nelson Mandela, after 27 years of imprisonment. Apartheid was banned in 1991, ending racial classification. Free elections were held in 1994, and Mandela became president. He formed a coalition government, including members of the former ruling party.

SWAZILAND H

Area: 6,706 sq.mi.(174,356 km²). **Population:** 970,000. **Capital:** Mbabane, 47,500. **Government:** Constitutional monarchy. **Language:** Siswati; English. **Religion:** Christianity 70%, indigenous 30%. **Exports:** Sugar, wood pulp, iron ore, asbestos, citrus. **Climate:** Mild; cooler in highlands. ☐ Swaziland is one of Africa's three remaining kingdoms (with Lesotho and Morocco). As a British protectorate until 1968, it remained independent of South Africa, which surrounds it on three sides. The nation has good land, climate, water, and mineral deposits. Descendants of white immigrants from South Africa own nearly half the major farms, the mines, the industries, and the forests they planted in the highlands. Swaziland is known for its popular vacation spots.

ZIMBABWE I

Area: 150,750 sq.mi.(390,452 km²). **Population:** 11,100,000. **Capital:** Harare, 800,000. **Government:** Republic. **Language:** Shona; Ndebele; English. **Religion:** Christianity 60%, indigenous 40%. **Exports:** Asbestos, chromium, gold, nickel, tobacco, food products. **Climate:** Mild; ample rainfall. ☐ Zimbabwe (zim bah' bway) is named after an early African kingdom, whose stone ruins are a major tourist attraction. Most of the land is a fertile, high plateau with ample water, mineral deposits, and hydroelectric power. Though landlocked, it has rail connections to ports in Mozambique and South Africa. Zimbabwe was a British colony (Southern Rhodesia) until ruling whites defied Britain and asserted their independence in 1965. Fourteen years later, they yielded to international sanctions and guerrilla pressure and transferred power to the 98% black majority. Many whites were persuaded to stay on in key positions. About 4,000 prospering white farmers still own about a third of the country. Land reform has failed but educational programs have been too successful—there are far more qualified citizens than there are jobs.

44

THE ARCTIC

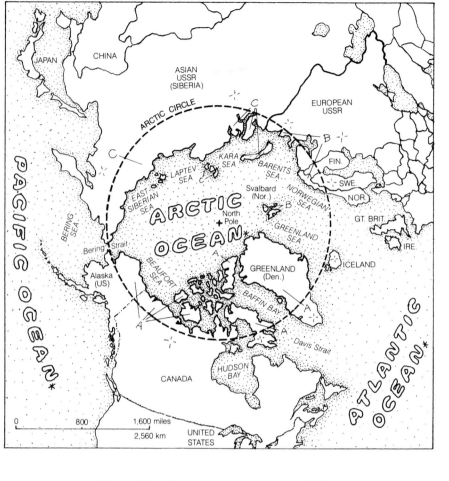

ARCTIC

NORTH AMERICA A
EUROPE B
ASIA C

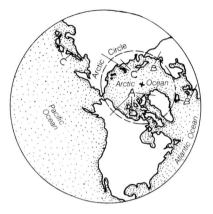

THE ANTARCTIC

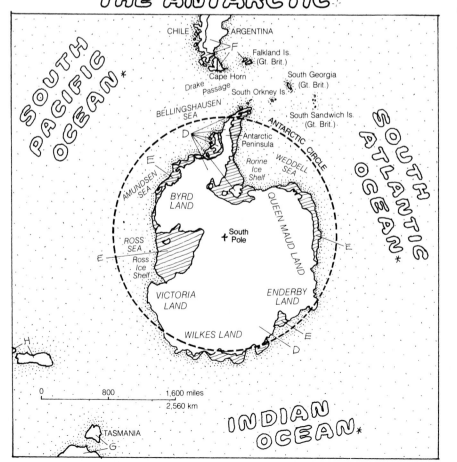

ANTARCTIC

ANTARCTICA D
ICE SHELF E
SOUTH AMERICA F
AUSTRALIA G
NEW ZEALAND H
AFRICA I

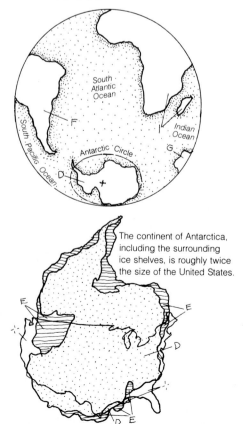

The continent of Antarctica, including the surrounding ice shelves, is roughly twice the size of the United States.

POLAR REGIONS

CN: (1) On the large upper map, color only the portions of the continents (A–C) that are within the Arctic Circle. On the global views, color the continents completely. (2) On the large lower map, use a very light color for Antarctica (D). (3) On the diagram below, use yellow to color the Sun (J) and the sunlight and daylit portions of the globes (J').

The polar regions are the lands bordered by the Arctic Circle in the Northern Hemisphere and the Antarctic Circle in the Southern. These circles mark the point at which the Sun stays above the horizon for at least one 24-hour period during the year. Many geographers and scientists prefer to define the Arctic as the region north of the "tree line" (the northernmost point at which trees will grow) or the "temperature line" (the line above which the average July temperature stays below 50° F/10° C). Both lines roughly correspond to the astronomically determined Arctic Circle.

Because of limited precipitation, the polar regions can be classified as deserts. Antarctica is the driest, with an average annual snowfall equal to 2 in.(10 cm) of rain. Because of the cold, even small amounts of precipitation do not melt, and are added to the ice cap. The Arctic receives somewhat more moisture (6–10 in. or 15–25 cm), and because of poor drainage and slow evaporation, the Arctic landscape remains marshy during the summer thaw.

Although the Arctic and Antarctic are comparable in size, have similar light and dark seasons, and receive very little precipitation, the two regions are in fact very different. The Arctic is basically an ocean surrounded by the northern regions of North America, Europe, and Asia. The Arctic Ocean, the world's smallest, covers an area of 5,500,000 sq.mi.(14,374,500 km²). The Antarctic region is just the opposite: it is a land mass that constitutes the continent of Antarctica (larger than Europe and Australia), surrounded by three oceans (the South Atlantic, South Pacific, and Indian).

Airliners taking the shorter "great circle routes" routinely fly over the Arctic, and nuclear submarines navigate under the ice at the North Pole. Yet it wasn't until 1909 that explorers reached the "the top of the world." Admiral Robert E. Peary was the first man to set foot on the North Pole. Two years later the Danish explorer Roald Amundsen was the victor by 34 days in the race to the South Pole against an Englishman, Robert F. Scott, who, with members of his team, perished on the trek back.

Because of warming ocean currents, the Arctic is not quite as cold as the interior of Siberia. But much of the Arctic Ocean stays covered by 10–15 ft.(3–4.5 m) of ice. The relative warmth of summer causes part of the ice to break into moving packs, and the tundra (treeless, permanently frozen land) thaws out enough to support colorful plant growth. The thaw involves the uppermost 6 in.(15 cm) of the 1,000 ft.(305 m) or more of permafrost (permanently frozen ground).

The Antarctic has no such dramatic change of season; it's on average 35° F (20° C) colder than the Arctic. Gale-force winds combine with the frigid cold to produce the Earth's fiercest weather. Winds in excess of 200 mph (320 kph) have been clocked, and the lowest temperature reading on record, –128° F (–89° C), was made during an Antarctic winter. Floating ice shelves attached to various parts of the Antarctic coastline considerably expand the size of the continent. In some places, the ice covering Antarctica is close to 3 mi.(4.8 km) thick. This ice cover contains 90% of the world's supply of fresh water. Most of Greenland (in the Arctic) is very similar to Antarctica because both lie under ice sheets of similar thickness. Icebergs break off from glaciers on both land masses to create hazards for shipping.

Over one million people live within the Arctic Circle. Most are of Mongoloid ancestry,

including the Inuit (Eskimos) of North America, the Lapps of Scandinavia, and the Chukchi and Samoyeds of Russia. The introduction of modern communication, transportation, and scientific and mining operations into these regions has changed the lifestyles of the natives. Snowmobiles are replacing dogsleds. The Arctic is sparsely populated, but it seems crowded when compared to Antarctica, which hasn't a single permanent resident. About 4,000 scientists from many nations limit their periods of residence to the summer months.

The world's most vacant continent, Antarctica, is also the highest. The average elevation is 6,000 ft.(1,830 m). Some of its mountain ranges are as tall as 15,000 ft. (4,570 m); the continent would be even higher if not for the weight of the thousands of feet of ice cover, which compresses land and mountains alike. The Antarctic Peninsula is an extension of the Andes Mountains of South America.

The Arctic region has a rich sampling of land animals: polar bears, reindeer (in Europe and Asia), caribou (the reindeer's cousins in North America), wolves, foxes, and numerous smaller creatures. Except for polar bears, seals, walruses, and some foxes, most of the animals migrate south during the winter. Antarctica, on the other hand, is even more devoid of land animals than it is of humans. The only land animal present is a tiny, wingless mosquito, about a tenth of an inch (2.5 mm) long. Bird and sea life are abundant in both regions. Antarctica is best known for its penguin population. The Emperor penguin, the largest of the species, stands 4 ft.(1.2 m) tall and is capable of surviving the brutal Antarctic winter.

Personnel in scientific stations and field laboratories are actively engaged in research throughout the polar lands. Meteorologists from both hemispheres gather data to assist in global weather forecasting. A variety of mineral deposits have been discovered, but mining operations, such as oil drilling on the north slopes of Alaska, are in progress only in the more accessible Arctic region. Scientists are concerned about the status of the ozone layer of the Earth's atmosphere. This barrier against ultraviolet radiation has been thinning rapidly because of certain contaminants released into the atmosphere by the industrialized nations. A hole has been discovered in the ozone layer above the South Pole, and there has been a decrease in the region's phytoplankton, the plant source that feeds the shrimp-like krill that are at the heart of the Antarctic's marine food chain.

Although there are no immediate prospects for the exploitation of Antarctica, 16 nations have established permanent bases, and 7 have staked out claims to ownership of the land. The United States and the rest of the world do not recognize these claims. The Antarctic Treaty, signed in 1959, grants to nations the right to pursue scientific investigations for peaceful purposes, but they must share all discoveries. Military activity, nuclear testing, and the dumping of toxic waste are prohibited. An unfortunate amount of dumping, burning of waste, and numerous fuel spills have already occurred, but efforts are being made to eliminate those practices. Nations are debating the merits of a new treaty, the Wellington Convention, that would allow mining (under the strictest supervision) in the Antarctic. Skeptical environmentalists fear the unbridled despoliation of the only remaining, truly wild continent. Along with a minority of nations, they favor a proposal that would make Antarctica a "World Park."

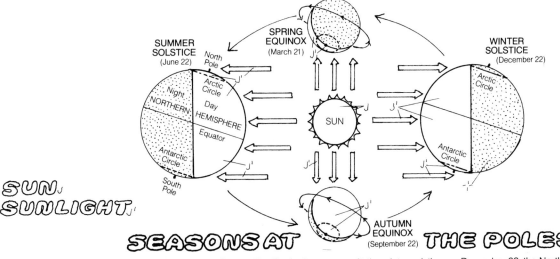

SUN SUNLIGHT

SEASONS AT THE POLES

The Earth revolves around the Sun once every 365¼ days. The Earth also revolves on its own axis (an imaginary line connecting the two poles) once every 24 hours, creating a night-and-day effect for most points on the planet. The Earth's axis is tilted at an angle of 23.5° to the plane of its orbit around the Sun. We have seasonal changes on Earth because of this tilt. Without the tilt, the same amount of sunlight would fall on a particular point on the planet every day, and climates would not vary throughout the year. At the summer solstice, around June 22, the Northern Hemisphere is tilted toward the sun and experiences its longest days and shortest nights. The diagram shows that on this day the entire region within the Arctic Circle receives 24 hours of sunshine. Simultaneously, the Antarctic region is in 24 hours of darkness, and winter begins in the Southern Hemisphere.

At the winter solstice, on December 22, the Northern Hemisphere is tilted furthest from the Sun, and the entire Arctic Circle is in darkness for the first 24 hours of winter. The spring and autumn equinoxes occur around March 21 and September 22. The equinoxes fall midway between the solstices. On the days of the equinoxes, the Earth's axis is perpendicular to the Sun's rays—the axis is neither toward nor away from the Sun (it might help to visualize the tilt as being "sideways" to the Sun). The sun is directly above the Equator, and days and nights are of equal length at all latitudes, in both hemispheres. Between the spring and autumn equinoxes, within the Arctic Circle there will be at least one day of 24-hour sunlight ("Midnight Sun"), up to a maximum of 6 months at the North Pole. During that period, an equivalent amount of darkness will prevail within the Antarctic Circle.

45

HISTORIC EMPIRES

CN: (1) Keep in mind that the maps on Plates 45-48 show the empires at their greatest point of expansion. Their actual size would have varied considerably over the period of their rule. (2) Use your lightest colors so that the underlying information on the maps remains visible. If you don't have enough light colors, feel free to repeat any of those already used in this section. (3) Color the names and dates with the same colors that you use on the areas within the dark outlines of the empires. (4) The names of the modern nations are printed in a light-face type; the historic names are printed in slightly darker type.

PERSIAN EMPIRE A

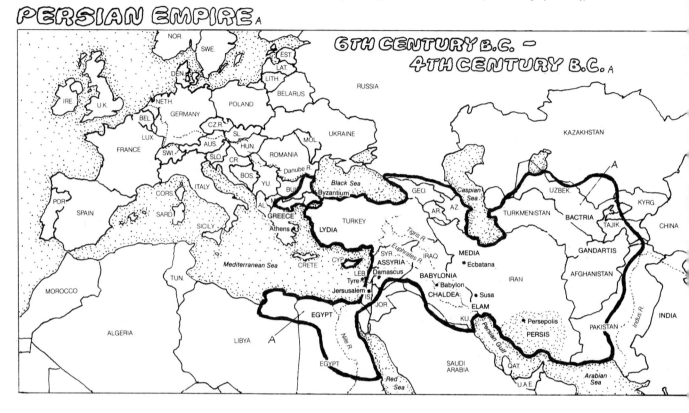

6TH CENTURY B.C. – 4TH CENTURY B.C. A

ALEXANDER THE GREAT B

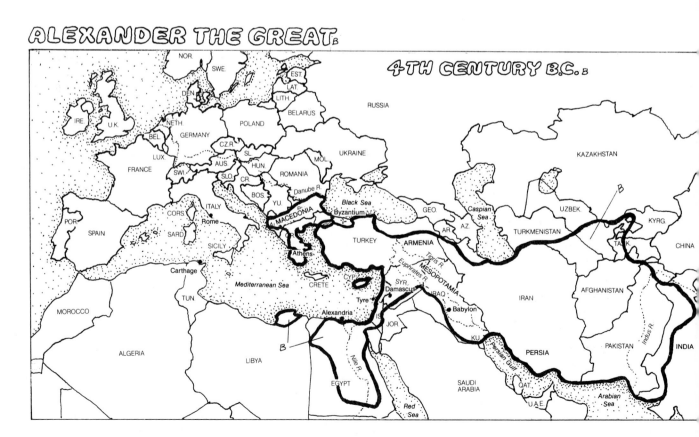

4TH CENTURY B.C. B

The Persian Empire originated in what is now Iran, which takes its name from the Aryans who settled there around 2,000 B.C. The Aryans were Indo-European, white-skinned people who migrated south and westward (to much of Europe) from Central Asia. Two powerful Aryan tribes competing for control of southwestern Asia were the Medes and the Persians. In the 6th century B.C., the Persians, under Cyrus the Great, overthrew the Median Kingdom and created the Persian Empire. The presence of this newly formed, powerful empire created great concern in the kingdoms of Egypt, Lydia, and Chaldea. The leaders of these regions formed a military alliance designed to eliminate the Persian threat. But before they could coordinate their efforts, Persian armies decisively defeated and captured Lydia and Chaldea. The fall of Chaldea brought about the unexpectedly passive surrender of the massively protected city of Babylon.

Cyrus called his Persian Empire the Achaemid Empire (after his ancestor, Achaemenes). Its rapid growth was halted only temporarily when Cyrus was killed in battle while capturing lands eastward toward India and northern Pakistan. His son, Cambyses, quickly took command and expanded the empire to the west by conquering Egypt and thereby making the Persian Empire the largest the world had ever seen. At its height, it controlled Egypt, the eastern and southern shores of the Mediterranean, all of the Anatolian Peninsula, the Middle East (north of Saudi Arabia), and most of Southwest Asia up to the border of India. The empire briefly occupied regions in Greece, southeastern Europe, and part of southern Russia.

It was during the reign of Darius I (in the latter part of the 6th century B.C.) that the empire enjoyed its greatest prosperity. In addition to Egypt and Babylonia, over which he personally ruled, Darius created twenty districts (called satraps) with four capitals: Susa, Ecbatana, Babylon, and Persepolis. These districts enjoyed self-rule and were allowed to retain their original laws and customs. Like so many empires that preceded and followed them, the Persians took away far more from their newly subjected people than they actually imposed upon them. This practice of conquerors, in effect being conquered by their subjects, has been especially true whenever culturally advanced societies have fallen to primitive invaders.

Darius seemed more concerned with maintaining the empire than in acquiring new territories. He managed to deter most dissent and rebellion by conducting an unusually rational and tolerant administration. Darius freed thousand of Jews held captive in Babylon and allowed them to rebuild their Great Temple in Jerusalem. The Persian policy of tolerance can be traced to their religion of Zoroastrianism and the prophet Zoroaster, who preached the necessity of doing good deeds in order to reach heaven.

Commerce and communication boomed in the empire when the Egyptian system of writing with ink on paper replaced the laborious method of cuneiform inscriptions on clay tablets, and also when minted coins became the medium of exchange. The Persians ingeniously converted vast regions of desert into fertile farmland by the building of narrow irrigation tunnels (called ghanats), which were able to transport water, underground, for miles, safe from the sun's evaporative heat. A large network of well-maintained roads facilitated the establishment of a fast and efficient postal system, operating in the manner of the American pony express.

Persian prosperity began to diminish at the beginning of the 5th century B.C., when captured Greek cities on the Aegean coast of the Anatolian Peninsula ("Anatolian Greeks") revolted, with the aid of ships furnished by Athens. Darius dispatched his army and navy both to punish the Athenians and to end their military capabilities. Thus began the series of Greek/Persian wars in which the Persians succeeded in destroying Athens but ultimately suffered humiliating defeats at the hands of greatly outnumbered but strategically superior Greek armies. The Persian invasion of Greece motivated the antagonistic Greek city-states to join forces to defeat the common enemy. For the next century, the Persian military remained a threat to Greece until Alexander the Great invaded Persia in 331 B.C. and dealt Darius III his final military defeat. Alexander quickly took control of most of the former Persian Empire, but he did not retaliate, as expected, for the Persian burning of the Athenian Acropolis. Instead, he once again demonstrated his unusual capacity for tact and restraint in dealing with captives by only symbolically setting fire to the Persian palace at Persepolis.

From the time he assumed command of his father's army at the age of twenty (in 336 B.C.) until his death just thirteen years later, Alexander the Great conquered more territory, in a shorter period of time, than any military leader before or after him. Ironically, this seemingly invincible man who amassed the largest empire the world had ever known was not felled by an enemy sword, but by the bite of a malaria-bearing mosquito.

No one was ever better prepared for a mission than was Alexander. Son of Philip, King of Macedonia (himself an accomplished military leader), Alexander received his combat training from a very early age, and when he was eighteen, he successfully commanded one of Philip's legions into battle.

Even more remarkable than his prodigious military skills were the elegance, grace, and understanding with which Alexander was to rule his conquered territories—an ability that in large measure was the result of his father's foresight in balancing the boy's military education with studies in philosophy, literature, science, and the arts. These studies were to foster a lifelong interest in and respect for learning—Alexander was known to have brought experts in various disciplines along on campaigns for the benefit of their advice. To find the best teacher for his son, and aware of Macedonia's cultural limitations, Philip looked to his eminent neighbor, Greece. He persuaded its greatest living philosopher, Aristotle, to become Alexander's personal tutor. Aristotle was to spend five years in Macedonia, teaching an extraordinarily receptive pupil.

As Alexander was being groomed for his future role as conqueror and ruler, Philip was expanding Macedonian rule northward to the Balkans and southward to Greece. As a young man, Philip had learned military strategy by studying the Greek army. He used their own tactics, along with improved weapons, to conquer the Greek city states. In 336 B.C., Philip was assassinated, and Alexander became Macedonia's King and military leader.

In a very short time, he was to show just how "ready" he was for his new role when Greek city states attempted to free themselves from Macedonian control. Alexander, intent on making a point and setting an example, was extremely brutal in suppressing the rebellion. His forces killed thousands in Thebes, and sold even more of its citizens into slavery. Alexander then leveled the entire city, except for the home of a famous poet. Neither his resolve nor ability was ever to be tested again, nor would he ever display such savagery against an enemy. If anything, he would be noted for the professional way in which he fought his battles, respected his opponents, and treated his subjects. His troops were never gratuitously brutal, nor were they allowed to pillage communities or physically harm the inhabitants.

Alexander had an extraordinary gift for gaining the cooperation of those he conquered. They were allowed to retain their languages, religions, and customs. Alexander would eagerly adopt the best that local cultures had to offer. He encouraged his men to intermarry (he himself married twice). He sought out the best soldiers from conquered armies and brought them into his forces. He added to his growing mystique by going to Egypt, where oracles proclaimed him a soon-to-be living God. While there, he founded the city of Alexandria at the mouth of the Nile River. It rapidly developed into one of the Mediterranean's leading seaports and the site of a magnificent library that would house the entire body of Greek learning. This was the first of many cities (16 with the name "Alexandria") he was to create. One city, "Bucephalia," was named after his favorite horse, which died in battle. The cities were laid out according to traditional Greek grid patterns and incorporated all of the usual amenities: civic centers, libraries, amphitheaters, baths, etc. These cities would spread Hellenistic culture throughout the ancient world; their remnants can be seen throughout Asia Minor.

When Alexander conquered the Persian Empire (and ended its threat to Greece), he was fulfilling an unrealized goal of his father. For two centuries, Persia presided over the world's largest and most advanced empire (Alexander's conquests were to barely exceed its boundaries). The Persian Empire was the first bridge between East and West, and Alexander had enormous respect for their accomplishments. He studied their laws and government, and married into the culture by taking as his bride the daughter of a Persian prince. When he began to dress in the lavish style of Persian royalty, his men felt he had gone too far. Their morale was already extremely low: they were war weary, they had been away from home for too long, and they resented the growing presence of former enemy soldiers in their ranks. After incurring heavy casualties in India—at the hands of elephant-led armies—they refused to go beyond the Ganges valley. The mutiny signaled the end of Alexander's eastern campaign. On his return to the west, Alexander caught fever and died in Babylon (his remains were later entombed in Alexandria). With his death, the empire dissolved almost as rapidly as it expanded. It was left to his generals to hold it together—a task well beyond their desire or capability.

46

HISTORIC EMPIRES.

ROMAN EMPIRE.

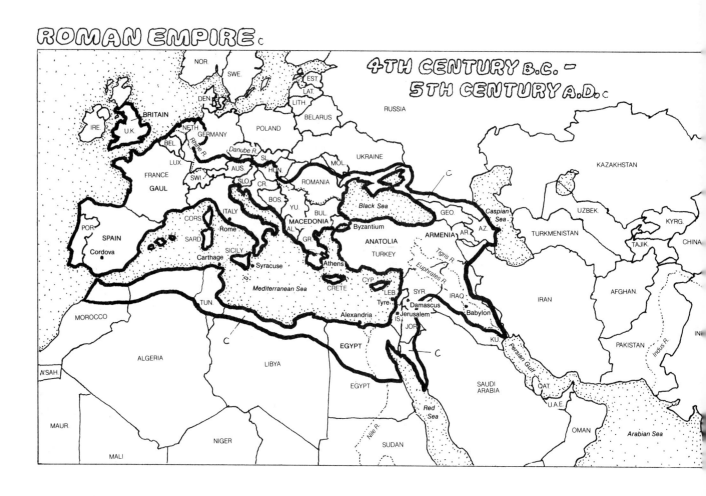

4TH CENTURY B.C. –
5TH CENTURY A.D.

BYZANTINE EMPIRE.

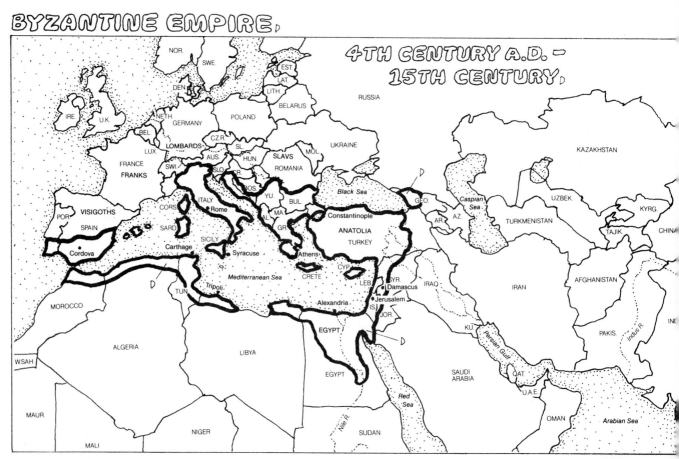

4TH CENTURY A.D. –
15TH CENTURY.

From its inception in the 8th century, the city state of Rome was ruled for 400 years by a succession of kingdoms. In the 4th century B.C., a republican form of government was established, but it represented the interests only of the landed elite. During the next 300 years, Rome, the most centrally located city in the Mediterranean, rapidly grew in power and influence. With wealth came external danger—a military defense was needed. The original army, composed of landowners (the only soldiers deemed trustworthy), had to give way to a larger, full-time military: "Roman Legions."

Carthage, the north African Mediterranean power, was Rome's chief threat. During the years of 264 to 118 B.C., the rivals clashed in the three Punic Wars. The first conflict, lasting 23 years, was fought primarily at sea, initially favoring Carthage's naval strength. But by war's end, Rome had built a fleet large enough to rule the Mediterranean. The second war was noted for the exploits of Hannibal, the Carthaginian general who left Spain with an elephant-led army and managed to march across the Alps into Italy. Although none of the elephants and only half of his men survived the journey, Hannibal revived his forces and defeated Rome's armies in southern Italy. Hannibal never attacked Rome itself, and eventually he had to return to Spain to fight against an invading Roman army. The third war ended with Rome in control of Carthage, Spain, and the islands of Sicily, Corsica, and Sardinia.

With its western empire firmly established, Rome fixed its sights on the eastern Mediterranean nations of Macedonia, Greece, Anatolia (Turkey), Syria, and Egypt. Around 50 B.C., Julius Caesar expanded the empire to include its most distant provinces: Britain and Gaul (France). Caesar was a gifted orator, writer, leader, and general. These enormous talents contributed to his death by a conspiracy of sixty Roman senators who feared that Caesar's growing dictatorial ambitions would end Rome's republican government. The armies of Caesar's successors would never go further north—Ireland, Scotland, and the the European continent east of the Rhine and north of the Danube rivers were to remain unconquered. Rome was reluctant to engage the powerful inhabitants of those regions. The "Barbarians" (the name for anyone not under Roman control) had a fearsome reputation, and Rome would not risk men and treasure for lands of dubious value. It had a better use for its military: when not on assignment, the army was put to work building roads, bridges, aqueducts, buildings, towns, and cities—public works that were central to the unification and prosperity of the empire.

Roman law was equally applied throughout the empire by Latin-speaking members of the privileged classes. Commoners, who were allowed to keep their regional languages and local customs, were still accorded rights as citizens. Slaves could expect to ultimately gain their freedom and become citizens. Usually taken as war booty, slaves were so numerous that Romans of even modest means could afford to own them. At one point, it was estimated that slaves represented nearly 40% of Rome's population. They practically ran the city—handling almost every job, from bureaucrat to house servant. The use of slaves was the ruination of Italy's masses, who were unable to find work because so much "free" labor was available. Peasants, conscripted for years of military service, would return to find that their neglected farms had been sold to wealthy landowners, who used slaves to operate their holdings. A growing class of poor citizens descended upon Rome demanding work. Instead, they were given free bread and admissions to the highly popular (and barbaric) public games. Little attention was paid to the plight of the Italian countryside until a serious effort was required to put down a two-year rebellion by an army of peasants led by Spartacus, an escaped gladiator.

As Italy withered, Rome's elite grew richer from taxes and investments, even as corruption and immorality were eating away at its leadership. The provinces had more trouble: Barbarians were becoming bolder and more successful in seizing outlying regions. Even the most efficient administrative system could not adequately rule so large an empire from a single capital. In 306 A.D., the empire split into a western half ruled by Constantine in Rome and an eastern division headed by the Emperor Licinius in Byzantium. In 330 A.D., the ambitious Constantine defeated his rival in the east, reunited the empire, and made Byzantium the Roman capital. Constantine beautified Byzantium with objects of art taken from cities across the empire. He changed the name of the capital to "Constantinople" (currently "Istanbul"), and made Christianity the official religion.

When problems in the west grew even worse, the empire was split again, but this time there was no forestalling the inevitable: in 476, when the last Emperor was forced off the throne, the Roman Empire ceased to exist. The richer and stronger Eastern Roman Empire would survive for another thousand years as the Byzantine Empire. Although Rome was gone, its law, language, religion, and social institutions continued to have a strong influence on the development of western Europe, and as a result of colonialism, Roman culture would be spread around the world.

In 306 A.D., the declining Roman Empire was split into two parts: a western half ruled from Rome and an eastern division whose capital was Byzantium (named after the ancient Greek colony that stood on that site). Byzantium faced the Straits of Bosporus, the last in a narrow network of water links between the Mediterranean and Black Seas. Today, a bridge spans this strait, linking the Turkish shores of southeastern Europe with Turkish Asia Minor.

In 324 A.D., the Western Roman Emperor, Constantine, defeated his rival in the East and took control of the entire empire. He moved the capital of Rome to Byzantium, which he then called "Constantinople" (to be renamed "Istanbul" by Turkey in 1930). With water facing three sides, and massive walls guarding its rear, Constantinople was to be the Byzantine Empire's impregnable capital for the next millenium. The Roman Empire was split again in 395 A.D.—a date generally regarded as the point in history when the Byzantine Empire grew out of the Eastern Roman Empire. In less than a century, the Western Roman Empire would be overrun by Germanic tribes and cease to exist, while the stronger, wealthier, and less corrupt Byzantine Empire was spared a similiar fate.

Although the new empire was beset by external dangers, it was able to survive, largely because of an efficient central government, with its powerful military, prosperous economies, and a strong, uncorrupted clergy. For 1,000 years, Constantinople continued to be the Empire's bulwark—except for the 57-year period following its destruction and capture in 1204 by the Fourth Crusade. The sacking of Constantinople by Western European Christians—on their way to free the Holy Land from Muslim control—was thought to have been motivated by Venetian merchants eager to eliminate their better-positioned rival for trade with Asia. The attack also may have had the blessing of the Papacy in Rome, angry over the secession of the Eastern Orthodox Church from the Roman Church.

The Eastern Orthodox Church played a major role in spreading Christianity throughout southeastern Europe and Russia. It also established its own view on how Byzantine art should depict religious themes. In contrast to its Western counterpart, Byzantine imagery of humans and animals tended to be less lifelike, more stylized, and often executed in complex mosaics of dazzling pieces of glass and stones. The dome-roofed design of Byzantine churches would strongly influence Russian architecture. Most Byzantine churches had simple, unadorned exteriors, concealing their highly ornate interiors (contrasting the blandness of everyday life with the glorious nature of the inner spirit). The most remarkable church of all was Istanbul's Hagia Sophia (Greek for "holy wisdom"), later rebuilt and enlarged by the famous Emperor Justinian in the 6th century. Ottoman Turks converted it to an Islamic mosque in 1453. Now a museum, this enormous building is surely one of the world's most enthralling structures.

Justinian was successful at recapturing much of the old Roman Empire's territory in North Africa, Spain, and Italy, but his goal of restoring the size and grandeur of the earlier empire was short-lived. Costly military campaigns and excessive public works programs (which made Constantinople the wonder of the world's cities) bankrupted the empire and prevented his successors from defending it against renewed attacks. Rather than military conquests, Justinian's most enduring achievement has turned out to be a book of laws: the Justinian Code. This concise compilation of the history of Roman law was the work of legal scholars who organized and codified all that had gone before. The Code was to become the legal foundation of Western societies.

The greatest accomplishment of the Byzantine Empire during its thousand-plus years of rule was its role as the cultural bridge between the ancient and modern worlds—it preserved the learning of classical Greece and Rome during the "Dark Ages" of medieval Europe. Although the privileged classes all spoke and wrote Classical Greek, Byzantines referred to themselves as "Romans." The masses resorted to a simpler form of Greek, and the cyrillic alphabet, derived from Greek letters, is still being used today in Russia, Bulgaria, Greece, and Yugoslavia.

Between the 7th and the 15th centuries, the empire's boundaries shrank or expanded, depending upon the strength of Byzantine leadership in fighting off recurring invasions from Arabs, Slavs, and Turks. Ironically, during that entire period, palatial Constantinople, the world's largest and most magnificent city, never once was harmed by its non-Christian enemies, but in 1204, Crusaders from Western Europe embarked on a rampage of violence, arson, theft, and destruction that stripped the Byzantine capital of virtually everything of value. Although the "Romans" recaptured their city in 1261, the Byzantine Empire never recovered its strength. The fall of Constantinople in 1453 to the Turks brought an end to the more than 1100 year-old empire.

ISLAMIC EMPIRE E

CN: The Byzantine Empire, which co-existed with the reign of the Islamic Empire, is shown in diagonal shading. Take note, but do not color it.

7TH CENTURY – 11TH CENTURY E

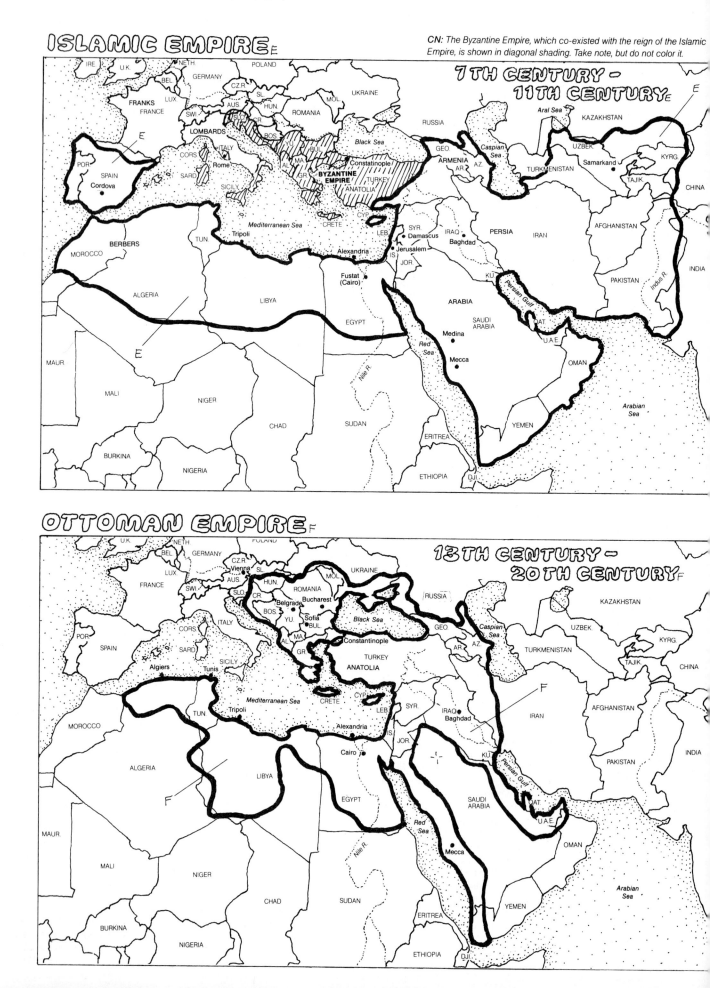

OTTOMAN EMPIRE F

13TH CENTURY – 20TH CENTURY F

At the beginning of the 7th century A.D., Islam, one of the world's great religions, emerged from the barren Arabian desert, and in less than 100 years, it would rule an area three times the size of the Roman Empire. At no time in history has one religion spread so far and so rapidly, dominated as much territory, and made as many converts as did Islam.

Islam means "Submission to the Will of Allah." Allah is the one true God and Muhammad is his prophet, according to the Koran, Islam's Holy Book. Within it can be found God's revelations to Muhammad. The Koran identifies Muhammad as the last in a line of 28 religious prophets sent to earth by God; it recognizes Jesus, Moses, and the other prophets of the Old Testament as predecessors of Muhammad. Parts of the Islamic religion are based on the teachings of Christianity and Judaism. Islam's view of the Day of Judgment is very similar, depicting an eternal life in a divine heaven or a horrific hell. Although more tolerant of other religions than it was in the past, Islam teaches that it alone contains the latest words of God.

Of the numerous rules that the Koran specifies for daily living, none are more important than the requirements to pray five times a day (while facing Mecca), to support the poor, to fast during daylight hours in the holy month of Ramadan, and to make a pilgrimage to Mecca at least once in one's lifetime.

Muhammad was born around 570 A.D. in Mecca, a trading city in western Arabia, close to the Red Sea. As a young adult, he began to receive divine revelations. After a period of seclusion in which he pondered God's teachings, he recognized his mission in life and went forward to make converts. Soon he was regarded by local officials as a threat to the established religion. They forced Muhammad and his followers to flee from Mecca. This flight to Medina in 622, which is known as the "Hegira," marks the beginning of the Muslim calendar. In Medina, Muhammad was well received, and within eight years he was to become the spiritual and political leader of the city and its surroundings. It was then that Muhammad and his followers returned to Mecca to defeat his enemies and destroy all their religious symbols except for the Kaaba (Islam's holiest shrine, which houses the sacred Black Stone).

At the time of his death in 632, Muhammad had unified Arabia under Islam. During the next 30 years, his successors dispatched Arab armies, endowed them with a fanatical fervor to spread the Islamic religion, and conquered parts of the Byzantine Empire (Egypt and Syria) and most of the Persian Empire (Iraq and Iran). As more non-Arabs were converted to Islam, only those choosing to speak the Arabic language were considered to be Arabs.

One may wonder how Arabia's small bands of fanatical horsemen could accomplish such a rapid and extensive spread of Islam. They were, in fact, greatly aided by the weakened condition of and high level of discontent in the adjacent empires. Many of the converted were oppressed people eager to accept a new regime that promoted its simple, comforting religion. Byzantine and Persian defenses, in disrepair and financially drained by prolonged warfare, were no match for the inspired Muslims. The Arabs, driven by their faith and a sense of mission, were immune to cultural influences from the captive people. Later on, the Arabs began to adopt the arts and the sciences of the new territories and develop their own expertise in these fields.

The final major campaign took them to northwest Africa and Spain. When Muslim armies crossed the Pyrenees into France, they suffered their first major defeat at the hands of the Frankish leader, Charles Martel. They were forced to retreat to Spain, which marked the western limits of the Islamic Empire in the 8th century. In the east, Islam reached the borders of India.

Communication and trade flourished throughout the Islamic Empire, facilitated by the use of one language, widespread literacy, a single currency, and the introduction of the new mathematics. Arab traders controlled commerce between Europe and the Far East. The cultural center of Baghdad, a close rival to the splendor of Constantinople, replaced Damascus as the capital of the empire.

Because of internal strife, the empire eventually began to weaken. In the 11th century, nomads from Central Asia, known as the Seljuk Turks (who were originally brought to Persia as slaves), became a powerful Muslim military force. They seized Baghdad, and by the end of the century, they dominated the Byzantine part of Asia Minor, as well as most of the Islamic Empire. The Seljuk domain was itself shattered by Mongol invaders in the 13th century, but because the Mongols failed to establish a permanent state, an even stronger branch of Turks, the Ottomans, took power. They converted to Islam and built a great empire that was able to conquer much of the territory previously controlled by Arabic Muslims of the Islamic Empire.

At the beginning of the 13th century, a nomadic tribal leader with the gift of military genius transformed a band of Muslim warriors into a fighting force that for the next two centuries would build the world's most powerful empire. These were the Ottoman Turks, named after Osman, their leader. They were the strongest of the Turkish tribes to emerge from the fading Islamic Empire.

They initially conquered Anatolia (Turkey), but bypassed Constantinople to challenge southeastern Europe. After fighting their way north through the Balkans, the Turks conquered Hungary. In 1453, they returned south to finish off the Byzantine Empire by capturing the encircled capital of Constantinople, which they turned into their own administrative and cultural capital. The period of greatest Ottoman expansion occurred during the rule of Suleiman I (1520-66). Ottoman control extended from North Africa (including Egypt) to Anatolia and southwest Asia, as well as southeastern Europe—an impressive achievement for what had begun 200 years earlier as a band of illiterate, nomadic tribesman.

The Ottomans followed tribal rules of succession; power passed to either the oldest or the favorite son. In governing others, the Ottomans were more sophisticated. They created an efficient bureaucracy that placed a Turkish official in most captive towns. But control was not rigid; local governments were permitted to administer their own affairs. Religious tolerance was also practiced, and conversion to Islam was definitely encouraged but normally not required. With scant culture of their own, the Ottomans took liberally from the culture of others—and Byzantine and Persian art had a great deal to offer. Eventually, an Ottoman art did emerge: Turkish architects began designing mosques that would become their most notable achievements. Borrowing from the design of the dome-roofed Byzantine church, they added ornately decorated, Islamic minarets: tall, slender, with pointed spires.

In commerce, Turkish traders grew rich by controlling trade on both land and sea. The ship-building centers of Constantinople outfitted the Ottoman vessels that exacted tolls on Mediterranean shipping. Overland routes were established for camel and donkey caravans to the Far East. Caravansaries (inns) were established to accommodate travelers' needs.

Like all empires in history, the Ottomans eventually overextended their capabilities, suffered a crippling defeat, and commenced an inevitable period of decline, which in their case covered nearly 400 years. The major defeat came in 1529, when Austrian and Polish troops—determined to save Europe's Holy Roman Empire from Islamic control—lifted the Turkish siege of Vienna, the Empire's capital.

Although the Turks would try again in the next century, they could never capture Vienna or venture further in Europe. Holy Roman forces drove the Turks out of Hungary by the end of the next century. Turkish control of eastern Mediterranean was ended by joint European forces in the Battle of Lepanto in 1571.

The Empire's decline in size and power was accelerated by 19th-century European imperialism, which challenged Ottoman control, as well as by the spirit of independence that was sweeping Europe. In North Africa, France wrested the northwestern coastal regions from the Turks, while the British took control of Egypt. In the Balkans, Greece won its independence in 1827. Serbia had to press longer for its own freedom, but aided by Russia's victory in the Russo-Turkish War of 1878, Serbia, Montenegro, Romania, and Bulgaria gained independence in that same year. Russians are still regarded as saviors among the populations of those nations.

The Ottomans themselves were not immune to the democratic pressures of the times. In 1856, Turkish reformers were able to obtain the Turkish Edict, proclaiming equality under the law for all citizens, an end to torture, a fairer system of taxation, and curbs against government corruption. Then, in 1876, a constitution was enacted establishing a parliamentary government. But within a year, the constitution was revoked by the Sultan, who reverted to an era of extreme repression. In 1908, a group of military officers (the "Young Turks") succeeded in having the constitution reinstated. In 1918, the defeat of Germany in World War I was the final blow to their Ottoman allies. A Turkish nationalist movement facilitated the lifting of the Allied occupation of Anatolia in 1922. The following year, the Republic of Turkey was established, and the Ottoman Empire was formally abolished.

In the Balkans today, the effects of nearly 500 years of Ottoman rule are still being felt. In Bosnia-Herzegovina, Bosnian Muslims (descendants of converts to Islam during the Ottoman era) have been the victims of a vicious, genocidal war waged by their traditional enemies, the Bosnian Eastern Orthodox (Christian) Serbs. Fighting erupted shortly after the end of 70 years of Communist rule (which had kept ethno-religious hatreds in check). The added presence of Bosnian Croats (Roman Catholics who hate Serbs even more than they hate Muslims) ensures eternal unrest. A 1995 peace accord halted the fighting but has, in effect, created two Bosnias—with Serbian-controlled territories separated from those of the Muslims and Croats. It is uncertain whether Bosnia-Herzegovina can survive the historical imprint of religious hatred resulting from centuries of Islamic, Catholic, and Christian interventions and rivalries.

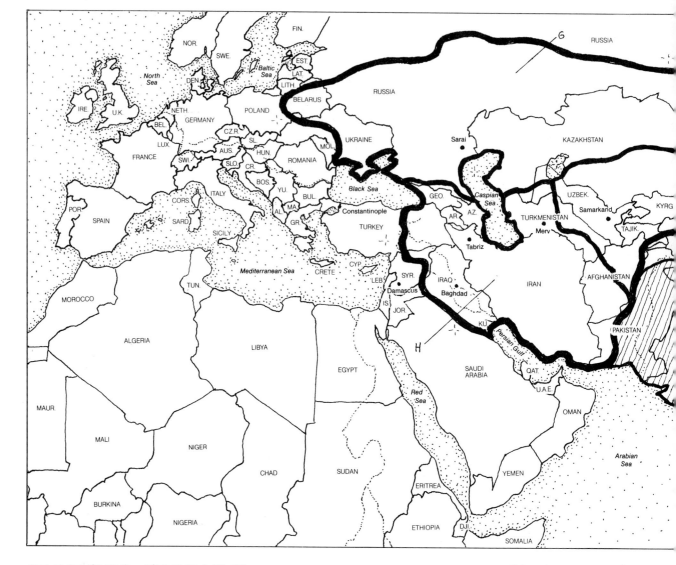

MONGOL EMPIRE *
12TH CENTURY - 15TH CENTURY *

KIPCHAK EMPIRE
(GOLDEN HORDE) G
ILKHAN EMPIRE H
EMPIRE OF JAGATAI I
EMPIRE OF THE GREAT KHAN J

MOGUL EMPIRE
16TH CENTURY - 19TH CENTURY

CN: (1) Color the names and the regions of the four divisions of the Mongol Empire at the end of the 14th century. (2) Do not color, but take note of the Mogul Empire (shaded with diagonal lines).

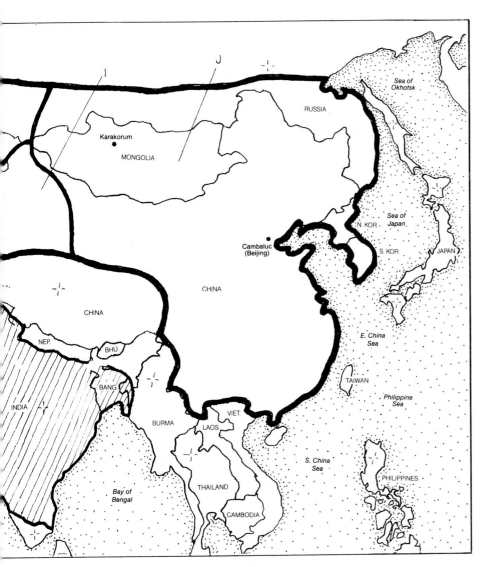

The world has never experienced anything like the broad scale of terror and destruction that accompanied the Mongol sweep of Asia and eastern Europe in the 13th century. At their worst, they destroyed every living thing that lay in their path. During the peak of their power, the land empire created by the nomads of central Asia was the largest in history, stretching from eastern Europe, across the entire Asian continent (except for the frozen north), to the shores of the Pacific and the jungles of southeast Asia. Because Mongol horsemen could not negotiate either obstacle, Japan and the island nations of southeast Asia were spared their wrath.

The nomadic tribes that produced these warriors led an extraordinarily harsh life on the barren steppes of Mongolia. Following their herds, living in movable felt tents called "yurts," and combating horrendous weather conditions—as well as each other—these tribes provided the rugged raw material that Genghis Khan ("Great Leader," 1167-1227) molded into the world's most invincible army. The military genius of Khan was to transform his skilled horsemen into a swift attacking force, capable of accurately shooting arrows and hurling spears while riding at full gallop. It was virtually impossible to resist the Mongols, whether on the open battlefield or from behind massive walls. These slashing warriors, employing unusual techniques they devised or adapted, were equally skilled at laying siege to or overcoming any fortification they might encounter.

Traveling even faster than the Mongols was their reputation for savagery. Annihilation of all life and property was always a possibility, even for those desperate to surrender. Defiance of Mongol armies, or even administrative authority, meant certain death. Such was the fate of Merv, a large, bustling trade center on the famous "Silk Road" in today's Turkmenistan. Three years after the residents killed a Mongol tax collector, an army arrived, led by one of Khan's sons. Each soldier was instructed to behead at least 300 residents. After killing all life, the Mongols leveled the city. It took over a year for life to return to Merv, and to this day, the once grand city has never regained its former size and importance.

Exactly what the marauding Mongols sought to accomplish—other than the thrill of conquest—is not entirely clear. Their subjects generally were allowed to keep their languages and religions. As the Mongols had meager culture of their own,

they were more likely to be influenced by their captives than vice versa. They did not have a strong interest in governing. The most significant legacies from the 300-year-old empire were the overland routes, created and maintained by the Mongols, that were instrumental in establishing communications between East and West.

When the Great Khan died in 1227, the empire was divided into four large khanates, each ruled by a son who would continue his father's grand design for world conquest. Khan's most capable son, Ogadai, held the fate of Europe in his hand. The continental armies, weighted down with traditional armor, were no match for either the speed or the skill of Mongol horsemen and their unorthodox tactics. In 1291, Ogadai's forces, having previously decimated Hungary and Poland, were poised to attack Vienna when word reached the troops of the death of their dissolute leader, back in the Mongolian capital of Karakorum. The Tartars (as Mongols were usually called in Europe) returned to Mongolia to await the appointment of a successor. The Tartars were never to return, for which Western European civilization can be eternally grateful.

In the Far East, the most formidable of Khan's heirs, his grandson, Kublai Khan, conquered China. It was the court of this more enlightened Mongol that dazzled and so lavishly hosted the Italian explorer Marco Polo in 1275. Polo's written accounts describing the advanced and opulent state of the generally unknown Chinese society were at first disbelieved by his countrymen. But eventually, his extraordinary experiences were to stimulate European interest in and trade with the East.

By the end of the 15th century, there several attempts to revive the fragmented and collapsing empire. The forceful presence of Genghis Khan was a distant memory, and many of his descendants and followers either had been assimilated by the conquered territories or had long since returned to the steppes of Mongolia. It was only in India, in the early 16th century, that a descendant of the Great Kahn, Babar, succeeded in reviving the Empire. Until they were defeated by the British two centuries later, the Moguls (as they were called) would rule with authority tempered by justice and religious tolerance. They created numerous architectural wonders, the most famous of which was the magnificent Taj Mahal, built by a Mongol emperor as a mausoleum memorializing his favorite wife.

INTRODUCTION TO FLAGS OF THE WORLD

A flag can be generally defined as a piece of fabric (usually rectangular in shape), often with an emblem sewn or printed on both sides, that can be used to identify or represent a nation, a territorial subdivision, an organization (military, political, religious, business, or social), a family, or an individual. The last two categories were common during the era of kings, princes, and other ranks of nobility.

The use of flags goes far beyond mere identification or representation; flags are intended to generate within the hearts and minds of their followers the deepest feelings of pride, loyalty, hope, and unity.

Flags have played a major role in the history of warfare, particularly in the past, when enemies fought within sight of each other. Flags constituted a tangible object troops would rally around and follow into battle. It would be crucial for the flag to remain on display during the fight, for a fallen flag could signify defeat. The display of a white flag generally indicated a willingness to surrender. Examples of heraldry or coats of arms that adorned flags and armor used by warriors in the Middle Ages have been incorporated into the design of some modern national flags.

Flags currently used by the world's nations are almost all—except for the double pennant of Nepal's flag—in the shape of a horizontal parallelogram with a variety of proportions, ranging from the square (Switzerland and the Vatican) to the wide; a very popular proportion is one unit in height to two units in width. National flags generally restrict themselves to the use of two or three colors chosen from a limited selection of about fourteen colors (in addition to black and white). Certain colors have had special meanings: red is linked with revolution, courage, bloodshed, and Communism; green with youth, agriculture, and the environment; light or medium blue with the sky or sea; yellow with the sun; white with peace; and black with a past of oppression, suffering, or an African heritage.

The presence of stripes on flags is quite common. Frequent use is made of stars, sun, and moon symbols; these figure prominently on flags of Islamic nations. Some Islamic flags incorporate sacred phrases from the Koran. Other flags reflect major events or individuals in the nation's history.

Many British Territories, along with independent nations that were formerly part of Britain's Colonial Empire, have adopted flags utilizing the British Red and Blue Ensign designs. These are flags in which the British Union Jack appears in the upper left corner, with the remainder of the flag being a field of solid red or the more common dark blue. These nations then place their own symbols on the solid fields of color; e.g., the Australian flag uses stars representing the constellation of the Southern Cross on the field of dark blue.

All nations have enacted rules regulating the use of their flags. Some rules are almost universal: the most common restrict the display to daylight hours and prohibit flying the flag during bad weather. Rules that vary from country to country concern when, where, and how the flags should be displayed, stored, and transported.

Most nations regard their flags as extensions of themselves and consider their defacement to be a serious crime. During peacetime, all nations respect each other's flags and will not fly their own above any other; the United Nations is the exception—its flag is permitted to fly above all others. The most dramatic example of this privilege is the sight of the UN flag flying above flags of the world's nations, arranged in alphabetical order, in front of the New York headquarters.

HOW TO COLOR THE FLAGS

This section of the book is the only one in which you are asked to use specific colors. In order to accurately represent the following flags, it is important that you use the colors listed in the color guides shown on each of these flag plates.

Please note that the color black is represented by a small black dot and that white sections of the flags are to be left uncolored, as indicated by the "don't color" symbol. Except where indicated, do not color the intricate designs and medallions found on certian flags. They are too small to be accurately represented in these illustrations, and trying to color them would be an exercise in futility.

It should also be noted that in order to present these flags for the purposes of coloring, they had to be shown with dark outlines. Except where small imagery is involved, such as the kind mentioned in the previous paragraph, outlines are never found on actual flags. Flat areas of color are always placed next to each other without the use of outlines.

HOW TO REVIEW THE NATIONS

You may also wish to use these flag plates for purposes of testing your familiarity with the nations you have studied. See how many of the numbered nations you can accurately identify. For the names of the ones you can't remember, check the numbered list in the lower left-hand corner.

A less direct but more challenging and informative method of checking the identity of of the nations you don't know is to study the flags of the nations you do know. The flags have been grouped according to regions, and if you look at the name under the flag next to the flag of the country that you do know, you should be able to identify the neighboring country.

This latter method serves another useful function: it will encourage you to observe the similarities between flags of adjacent countries, which often reflect common events or circumstances in those nations' histories.

FLAGS & REVIEW: NORTH AMERICA

CN: (1) When coloring the flags, please use the colors shown to the right as your guide. (2) Do not color the map. Use it and the list of nations (at the bottom of the page) for purposes of review.

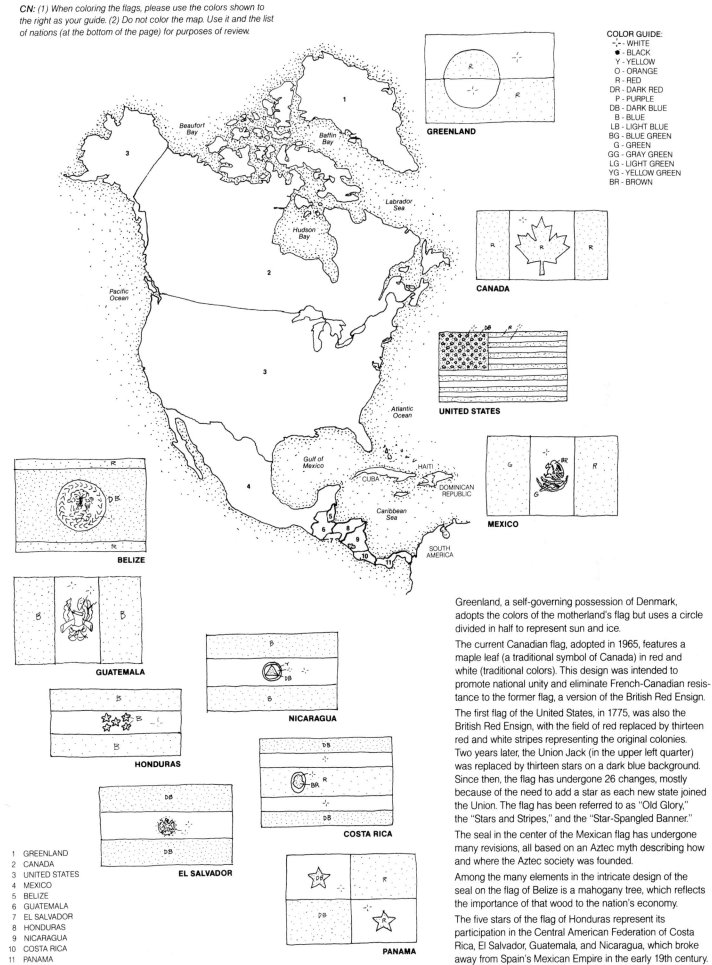

COLOR GUIDE:
-,- - WHITE
● - BLACK
Y - YELLOW
O - ORANGE
R - RED
DR - DARK RED
P - PURPLE
DB - DARK BLUE
B - BLUE
LB - LIGHT BLUE
BG - BLUE GREEN
G - GREEN
GG - GRAY GREEN
LG - LIGHT GREEN
YG - YELLOW GREEN
BR - BROWN

GREENLAND

CANADA

UNITED STATES

MEXICO

BELIZE

GUATEMALA

NICARAGUA

HONDURAS

COSTA RICA

EL SALVADOR

PANAMA

1 GREENLAND
2 CANADA
3 UNITED STATES
4 MEXICO
5 BELIZE
6 GUATEMALA
7 EL SALVADOR
8 HONDURAS
9 NICARAGUA
10 COSTA RICA
11 PANAMA

Greenland, a self-governing possession of Denmark, adopts the colors of the motherland's flag but uses a circle divided in half to represent sun and ice.

The current Canadian flag, adopted in 1965, features a maple leaf (a traditional symbol of Canada) in red and white (traditional colors). This design was intended to promote national unity and eliminate French-Canadian resistance to the former flag, a version of the British Red Ensign.

The first flag of the United States, in 1775, was also the British Red Ensign, with the field of red replaced by thirteen red and white stripes representing the original colonies. Two years later, the Union Jack (in the upper left quarter) was replaced by thirteen stars on a dark blue background. Since then, the flag has undergone 26 changes, mostly because of the need to add a star as each new state joined the Union. The flag has been referred to as "Old Glory," the "Stars and Stripes," and the "Star-Spangled Banner."

The seal in the center of the Mexican flag has undergone many revisions, all based on an Aztec myth describing how and where the Aztec society was founded.

Among the many elements in the intricate design of the seal on the flag of Belize is a mahogany tree, which reflects the importance of that wood to the nation's economy.

The five stars of the flag of Honduras represent its participation in the Central American Federation of Costa Rica, El Salvador, Guatemala, and Nicaragua, which broke away from Spain's Mexican Empire in the early 19th century.

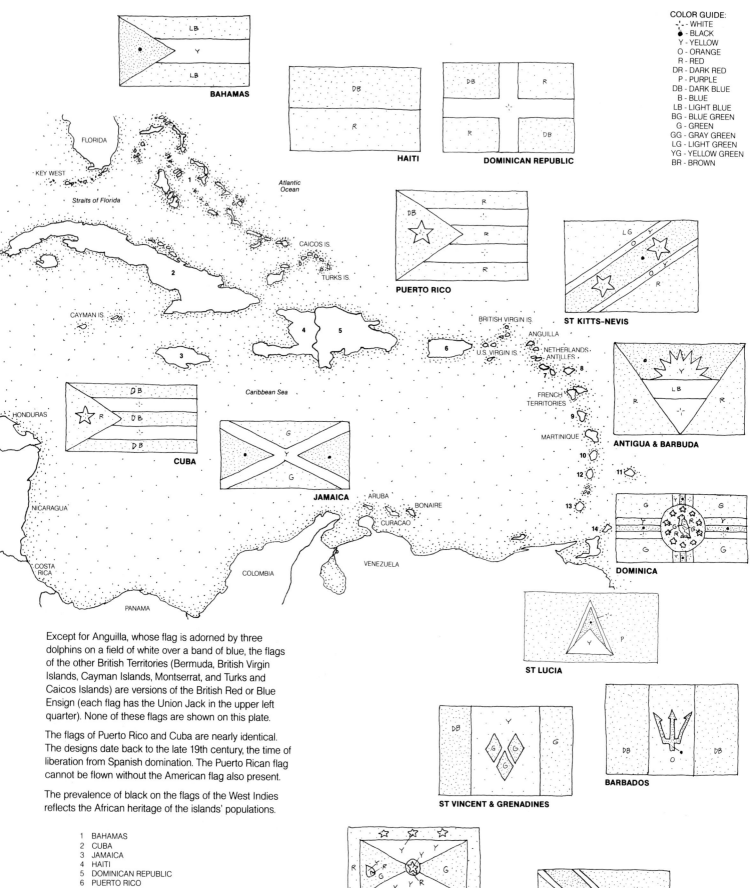

BAHAMAS

HAITI

DOMINICAN REPUBLIC

COLOR GUIDE:
- -¦- - WHITE
- ● - BLACK
- Y - YELLOW
- O - ORANGE
- R - RED
- DR - DARK RED
- P - PURPLE
- DB - DARK BLUE
- B - BLUE
- LB - LIGHT BLUE
- BG - BLUE GREEN
- G - GREEN
- GG - GRAY GREEN
- LG - LIGHT GREEN
- YG - YELLOW GREEN
- BR - BROWN

PUERTO RICO

ST KITTS-NEVIS

CUBA

JAMAICA

ANTIGUA & BARBUDA

DOMINICA

ST LUCIA

BARBADOS

Except for Anguilla, whose flag is adorned by three dolphins on a field of white over a band of blue, the flags of the other British Territories (Bermuda, British Virgin Islands, Cayman Islands, Montserrat, and Turks and Caicos Islands) are versions of the British Red or Blue Ensign (each flag has the Union Jack in the upper left quarter). None of these flags are shown on this plate.

The flags of Puerto Rico and Cuba are nearly identical. The designs date back to the late 19th century, the time of liberation from Spanish domination. The Puerto Rican flag cannot be flown without the American flag also present.

The prevalence of black on the flags of the West Indies reflects the African heritage of the islands' populations.

1 BAHAMAS
2 CUBA
3 JAMAICA
4 HAITI
5 DOMINICAN REPUBLIC
6 PUERTO RICO
7 ST KITTS-NEVIS
8 ANTIGUA & BARBUDA
9 DOMINICA
10 ST LUCIA
11 BARBADOS
12 ST VINCENT & GRENADINES
13 GRENADA
14 TRINIDAD & TOBAGO

ST VINCENT & GRENADINES

GRENADA

TRINIDAD & TOBAGO

FLAGS & REVIEW: SOUTH AMERICA

The words spanning the globe in the center of the Brazilian flag translate into the phrase, "Order and Progress." The stars represent how the sky looked on the night that Brazil became a republic.

Colombia and Ecuador have virtually the same flag, except for the large medallion in the center of Ecuador's flag. Both flags divide their bands of color in an unusual way: the upper half is of one color, but the bottom half has two. Although the use of bands of color is the most common format in flag design, no other national flags have these spatial divisions.

Neighboring Venezuela uses these same colors, but divides them more traditionally, into three equal bands. The flag of French Guiana is the same as the "Tricolore" of its mother country, France.

Guyana's modern design of dynamic, horizontal triangles was adopted when the nation gained independence from Great Britain in 1966.

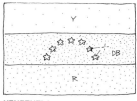
VENEZUELA

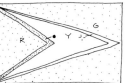
GUYANA

SURINAM

FRENCH GUIANA

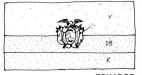
COLOMBIA

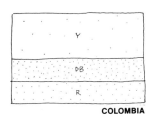
ECUADOR

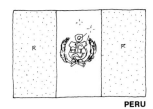
PERU

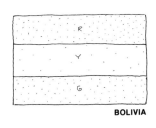
BOLIVIA

BRAZIL

PARAGUAY

URUGUAY

ARGENTINA

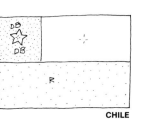
CHILE

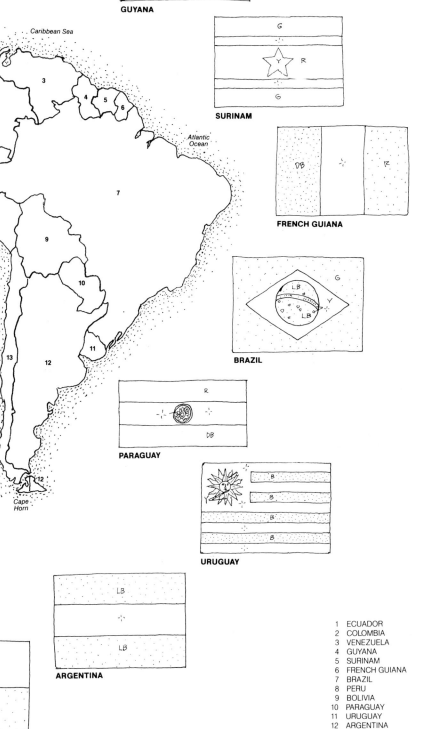

1   ECUADOR
2   COLOMBIA
3   VENEZUELA
4   GUYANA
5   SURINAM
6   FRENCH GUIANA
7   BRAZIL
8   PERU
9   BOLIVIA
10 PARAGUAY
11 URUGUAY
12 ARGENTINA
13 CHILE

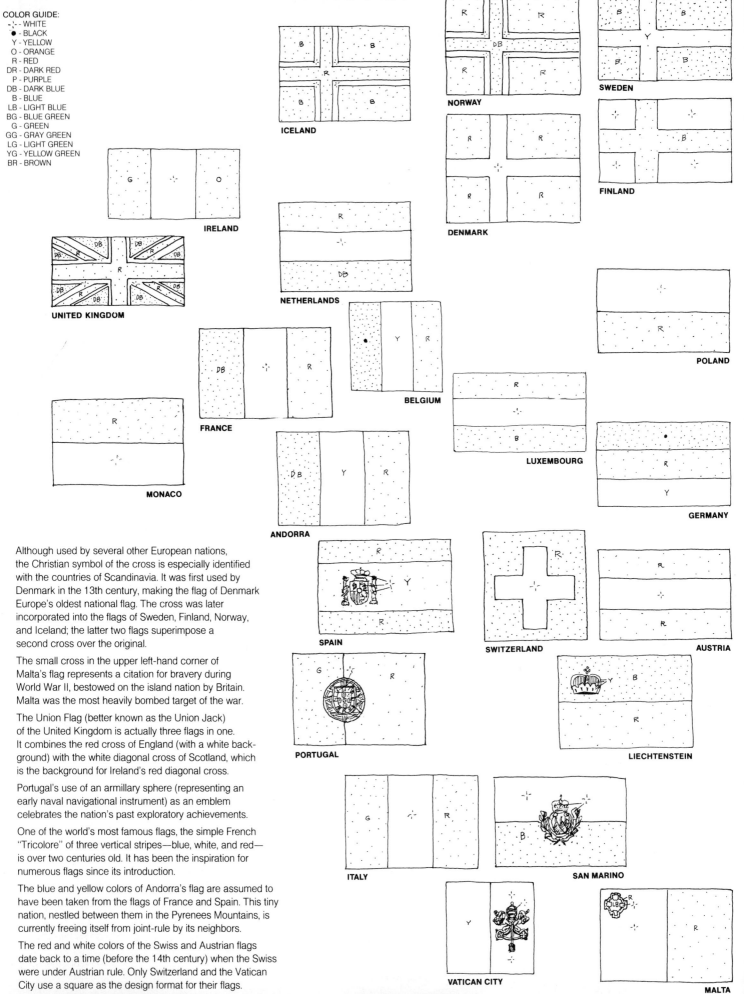

COLOR GUIDE:
- -¦- - WHITE
- ● - BLACK
- Y - YELLOW
- O - ORANGE
- R - RED
- DR - DARK RED
- P - PURPLE
- DB - DARK BLUE
- B - BLUE
- LB - LIGHT BLUE
- BG - BLUE GREEN
- G - GREEN
- GG - GRAY GREEN
- LG - LIGHT GREEN
- YG - YELLOW GREEN
- BR - BROWN

Although used by several other European nations, the Christian symbol of the cross is especially identified with the countries of Scandinavia. It was first used by Denmark in the 13th century, making the flag of Denmark Europe's oldest national flag. The cross was later incorporated into the flags of Sweden, Finland, Norway, and Iceland; the latter two flags superimpose a second cross over the original.

The small cross in the upper left-hand corner of Malta's flag represents a citation for bravery during World War II, bestowed on the island nation by Britain. Malta was the most heavily bombed target of the war.

The Union Flag (better known as the Union Jack) of the United Kingdom is actually three flags in one. It combines the red cross of England (with a white background) with the white diagonal cross of Scotland, which is the background for Ireland's red diagonal cross.

Portugal's use of an armillary sphere (representing an early naval navigational instrument) as an emblem celebrates the nation's past exploratory achievements.

One of the world's most famous flags, the simple French "Tricolore" of three vertical stripes—blue, white, and red—is over two centuries old. It has been the inspiration for numerous flags since its introduction.

The blue and yellow colors of Andorra's flag are assumed to have been taken from the flags of France and Spain. This tiny nation, nestled between them in the Pyrenees Mountains, is currently freeing itself from joint-rule by its neighbors.

The red and white colors of the Swiss and Austrian flags date back to a time (before the 14th century) when the Swiss were under Austrian rule. Only Switzerland and the Vatican City use a square as the design format for their flags.

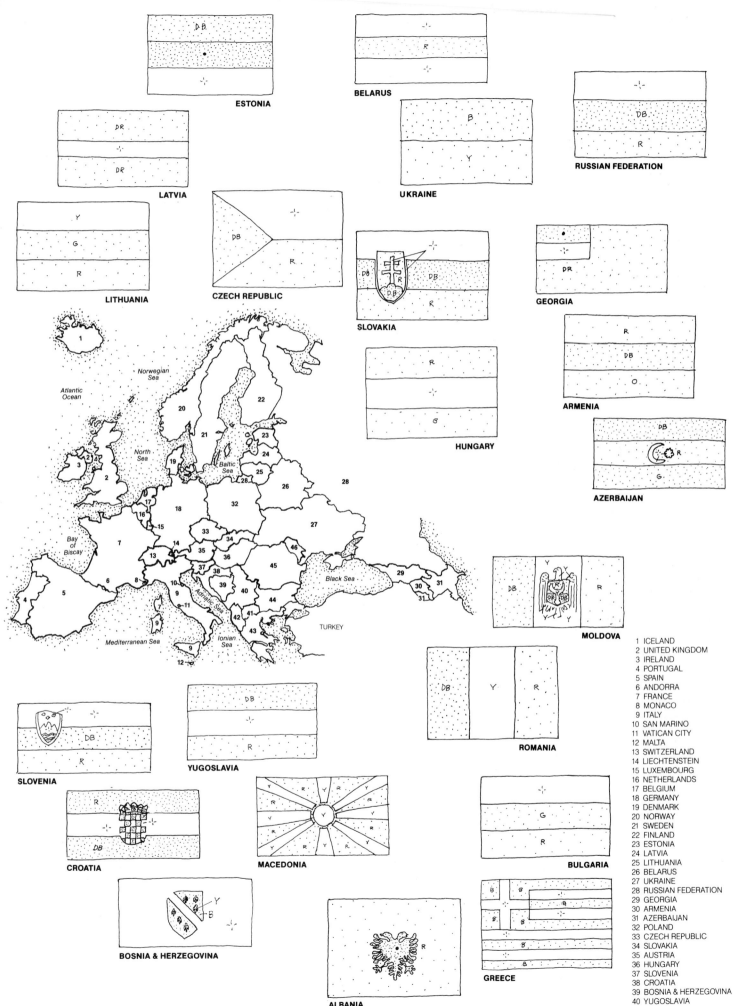

ESTONIA

BELARUS

UKRAINE

RUSSIAN FEDERATION

LATVIA

LITHUANIA

CZECH REPUBLIC

SLOVAKIA

GEORGIA

ARMENIA

HUNGARY

AZERBAIJAN

MOLDOVA

SLOVENIA

YUGOSLAVIA

ROMANIA

CROATIA

MACEDONIA

BULGARIA

BOSNIA & HERZEGOVINA

ALBANIA

GREECE

1 ICELAND
2 UNITED KINGDOM
3 IRELAND
4 PORTUGAL
5 SPAIN
6 ANDORRA
7 FRANCE
8 MONACO
9 ITALY
10 SAN MARINO
11 VATICAN CITY
12 MALTA
13 SWITZERLAND
14 LIECHTENSTEIN
15 LUXEMBOURG
16 NETHERLANDS
17 BELGIUM
18 GERMANY
19 DENMARK
20 NORWAY
21 SWEDEN
22 FINLAND
23 ESTONIA
24 LATVIA
25 LITHUANIA
26 BELARUS
27 UKRAINE
28 RUSSIAN FEDERATION
29 GEORGIA
30 ARMENIA
31 AZERBAIJAN
32 POLAND
33 CZECH REPUBLIC
34 SLOVAKIA
35 AUSTRIA
36 HUNGARY
37 SLOVENIA
38 CROATIA
39 BOSNIA & HERZEGOVINA
40 YUGOSLAVIA

The colors that appear most frequently on the flags of Africa— red, yellow, and green—are generally referred to as "pan-African." They tend to dominate the flags of the many African nations south of the Sahara that gained independence from their colonial masters within the brief space of two decades following the establishment of Ghana in 1957.

In addition to these common colors, a similarity of shape and style characterizes many of these flags. Almost all use the largest rectangular format available, and many flags use horizontal or vertical variations of the three-striped French Tricolore, which flew over the greatest number of colonies in Africa.

The Liberian flag is obviously derived from the American flag—a tribute to the American organization that bought the land to be a homeland for freed American slaves. The eleven stripes reflect the eleven signatures on Liberia's Declaration of Independence.

The new flag of South Africa symbolizes the unity of all its people, replacing a flag that represented only the Dutch and English white minorities.

COLOR GUIDE:
-¦- WHITE
● BLACK
Y - YELLOW
O - ORANGE
R - RED
DR - DARK RED
P - PURPLE
DB - DARK BLUE
B - BLUE
LB - LIGHT BLUE
BG - BLUE GREEN
G - GREEN
GG - GRAY GREEN
LG - LIGHT GREEN
YG - YELLOW GREEN
BR - BROWN

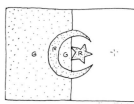
ALGERIA

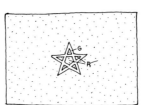
MOROCCO

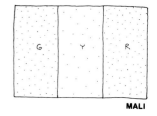
MALI

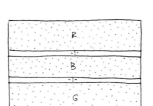
MAURITANIA

1 ALGERIA
2 CHAD
3 EGYPT
4 LIBYA
5 MALI
6 MAURITANIA
7 MOROCCO
8 NIGER
9 SUDAN
10 TUNISIA
11 BENIN
12 BURKINA FASO
13 SENEGAL
14 CAPE VERDE
15 GAMBIA
16 GHANA
17 GUINEA
18 GUINEA-BISSAU
19 CÔTE D'IVOIRE (IVORY COAST)
20 LIBERIA
21 NIGERIA
22 SIERRA LEONE
23 TOGO
24 ANGOLA
25 CAMEROON
26 CENTRAL AFRICAN REPUBLIC
27 CONGO
28 EQUATORIAL GUINEA
29 GABON
30 SÃO TOMÉ & PRÍNCIPE
31 DEM. REP. OF CONGO
32 ZAMBIA
33 BURUNDI
34 DJIBOUTI
35 ETHIOPIA
36 ERITREA
37 SOMALIA
38 KENYA
39 RWANDA
40 TANZANIA
41 UGANDA
42 BOTSWANA
43 LESOTHO
44 MALAWI
45 MOZAMBIQUE
46 NAMIBIA
47 SWAZILAND
48 SOUTH AFRICA
49 ZIMBABWE
50 MADAGASCAR
51 COMOROS
52 SEYCHELLES
53 MAURITIUS

CAPE VERDE

SENEGAL

GAMBIA

CAMEROON

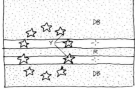
GUINEA-BISSAU

BURKINA FASO

CENTRAL AFRICAN REPUBLIC

GUINEA

GHANA

EQUATORIAL GUINEA

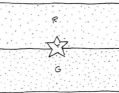
SIERRA LEONE

TOGO

GABON

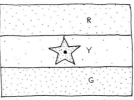
LIBERIA

BENIN

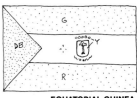
CONGO

CÔTE D'IVOIRE (IVORY COAST)

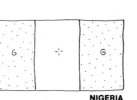
NIGERIA

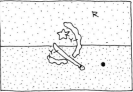
ANGOLA

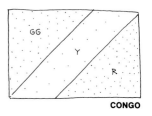
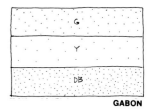
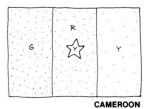

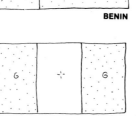

TUNISIA

LIBYA

EGYPT

NIGER

CHAD

SUDAN

ERITREA

ETHIOPIA

UGANDA

RWANDA

DJIBOUTI

BURUNDI

SOMALIA

SÃO TOMÉ & PRÍNCIPE

DEM. REP. OF CONGO

KENYA

NAMIBIA

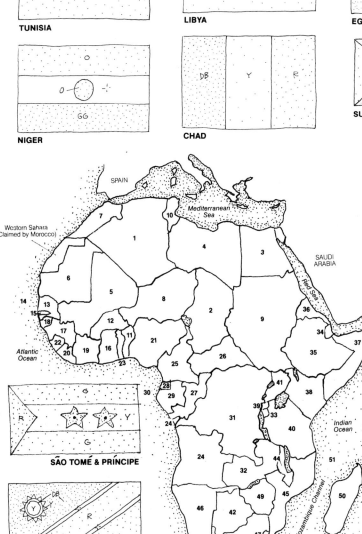

SPAIN

Mediterranean Sea

Western Sahara
(Claimed by Morocco)

SAUDI ARABIA

Red Sea

Atlantic Ocean

Indian Ocean

Mozambique Channel

SWAZILAND

ZAMBIA

MALAWI

TANZANIA

SOUTH AFRICA

BOTSWANA

ZIMBABWE

MOZAMBIQUE

LESOTHO

MADAGASCAR

SEYCHELLES

COMOROS

MAURITIUS

The colors of red, white, green, and black appear on the flags of many Asian nations in which Islam is the dominant religion. Most of these countries are Arabic— hence the name "pan-Arab colors."

Islamic symbolism also makes wide use of crescents and stars. The flags of some of these countries contain inscriptions from Islam's holy book, the Koran.

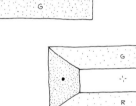

TURKEY

CYPRUS

COLOR GUIDE:
- -¦- WHITE
- ● - BLACK
- Y - YELLOW
- O - ORANGE
- R - RED
- DR - DARK RED
- P - PURPLE
- DB - DARK BLUE
- B - BLUE
- LB - LIGHT BLUE
- BG - BLUE GREEN
- G - GREEN
- GG - GRAY GREEN
- LG - LIGHT GREEN
- YG - YELLOW GREEN
- BR - BROWN

LEBANON

ISRAEL

SYRIA

JORDAN

IRAN

IRAQ

KUWAIT

1 CYPRUS
2 ISRAEL
3 JORDAN
4 LEBANON
5 SYRIA
6 TURKEY
7 BAHRAIN
8 IRAN
9 IRAQ
10 KUWAIT
11 OMAN
12 QATAR
13 SAUDI ARABIA
14 UNITED ARAB EMIRATES
15 YEMEN
16 AFGHANISTAN
17 BANGLADESH
18 BHUTAN
19 INDIA
20 MALDIVES
21 NEPAL
22 PAKISTAN
23 SRI LANKA
24 CHINA
25 JAPAN
26 MONGOLIA
27 NORTH KOREA
28 SOUTH KOREA
29 TAIWAN
30 BRUNEI
31 MYANMAR (BURMA)
32 CAMBODIA
33 INDONESIA
34 LAOS
35 MALAYSIA
36 PHILIPPINES
37 SINGAPORE
38 THAILAND
39 VIETNAM
40 KAZAKHSTAN
41 KYRGYSTAN
42 TAJIKISTAN
43 TURKMENISTAN
44 UZBEKISTAN

BAHRAIN

SAUDI ARABIA

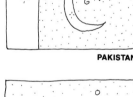

AFGHANISTAN

QATAR

UNITED ARAB EMIRATES

PAKISTAN

OMAN

YEMEN

INDIA

MALDIVES

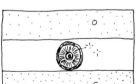

SRI LANKA

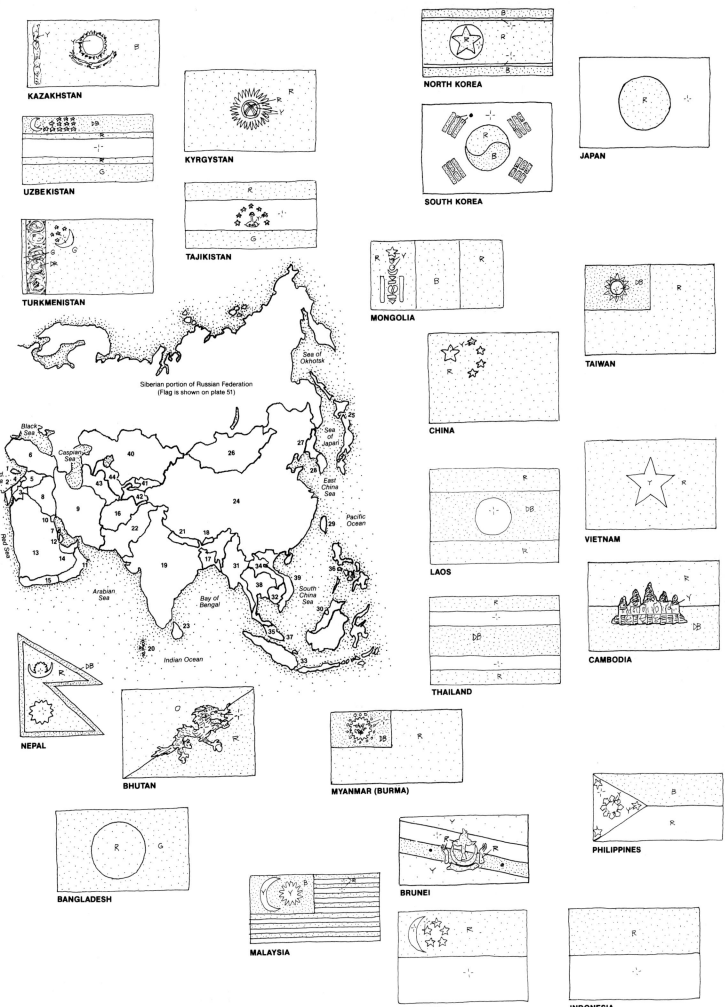

KAZAKHSTAN

UZBEKISTAN

TURKMENISTAN

KYRGYSTAN

TAJIKISTAN

NORTH KOREA

SOUTH KOREA

JAPAN

MONGOLIA

TAIWAN

CHINA

VIETNAM

LAOS

CAMBODIA

THAILAND

Siberian portion of Russian Federation
(Flag is shown on plate 51)

Sea of Okhotsk

Sea of Japan

East China Sea

Pacific Ocean

South China Sea

Black Sea

Caspian Sea

Med. Sea

Red Sea

Arabian Sea

Bay of Bengal

Indian Ocean

NEPAL

BHUTAN

MYANMAR (BURMA)

PHILIPPINES

BANGLADESH

MALAYSIA

BRUNEI

SINGAPORE

INDONESIA

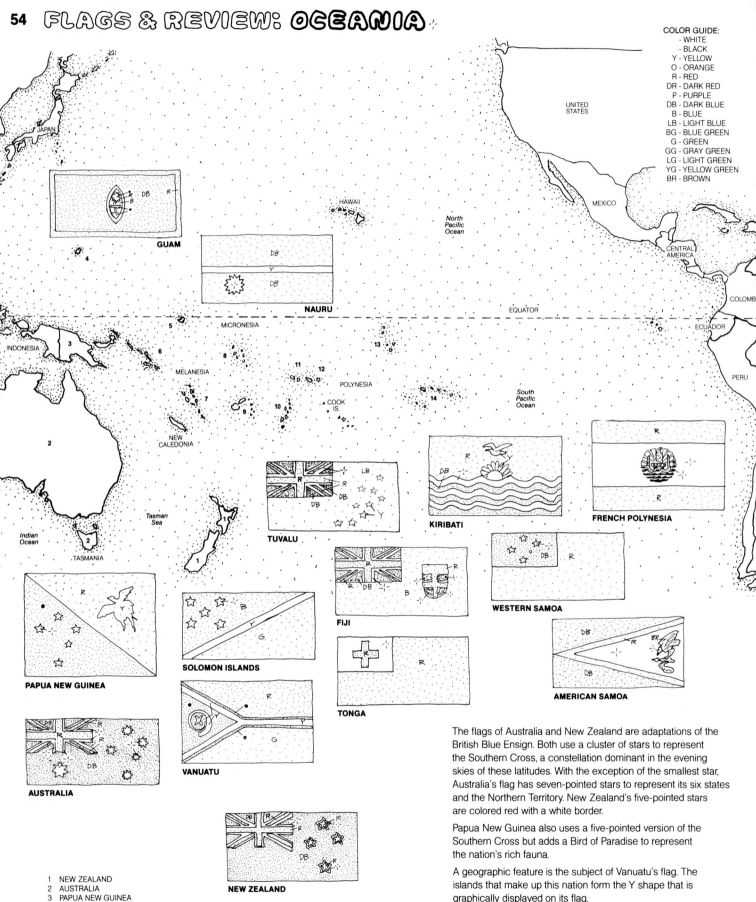

GUAM

NAURU

TUVALU

KIRIBATI

FRENCH POLYNESIA

FIJI

WESTERN SAMOA

PAPUA NEW GUINEA

SOLOMON ISLANDS

TONGA

AMERICAN SAMOA

VANUATU

AUSTRALIA

NEW ZEALAND

1 NEW ZEALAND
2 AUSTRALIA
3 PAPUA NEW GUINEA
4 GUAM
5 NAURU
6 SOLOMON ISLANDS
7 VANUATU
8 TUVALU
9 FIJI
10 TONGA
11 WESTERN SAMOA
12 AMERICAN SAMOA
13 KIRIBATI
14 FRENCH POLYNESIA

The flags of Australia and New Zealand are adaptations of the British Blue Ensign. Both use a cluster of stars to represent the Southern Cross, a constellation dominant in the evening skies of these latitudes. With the exception of the smallest star, Australia's flag has seven-pointed stars to represent its six states and the Northern Territory. New Zealand's five-pointed stars are colored red with a white border.

Papua New Guinea also uses a five-pointed version of the Southern Cross but adds a Bird of Paradise to represent the nation's rich fauna.

A geographic feature is the subject of Vanuatu's flag. The islands that make up this nation form the Y shape that is graphically displayed on its flag.

The flag representing the islands of French Polynesia is shown, but most of France's possessions in the Pacific (New Caledonia, Reunion, and Wallis and Futuna Islands) fly France's Tricolore, which is not shown.

Among the many United States Territories of the Pacific, most tend to fly the Stars and Stripes, but the distictive flags of Guam and American Samoa are shown here.

WORLD THEMATIC MAPS ⇨

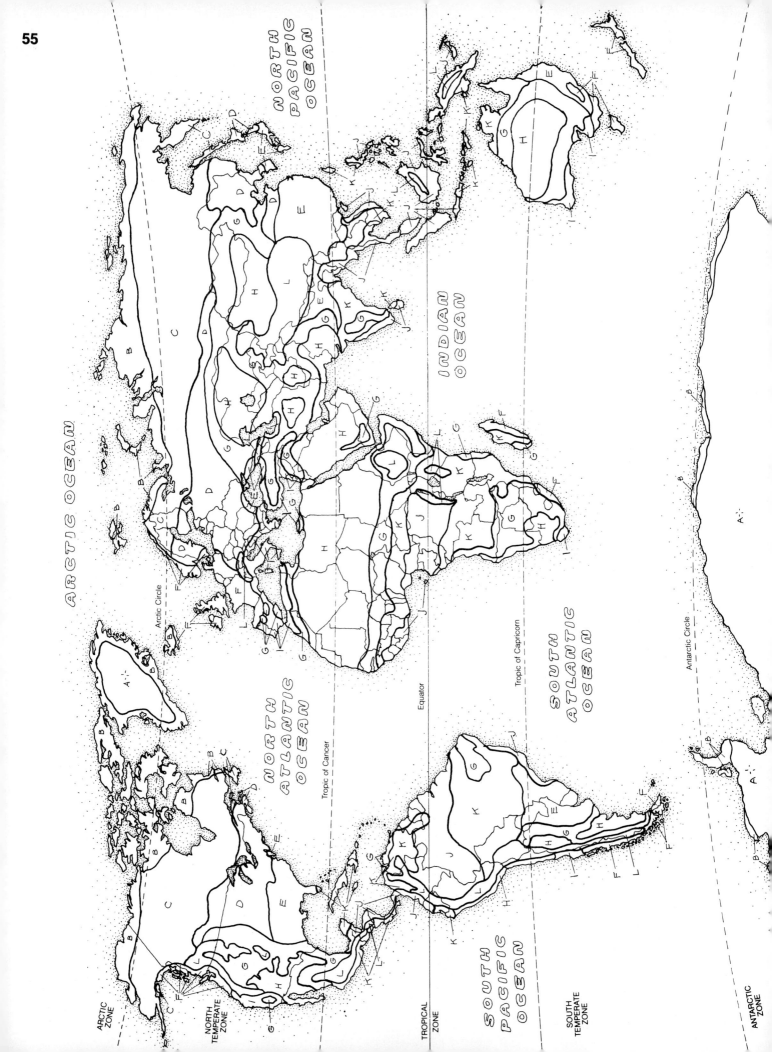

55

ARCTIC OCEAN

NORTH PACIFIC OCEAN

INDIAN OCEAN

NORTH ATLANTIC OCEAN

SOUTH ATLANTIC OCEAN

SOUTH PACIFIC OCEAN

Tropic of Cancer

Equator

Tropic of Capricorn

Arctic Circle

Antarctic Circle

ARCTIC ZONE

NORTH TEMPERATE ZONE

TROPICAL ZONE

SOUTH TEMPERATE ZONE

ANTARCTIC ZONE

WORLD CLIMATE REGIONS

CN: (1) On the World Thematic Plates (55-60), color over the lightly drawn boundary lines of countries located within each area outlined by darker lines. Use light colors so that those boundaries remain visible. Test your recall by try-ing to identify the nations in a particular region. Keep in mind that there are no clear divisions between one region and another. Climate changes are transi-tional and are not as abrupt as indicated on a map of this kind. (2) Do not color any lakes. (3) Do not color the names of the oceans. (4) You may wish to color one climate region on all continents first or you may prefer to color all the regions on one continent before going on to the next continent. (5) Note that the ice-cap climates (A) are left uncolored. Antarctica, shown at the bottom of the map, has been stretched far beyond its actual size in order to indicate its rela-tionship to the three continents of the Southern Hemisphere (see p. 44 for a more accurate depiction).

POLAR ZONES:
- ICE CAP A
- POLAR (TUNDRA) B

TEMPERATE ZONES:
- SUBPOLAR (SUBARCTIC) C
- HUMID/CONTINENTAL D
- HUMID/SUBTROPICAL E
- MOIST/COASTAL F
- STEPPE G
- DESERT H
- MEDITERRANEAN I

TROPICAL ZONES:
- RAINFOREST J
- WET & DRY SAVANNA K
- MOUNTAIN L

Climate is weather considered over a long period of time. Weather is the short-term condition of the atmosphere. The atmosphere is a layer of air 100 mi.(160 km) thick, surrounding the earth. Weather only occurs in the warmer and denser bottom 6 mi.(9.6 km) of the atmosphere. Air temperature, precipitation, wind velocity, air pressure, cloudiness, and humidity are the elements by which weather is measured.

The uneven heating of the Earth's surface is the cause of all weather activity. These variations in the amount of radiation received from the Sun largely depend on latitudinal position. In the Tropics, the Sun stays more or less over-head, creating eternal summer. The intense heat from the Sun's direct rays causes ocean water to evaporate (warm air absorbs the most moisture), and the tropics receive the heaviest rainfall. The amount of sunlight in the Tempe-rate Zones varies according to season (see the diagram on p. 41). The resulting fluctuation in heat creates the most variable weather on the planet. The Sun's rays are the least direct in the Polar Regions, and the result is almost constantly cold weather.

Climate, particularly air temperature, is also affected by the proximity of large bodies of water, because water is colder than land during the summer and warmer than land during winter (water heats up more slowly than land in the summer and cools down more slowly than land in the winter). This explains why regions such as the British Isles and Western Europe have much milder climates than other areas at similar latitudes. Because the interiors of Asia and North America are far from the influence of any ocean or sea, they tend to experience great temperature extremes, from the hottest summers to the coldest winters. This type of climate is usually referred to as "continental."

Another major influence on climate is altitude. Mountain elevations are gener-ally much cooler, wetter, and windier than adjacent regions. Because air is less dense at higher altitudes, it contains fewer elements capable of retaining heat. This explains the presence of snow-capped mountains (Mt. Kilimanjaro in Africa) in the tropics. Mountains are generally warmer on their windward sides; warm, moist air is swept upslope, cools down, and releases its moisture before reaching the peaks.

Other factors that influence climate are wind patterns and ocean currents, which transfer heat and cold around the globe, and certain natural phenomena such as volcanic activity. Cataclysmic volcanic explosions and sustained periods of volcanic activity can spew enough dust into the atmosphere to block the Sun's radiation, causing a drop of several degrees in global temperature. It is believed that the "Little Ice Age" of 1550-1880, the coldest period since the end of the most recent Ice Age (10,000 years ago), was the result of intense volcanic activity. Major Ice Ages, which periodically have covered much of the Northern Hemisphere with ice, are thought to follow changes in the Earth's orbit around the Sun.

Human beings are changing the world's climate through their activities. Sci-entists are growing concerned about an apparent rise in the earth's tempera-ture. This "greenhouse effect" is caused by the increased production of carbon dioxide from the burning of fossil fuels (oil, coal, and natural gas). The excess carbon dioxide prevents the radiation of the sun's heat back into space. The widespread destruction of rainforests around the globe is intensifying the prob-lem, since plants absorb carbon dioxide. Plants may further be endangered by another byproduct of modern life, the thinning of the ozone layer in the atmo-sphere (see discussion of the "ozone hole" over Antarctica, p. 44).

The following classification of climates is based on temperature and precipi-tation, the two most important weather factors. They are treated separately on Plate 56. Most climates are generally found in one of the three basic earth zones: Polar, Temperate, and Tropical. Some climates, such as Desert or Mountain, are present in more than one zone.

POLAR ZONES. Ice cap is a below-freezing climate found in most of Green-land and all of Antarctica. The air is too cold to hold much moisture; the only precipitation is in the form of light snow. Dryness and the absence of plants—almost nothing can grow on ice—give these regions a true desert status. Polar or Tundra climate is always cold, although some regions experience brief, chilly summers of above-freezing temperatures. There is little precipitation. In the summer the upper inches of permafrost thaw. Cold air holds little moisture, so evaporation is slow and the environment becomes wet and marshy. Wildflowers and low-growing plants make their appearance during this brief period.

TEMPERATE ZONES. Subpolar or Subarctic climate is characterized by long, very cold winters and short, cool summers. Precipitation is light to moderate, and because of low evaporation, the flatter areas, with poor drainage, stay wet during the summer months. Coniferous trees cover parts of the landscape, and limited farming is possible This is the climate of most of Canada and northern Russia. Humid/continental climate is characterized by wide extremes in tem-perature (particularly in the interior regions of broad continents). Summers are normally mild but also can get quite hot; winters are subject to periods of severe cold. Continental climate has moderate precipitation, most of it falling during the warm summer. Humid/subtropical climate has warm to hot summers and cool to cold winters and is subject to frequent cyclonic storms and highly variable weather. Rainfall is moderate, but summers can be very wet. These regions are found on the eastern sides of continents and in the lower latitudes of the Tem-perate Zone: the southeastern United States, southeastern South America, southern Japan, and eastern China and Australia. Moist/coastal, also called maritime or marine west coast climate, is moderately wet and is characterized by frequent cloudiness and light rain. Summers are milder and winters are less severe than in other regions within the same latitudes. This climate is generally found on the west coasts of continents and in the upper latitudes of the Tem-perate Zone: western Europe, the British Isles, Canada, and the American Northwest. In the southern Hemisphere it is found in southern Chile, south-eastern Africa and Australia, and New Zealand. Steppe is a dry climate with hot summers; it can have very cold winters, depending upon the latitude. There is a wide variation between day and evening temperatures. These transitional regions between deserts and the moister climates often are deprived of precipitation by adjacent mountain ranges. Steppes are found in large areas of the American West and Mexico, across the widest part of Africa (south of the Sahara), in southcentral Asia, and encircling the western desert in Australia. Desert climates have very limited precipitation, which is likely to fall in isolated downpours followed by long dry periods. The deserts of the higher temperate latitudes can experience very cold winters; those further to the south, such as the enormous Sahara, are hot all year long. A desert is a barren region with little or no rainfall. It is not necessarily sandy—only 20% of the Sahara is sandy. Some of the tropical deserts, such as those along the coasts of Peru, Chile, and Namibia, can go for many years without measurable rainfall. But since they are adjacent to the coast, these unusual deserts are often shrouded in fog. They are deprived of rain by cold ocean currents that cool the atmosphere, wringing moisture from the clouds before they can reach land. Mediterranean regions take their name from the climate in lands surrounding the Mediterranean Sea, which have very warm, dry summers and mild, wet winters. This climate is also found along parts of the west coasts of continents in the lower temperate lati-tudes: Central and Southern California, central Chile, the Cape Town region of South Africa, and the southern coast of Australia. These climates of moderate temperatures, low humidity, and plentiful sunshine are generally viewed as very desirable places to live. Native trees and shrubs in these regions can survive long dry periods.

TROPICAL ZONE. Rainforest temperatures are uniformly warm throughout the year. In the very humid rainforest climate, precipitation is heavy, varying from the Amazon Basin's almost daily afternoon downpours to the seasonal monsoons of Southeast Asia. Other wet Equatorial areas are the Caribbean coast of Central America and the west coast of Africa. This hot and wet environment creates the lushest vegetation on earth. Wet and dry savanna climates are found in the tropics and are at times hotter than the rainforest. Rainfall is heavy only during the brief wet season. For the remainder of the year the savanna is dry. This climate characterizes large regions surrounding the rainforests of central Africa and the Amazon Basin in South America. Mountain climates can be found in any latitude. They are the result of cold or cool tem-peratures found in high altitudes. Mountains are generally wetter and windier than surrounding environments, and many are permanently covered by snow and ice. Mountain climates are found in northwestern North America, central Mexico, the Andes in South America, the Tibetan Plateau and central Asia, and regions of Ethiopia and Eastern Africa.

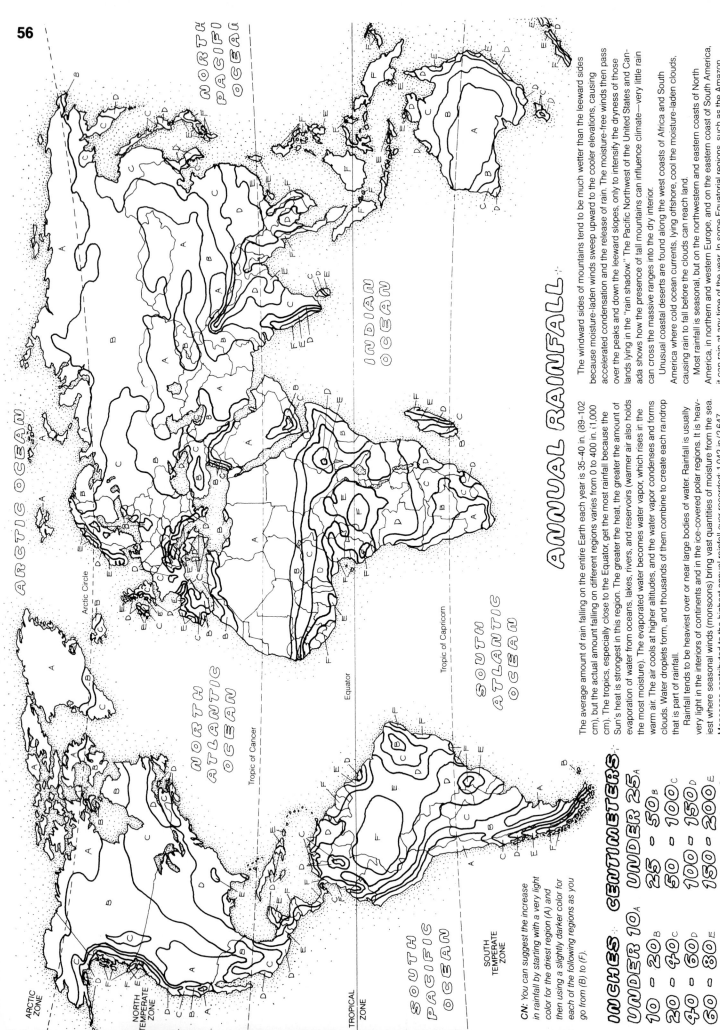

56

ARCTIC OCEAN

NORTH PACIFIC OCEAN

INDIAN OCEAN

NORTH ATLANTIC OCEAN

Tropic of Cancer

Equator

Tropic of Capricorn

SOUTH ATLANTIC OCEAN

SOUTH PACIFIC OCEAN

Arctic Circle

ARCTIC ZONE

NORTH TEMPERATE ZONE

TROPICAL ZONE

SOUTH TEMPERATE ZONE

CN: You can suggest the increase in rainfall by starting with a very light color for the driest region (A) and then using a slightly darker color for each of the following regions as you go from (B) to (F).

INCHES   CENTIMETERS

UNDER 10ₐ   **UNDER 25**ₐ

10 – 20ᵦ   **25 – 50**ᵦ

20 – 40c   **50 – 100**c

40 – 60ᴅ   **100 – 150**ᴅ

60 – 80ₑ   **150 – 200**ₑ

OVER 80F   **OVER 200**F

ANNUAL RAINFALL

The average amount of rain falling on the entire Earth each year is 35–40 in. (89–102 cm), but the actual amount falling on different regions varies from 0 to 400 in. (1,000 cm). The tropics, especially close to the Equator, get the most rainfall because the Sun's heat is strongest in this region. The greater the heat, the greater the amount of evaporation of water from oceans, lakes, rivers, and reservoirs (warmer air also holds the most moisture). The evaporated water becomes water vapor, which rises in the warm air. The air cools at higher altitudes, and the water vapor condenses and forms clouds. Water droplets form, and thousands of them combine to create each raindrop that is part of rainfall.

Rainfall tends to be heaviest over or near large bodies of water. Rainfall is usually very light in the interiors of continents and in the ice-covered polar regions. It is heaviest where seasonal winds (monsoons) bring vast quantities of moisture from the sea. Monsoons contributed to the highest annual rainfall ever recorded: 1,042 in. (2,647 cm) in Cherrapunji, India, north of the Bangladesh border.

The windward sides of mountains tend to be much wetter than the leeward sides because moisture-laden winds sweep upward to the cooler elevations, causing accelerated condensation and the release of rain. The moisture-free winds then pass over the peaks and down the leeward slopes, only to intensify the dryness of those lands lying in the "rain shadow." The Pacific Northwest of the United States and Canada shows how the presence of tall mountains can influence climate—very little rain can cross the massive ranges into the dry interior.

Unusual coastal deserts are found along the west coasts of Africa and South America where cold ocean currents, lying offshore, cool the moisture-laden clouds, causing rain to fall before the clouds can reach land.

Most rainfall is seasonal, but on the northwestern and eastern coasts of North America, in northern and western Europe, and on the eastern coast of South America, it can rain at any time of the year. In some Equatorial regions, such as the Amazon Basin, it usually rains every day of the year.

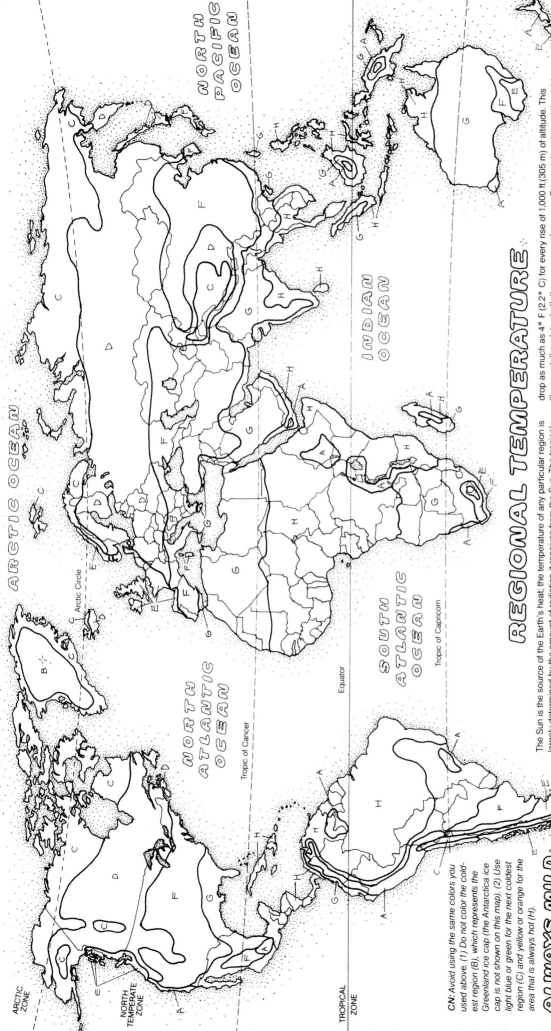

ARCTIC ZONE

NORTH TEMPERATE ZONE

CN: *Avoid using the same colors you used above. (1) Do not color the coldest region (B), which represents the Greenland ice cap (the Antarctica ice cap is not shown on this map). (2) Use light blue or green for the next coldest region (C) and yellow or orange for the area that is always hot (H).*

ALWAYS MILDₐ
ALWAYS COLDᵦ

SUMMER: **WINTER**:
COOLᴄ   **VERY COLD**ᴅ
MILDᴅ   **COLD**ᴅ
MILDᴇ   **COOL**ᴇ
HOTꜰ   **COLD**ꜰ
VERY HOTɢ   **MILD**ɢ

ALWAYS HOTₕ

TROPICAL ZONE

ARCTIC OCEAN

NORTH PACIFIC OCEAN

NORTH ATLANTIC OCEAN

Tropic of Cancer

INDIAN OCEAN

Arctic Circle

Equator

SOUTH ATLANTIC OCEAN

Tropic of Capricorn

REGIONAL TEMPERATURE

The Sun is the source of the Earth's heat; the temperature of any particular region is largely determined by the amount of radiation it receives from the Sun. The tropical region is consistently the hottest because it receives the most radiation from the Sun, which is directly overhead most of the year. It follows that the polar regions are the coldest because there the Sun's rays are the most indirect; even in summer the polar Sun stays close to the horizon. The Temperate Zones, in which the angle of the sun changes throughout the year (Pls. 1, 44, 55), are the only regions to experience seasonal changes in temperature.

Other factors that influence temperature are proximity to large bodies of water, altitude, and prevailing wind conditions. Because water is much slower to heat up or cool down than land is, the summer ocean still has some winter coolness and the winter ocean has not lost all of its summer warmth. The water temperature affects the ocean air, which exerts a moderating influence on the air above the coastal lands. Bodies of water are also heated or cooled by the flow of ocean currents. The coast of Norway is much warmer than regions at comparable latitudes in Canada or Asia, due to the presence of the North Atlantic Drift (Gulf Stream).

Altitude also influences temperature. As air rises, it becomes thinner and loses its chief heat-retaining constituents: water vapor and carbon dioxide. Temperatures can drop as much as 4° F (2.2° C) for every rise of 1,000 ft (305 m) of altitude. This "lapse rate" explains both the presence of snow on Equatorial peaks and the mild temperatures experienced by some tropical areas.

Wind patterns and ocean currents around the globe do a great deal to transfer heat from the tropics. Without this air and ocean activity (created by temperature differences), the tropics would be much hotter and the rest of the globe much colder. As it is, temperature differences still can be immense: the highest temperature ever recorded was 136° F (58° C) in summer shade at Al Azīzīyah, Libya, and the coldest was -128° F (-89° C) during winter in Antarctica.

For millions of years, the Earth's overall temperature has remained relatively constant because of the "heat balance" between the amount of solar radiation absorbed by the Earth and the radiation sent back to space. But carbon dioxide is building up in the Earth's atmosphere because of the burning of fossil fuels, and a "greenhouse effect" is being created as radiation is being retained by the Earth. Most scientists believe that a global warming trend has begun, but some insist that the Earth is due for another Ice Age in which a temperature drop of only a few degrees will bring back the glaciers over much of the Northern Hemisphere. Ice Ages have coincided with periodic variations in the Earth's orbit around the Sun.

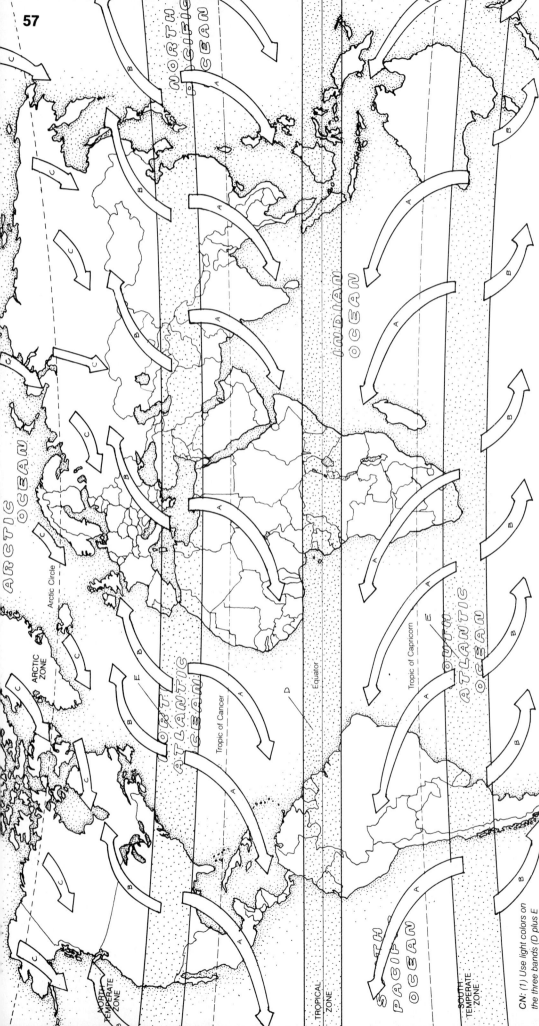

57

Weather is generally the effect of large masses of air—warm or cold, wet or dry—moving around the planet. These masses originate in "source regions" where relatively stagnant air assumes a uniformity of moisture and temperature before beginning to spread its influence over great distances as a weather front. Winds are created when air moves from high-pressure areas toward low-pressure areas. Pressure gradients are caused by the uneven heating of the Earth's atmosphere. Warm air around the Equator rises from a low-pressure trough called the *Doldrums* (which spans the Equator between 5° N and 5° S latitude), and then flows north or south. When this air reaches a subtropical belt of high pressure in the Horse Latitudes, between 30° and 40° (N and S), it descends and becomes the *Trade Winds*, which blow toward the Equator, and the *Westerlies*, which blow toward the polar regions. Because of the rotation of the Earth, these winds do not blow directly north or south, but instead blow diagonally. Of the two "prevailing winds" (winds that blow most frequently in a particular region), the *Trades* are the steadiest in force and direction. But the *Westerlies* are generally stronger, especially in a 10° band of latitude just below the south *Horse Latitudes*. Here, without any continental obstructions, winds can build up great force—both the winds and the latitudes are called the "Roaring Forties." The least predictable prevailing winds are the *Polar Easterlies*, which arise in subpolar regions (outside of the polar circles). Antarctica

(not shown) has the most ferocious winds.

The upward movement of warm air in the Doldrums and the downward movement of cool air in the Horse Latitudes discourage any major horizontal movements, and some of the time these regions are without any winds. The Horse Latitudes were supposedly so named because horses on sailing ships had to be killed when water ran short during periods of tropical stagnation. The name "Doldrums" is self-evident. Tropical oceans spawn great windstorms that go by different names according to the part of the world in which they occur: "hurricanes" in the Caribbean, "typhoons" in the southwest Pacific, and "cyclones" in the Indian Ocean.

Regional winds (often seasonal) can bring excessive moisture (the monsoons of Southeast Asia), searing heat (the sirocco from northern Africa to southern Europe), or unseasonable cold (from northern to southern Europe). Local winds occur mostly in coastal areas. During the afternoon, cool ocean air rushes in to replace the heated air rising above the warm landscape. At night, the situation is reversed, but winds heading toward the ocean are not as strong as the ocean breezes. In the interiors, winds are likely to ascend hillsides during the day and descend at night.

At much higher altitudes (5–12 mi., 8–20 km), the winds are dominated by Westerlies; within them flow the jet streams at speeds up to 250 mph (400 kph).

CN: (1) Use light colors on the three bands (D plus E twice). (2) Color the three types of winds (A–C). The Polar Easterlies in the Southern Hemisphere are not shown.

PREVAILING WINDS

TRADE WINDS_A
WESTERLIES_B
POLAR EASTERLIES_C

DOLDRUMS_D
HORSE LATITUDES_E

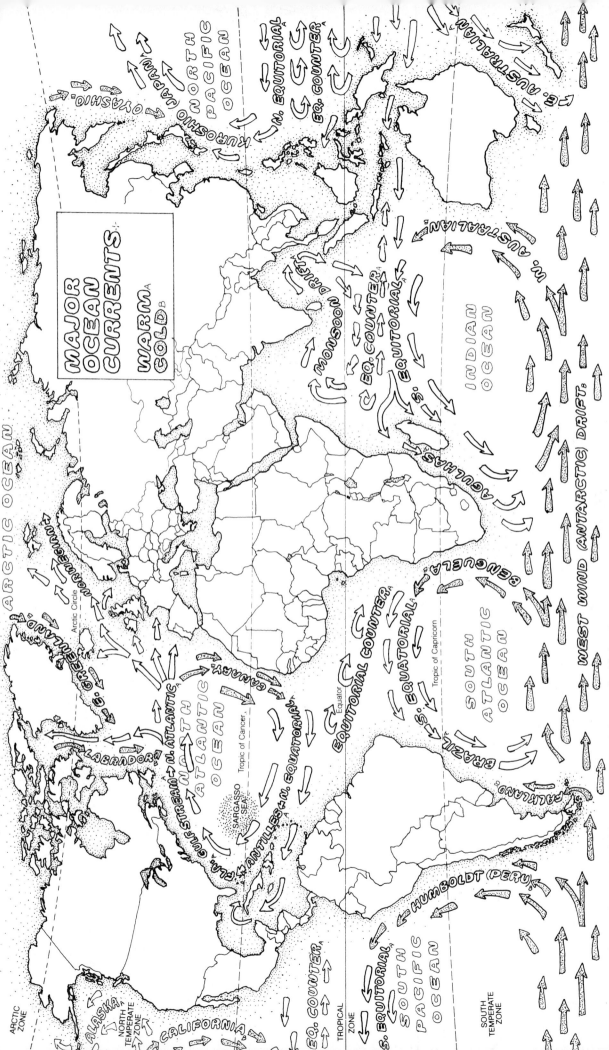

CN: Use a warm light color for the names of all the warm currents and their directional arrows (A), and a cool light color for the cold currents and their dotted arrows (B). Because this world view shows only the edges of the Pacific Ocean, the important North Pacific Current—which is an extension of the warm Kuroshio (Japan) current—is not shown.

Across the oceans stream broad "rivers" of warm and cool water. These "surface currents" are propelled mostly by prevailing winds. These winds create a general movement of water westward near the Equator and eastward in the higher latitudes (note the similarity to wind patterns shown on the upper map). Surface currents range in depth to 2,500 ft.(760m) and in speed from 0.5 to 5 mph (0.8 to 9 kph), depending upon the strength and duration of the wind. The "coriolis force," created by the rotation of the Earth, exerts an influence on surface currents, causing them to veer toward the right (clockwise) from the wind's direction in the Northern Hemisphere and toward the left (counter-clockwise) in the Southern Hemisphere. This force produces five major "gyres" (circular patterns)—two in the northern oceans and three in the southern. The seasonal changes in wind directions similarly affect surface currents. The Monsoon Drift, shown below the Indian subcontinent, alternates in direction as winds blow from land in the winter and bring rain from the sea in the summer.

Beneath the surface currents and moving at a much slower pace (about 1 mile per day), the cold, dense "deep currents" (not shown) move from the polar regions toward the warm and less dense waters of the Equator. Water from deep currents often rises along the coasts of continents to replace warm surface currents that have been driven away from shore by the coriolis force. These "upwellings" bring valuable nutrients to the surface waters and attract and support large fish populations.

Surface currents play an important role in moderating coastal climates by moving vast quantities of warm or cold water. The Gulf Stream, which is actually a system of different currents (North Equatorial, Antilles, Florida, Gulf Stream, and North Atlantic Drift), raises the average winter temperature of the British Isles and the west coast of Europe by 20° F (11° C). The southwestern coasts of continents in the Southern Hemisphere are cooled by currents spawned by the West Wind Drift, which encircles Antarctica.

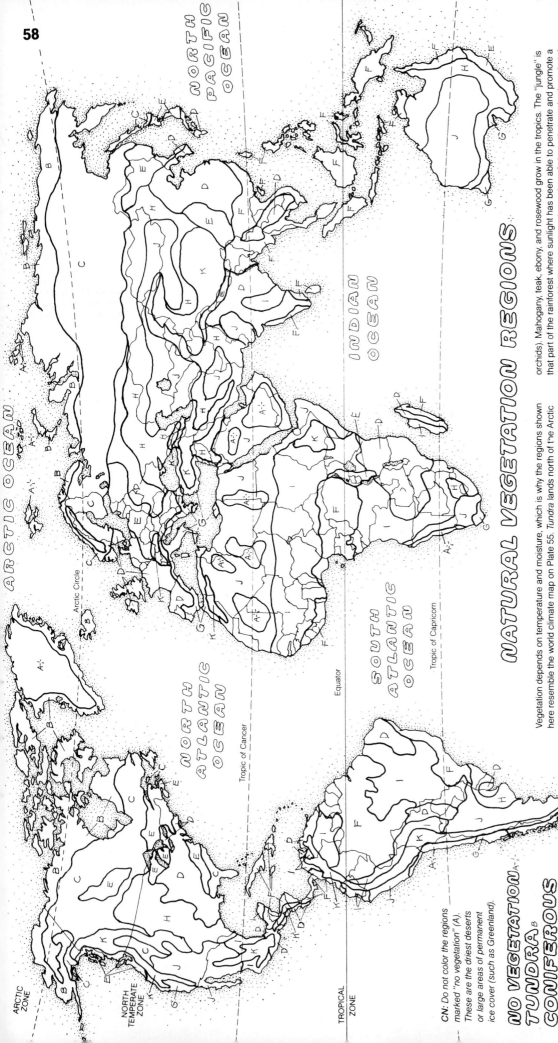

58

NATURAL VEGETATION REGIONS

Vegetation depends on temperature and moisture, which is why the regions shown here resemble the world climate map on Plate 55. *Tundra* lands north of the Arctic Circle have limited plant growth during the brief, cool summer season. Mosses, lichens, wildflowers, and stunted shrubs appear when the snow cover melts and the permafrost begins to thaw. *Coniferous forests* are found in the cold regions of the Northern Hemisphere. These softwood forests are called "taiga" in Asia and "boreal forest" in North America and Europe. Their conical shapes are able to withstand heavy snow buildups. The evergreens, with small needle-like leaves, are the familiar "Christmas trees." The limited number of species include fir, spruce, larch, and pine. *Deciduous forests* of broadleaf hardwood trees once covered western Europe and the eastern half of the United States. Most of these trees in the Temperate Zone lose their leaves in the fall. Among the many varieties are oak, ash, beech, maple, hickory, and chestnut. *Mixed coniferous and deciduous forests* are transitional regions, comprising both kinds of trees. *Tropical rainforests* contain an enormous number of species of broadleaf evergreen trees, packed into wet Equatorial regions. Tall trees with smooth, unbranched trunks form solid canopies that darken the forest floor. The trees are surrounded by light-seeking plants: "lianas" (thick climbing vines) and "epiphytes" (non-parasitic plants that grow on tree branches, such as mosses and orchids). Mahogany, teak, ebony, and rosewood grow in the tropics. The "jungle" is that part of the rainforest where sunlight has been able to penetrate and promote a dense undergrowth. *Mediterranean scrub* consists of plants that survive long, hot dry periods following mild, wet winters. Trees are generally low, with leathery leaves. The Mediterranean region, because of centuries of human activity, has lost nearly all its native trees. *Temperate grasslands* are called "prairies" in the U.S. Midwest, "pampas" in Argentina, and "steppes" in Russia. All are major wheat-producing areas, with soil too dry to support the growth of trees. *Tropical grasslands* are called "savannas" in Africa (where they surround the rainforest and cover over a third of the continent), "llanos" in Venezuela, and "campos" in Brazil. Because *tropical grasslands* are subject to occasional seasons of heavy rain, they are likely to have scattered trees and very thick grass. *Deserts* actually have very few sand dunes, and often have some vegetation in the form of drought-resistant shrubs or scrubby plants and spiny succulents. The cactus is a familiar sight in the American Southwest. Deserts are generally in the western part of a continent, located between 25° and 30° latitude. *Mountains* usually have more than one vegetative zone, because temperature drops at higher elevations. A lofty peak in the tropics can emerge from a rainforest and have deciduous, coniferous, and tundra zones, ending in an eternal ice cap.

NO VEGETATION_A
TUNDRA_B
CONIFEROUS FOREST_C
DECIDUOUS FOREST_D
MIXED CONIF & DECID_E
TROPICAL RAIN FOREST_F
MEDITERRANEAN SCRUB_G
TEMPERATE GRASSLAND_H
TROPICAL GRASSLAND_I
DESERT SHRUB_J
MOUNTAIN_K

CN: Do not color the regions marked "no vegetation" (A). These are the driest deserts or large areas of permanent ice cover (such as Greenland).

ARCTIC ZONE

NORTH TEMPERATE ZONE

TROPICAL ZONE

Arctic Circle

Tropic of Cancer

Equator

Tropic of Capricorn

ARCTIC OCEAN

NORTH PACIFIC OCEAN

NORTH ATLANTIC OCEAN

SOUTH ATLANTIC OCEAN

INDIAN OCEAN

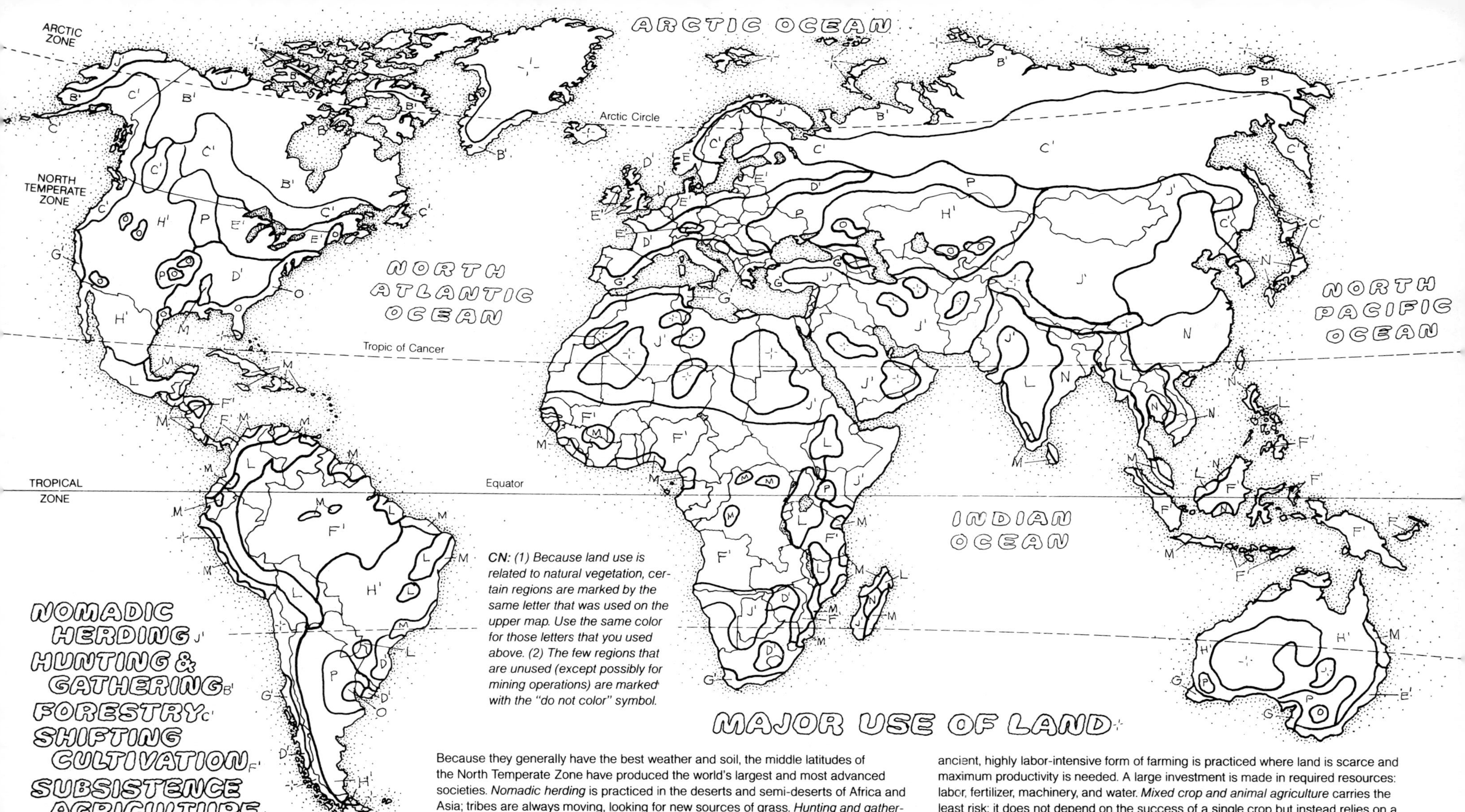
ARCTIC OCEAN
NORTH ATLANTIC OCEAN
NORTH PACIFIC OCEAN
INDIAN OCEAN
ARCTIC ZONE
NORTH TEMPERATE ZONE
TROPICAL ZONE
Arctic Circle
Tropic of Cancer
Equator

MAJOR USE OF LAND

NOMADIC HERDING J'
HUNTING & GATHERING B'
FORESTRY C'
SHIFTING CULTIVATION F'
SUBSISTENCE AGRICULTURE L
PLANTATIONS M
RICE PADDIES N
MIXED CROP & ANIMAL D'
MEDITERRANEAN G'
SPECIALIZED FARMING O
DAIRYING E'
LIVESTOCK RANCHING H
COMMERCIAL GRAIN P

CN: *(1) Because land use is related to natural vegetation, certain regions are marked by the same letter that was used on the upper map. Use the same color for those letters that you used above. (2) The few regions that are unused (except possibly for mining operations) are marked with the "do not color" symbol.*

Because they generally have the best weather and soil, the middle latitudes of the North Temperate Zone have produced the world's largest and most advanced societies. *Nomadic herding* is practiced in the deserts and semi-deserts of Africa and Asia; tribes are always moving, looking for new sources of grass. *Hunting and gathering* is the way of life for Eskimos in the Arctic and the Bushmen in southern Africa's Kalahari Desert. Like nomads, hunter-gatherers are declining in number. *Forestry* (managing and replenishing timber) has become standard practice in regions of the Temperate Zones, but not in the tropics, where slow-growing hardwood forests are rapidly being destroyed. *Shifting cultivation* ("slash-and-burn") is an ancient agricultural practice. Tropical trees and plants are hacked down, allowed to dry, and then burned to the ground, creating farmland that is barely fertile—good for 2 or 3 years, at the most. *Subsistence agriculture* methods just about meet the needs of farmers who must use extensive human labor and very simple equipment to raise crops and cattle. *Plantations* are large tropical landholdings that specialize in growing and partially processing a single cash crop, such as coffee, bananas, sugar, tea, cacao, spices, or tobacco. With the use of slave labor, early plantations made great fortunes for their colonial owners. *Rice paddies* in the fields and hillside terraces of Southeast Asia are flooded by monsoon rains and irrigation networks to grow the region's food staple. This ancient, highly labor-intensive form of farming is practiced where land is scarce and maximum productivity is needed. A large investment is made in required resources: labor, fertilizer, machinery, and water. *Mixed crop and animal agriculture* carries the least risk; it does not depend on the success of a single crop but instead relies on a wide variety for both human and animal consumption. North American and European growers use their land "intensively"; Asian and Latin American farmers work their land "extensively" (less input of resources and lower productivity). *Mediterranean farming* specializes in drought-tolerant vines and trees (grapes, olives, and figs) that survive long, dry summers, and in grain crops that grow during mild, wet winters. Livestock is raised, and with the use of irrigation, citrus trees are grown. *Specialized farming*, also called "truck farming," is a highly intensive method, used mostly in temperate regions, that produces large amounts of fruit and vegetables for adjacent urban areas. *Dairying* in many areas takes place in highly automated "milk factories." In nations far from export markets, liquid milk is converted to processed milk, butter, or cheese. *Livestock ranching* relies on plentiful land so that herds are able to find enough food even where grass is sparse. *Commercial grain production* is the most mechanized form of food production; giant farms are owned by absentee corporations whose employees are generally present only during the planting and harvesting seasons.

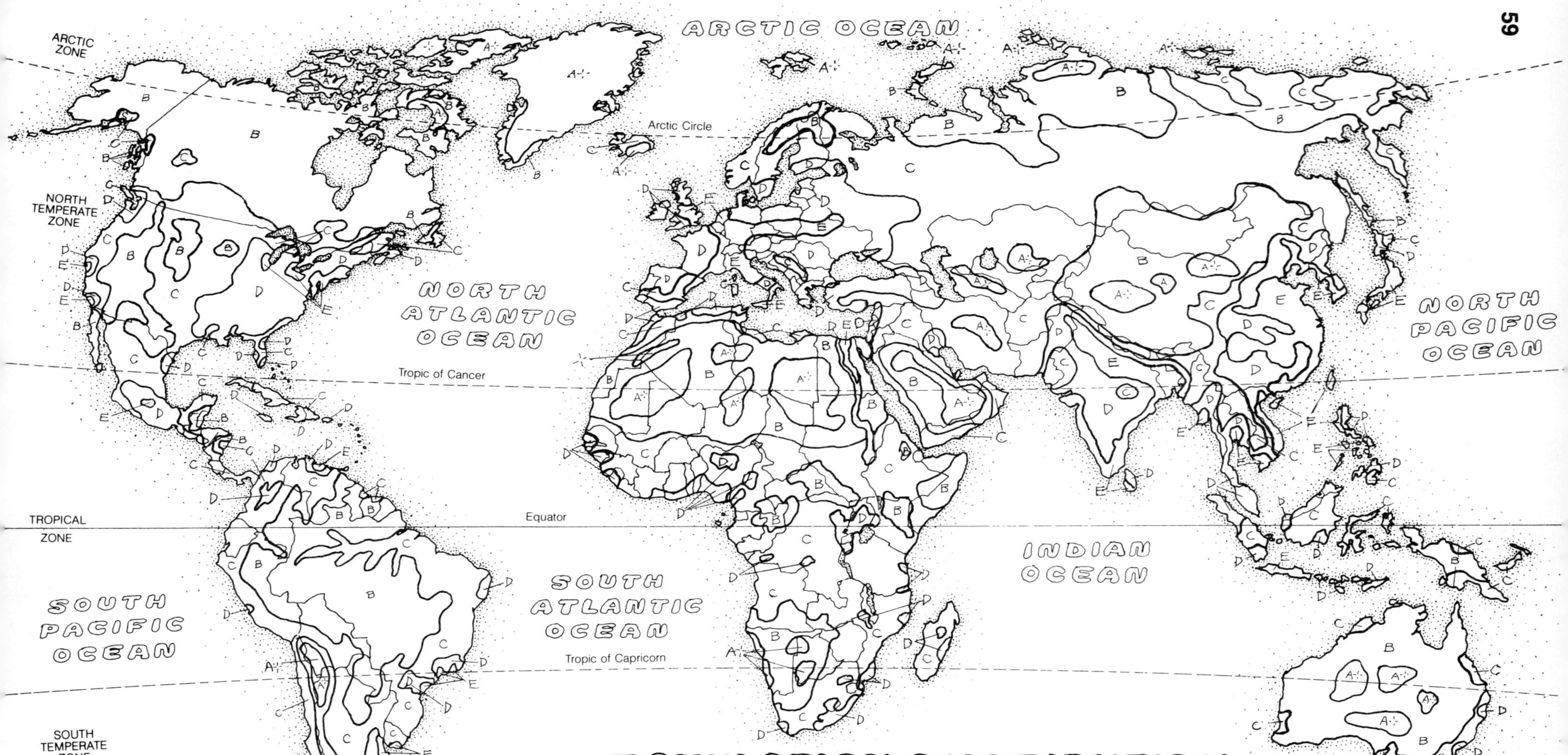

59

POPULATION DISTRIBUTION

Although the overall rate of population growth has been declining since 1963's peak of 2.2% per year, the number of people on the planet continues to increase—a cause of great concern around the world. Much of the earth's landscape is inhospitable and thus thinly populated; the distribution of 5.8 billion people is extremely uneven. Eighty percent live in one of three clusters: (1) eastern, southeastern, and southern Asia, (2) Europe, or (3) central and eastern North America. The majority of these populations are located in the middle latitudes of the North Temperate Zone, a region generally favored by good climate and fertile soil.

Although many countries are densely populated, not all are "overpopulated," a condition in which there are more people than an area can support. For example, because of its high standard of living, the Netherlands, one of the most densely populated nations, is not considered overpopulated. Uneven population distribution exists within nations themselves. The greatest concentrations are found in the job-producing urban areas ("urban" means having over 20,000 residents). Forty percent of the world's population now resides in or near cities, and the trend is continuing. In the United States, the figure is over 75 percent. In the past, people congregated only in food-producing areas, and this is still true in much of Asia, Africa, and Latin America.

In Europe, there is almost an even balance between rural and urban populations. When a nation's birthrate exceeds its death rate, a "natural increase" in population occurs. Immigration and emigration generally play a small role (except in the U.S., which has grown from a significant number of Asian and Latin-American immigrants). In most of the wealthier, industrialized nations of Europe, a low birthrate is causing a decrease in population. In Asia, Africa, and Latin America, an increasing birthrate (due to an increase in food production) and a declining death rate (due to medical intervention) has produced a population explosion, but even in those nations the rate of increase is declining. They all have "young" populations, with half the people under the age of 15. Farmers in the poorer nations depend upon their children to perform free labor and provide old-age assistance. Parents in industrialized nations do not normally need their children's labor, and raising a family can be quite expensive (especially the cost of higher education). The elderly are taken care of by pensions and social security. Until the developing nations raise their standard of living—which ironically depends on reducing population growth—poor people are unlikely to reduce the size of their families. China's rigid insistence on a one-child-per-family policy is a notable exception.

CN: Do not color the uninhabited areas (A). Color regions B-E with progressively darker colors to suggest an increasing population density.

PEOPLE PER SQ. MI. / KM²

UNINHABITED _A⁻_
UNDER 2 / 1 _B_
2 - 60 / 1 - 25 _C_
60 - 250 / 25 - 100 _D_
OVER 250 / 100 _E_

ARCTIC OCEAN

NORTH PACIFIC OCEAN

INDIAN OCEAN

NORTH ATLANTIC OCEAN

SOUTH ATLANTIC OCEAN

SOUTH PACIFIC OCEAN

ARCTIC ZONE

NORTH TEMPERATE ZONE

TROPICAL ZONE

SOUTH TEMPERATE ZONE

Arctic Circle

Tropic of Cancer

Equator

Tropic of Capricorn

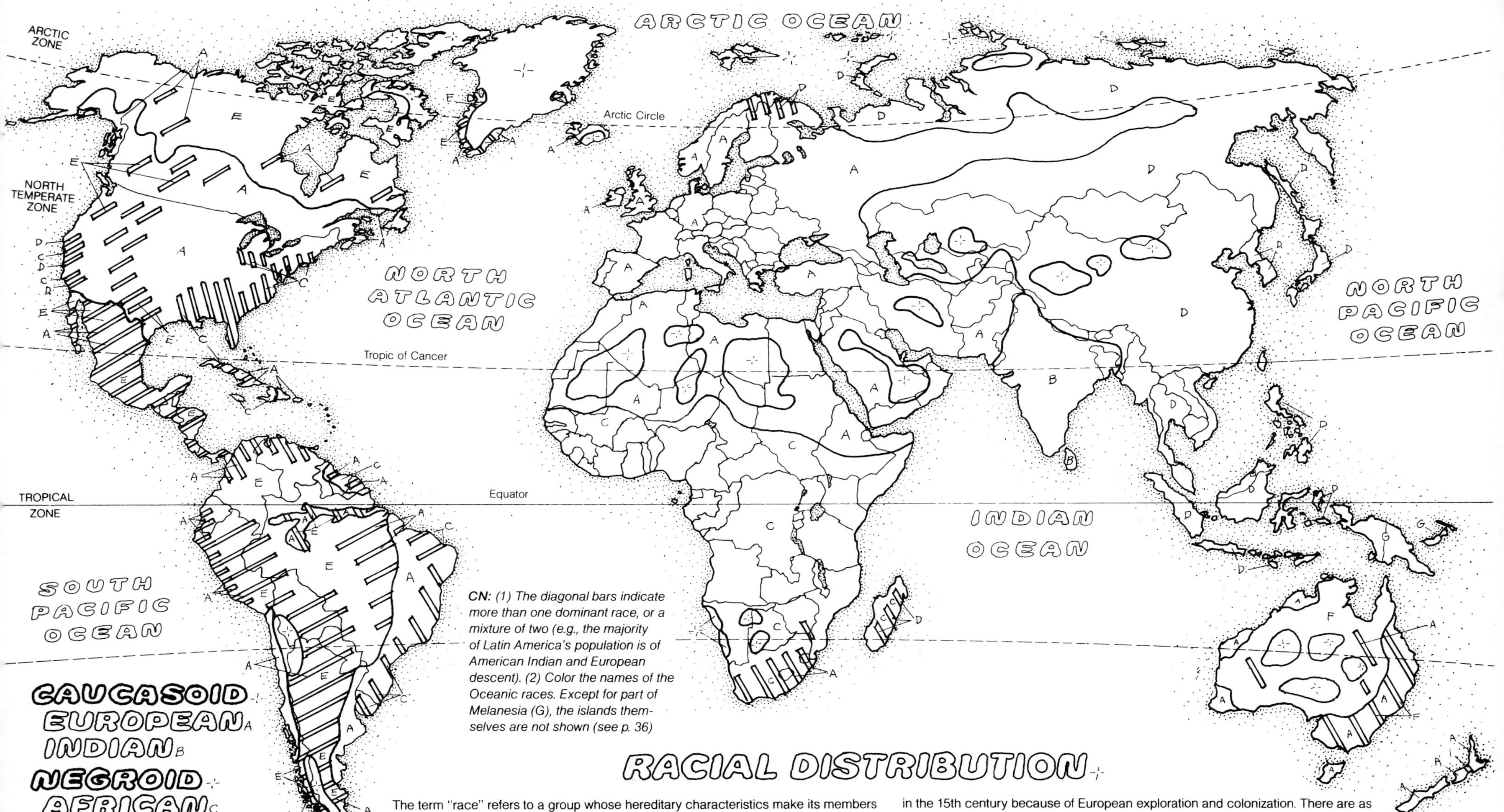

ARCTIC OCEAN

NORTH PACIFIC OCEAN

INDIAN OCEAN

NORTH ATLANTIC OCEAN

ARCTIC ZONE

NORTH TEMPERATE ZONE

Arctic Circle

Tropic of Cancer

Equator

TROPICAL ZONE

SOUTH PACIFIC OCEAN

CN: (1) The diagonal bars indicate more than one dominant race, or a mixture of two (e.g., the majority of Latin America's population is of American Indian and European descent). (2) Color the names of the Oceanic races. Except for part of Melanesia (G), the islands themselves are not shown (see p. 36)

CAUCASOID
EUROPEAN_A
INDIAN_B

NEGROID
AFRICAN_C

MONGOLOID
ASIATIC_D
AMERINDIAN_E

OCEANIC
AUSTRALIAN_F
MELANESIAN_G
MICRONESIAN_H
POLYNESIAN_I

RACIAL DISTRIBUTION

The term "race" refers to a group whose hereditary characteristics make its members resemble each other more than individuals of any other group. But "race" is an artificial classification that has no basis in biology. There is really only one race, the human species: Homo sapiens. Most anthropologists agree that differences among humans are too minor (and too recently acquired) to represent truly different biological races. It is generally believed that we all evolved from the early inhabitants of eastern Africa. As groups emigrated to other parts of the world, their adaptation to different environments may have played a major role in determining variations in appearance. There is no single physical characteristic that is exclusive to any one race. The traditional classification of Caucasoid, Mongoloid, and Negroid was based solely upon physical characteristics (skin color; hair texture; bone structure; shape of eyes, nose, and lips, etc.). On this page you will be coloring a newer classification, based upon geographical origins, which defines nine major racial groups (with hundreds of sub-groups). For centuries, these races lived in virtual isolation because physical barriers separated their homelands. Racial "purity" began to disintegrate

in the 15th century because of European exploration and colonization. There are as many people in the world with some European ancestry in their blood as there are living in Europe. The trend toward the merging of races—creating groups with more variabilities—is accelerating.

There is now greater diversity within each race. The "white" Caucasoid race ranges from fair-skinned Scandinavians to dark-skinned North Africans and still darker Ethiopians. The dark-skinned residents of India are also considered Caucasoid. Blacks living in the Americas are descended from the Negroid races of sub-Saharan Africa. The Mongoloid race includes American Indians and Asiatics. The Americas were originally populated by "red-" and "brown-skinned" people who migrated from Asia across the Bering Strait, as did the Eskimos at a later date. "Yellow-skinned" Asiatics (Chinese, Japanese, Koreans, and Southeast Asians) share similar eye and facial characteristics. The Oceanic group ranges from very dark-skinned Australian Aborigines to the progressively lighter Pacific island races: Melanesians, lighter Micronesians, and even lighter Polynesians (p. 36).

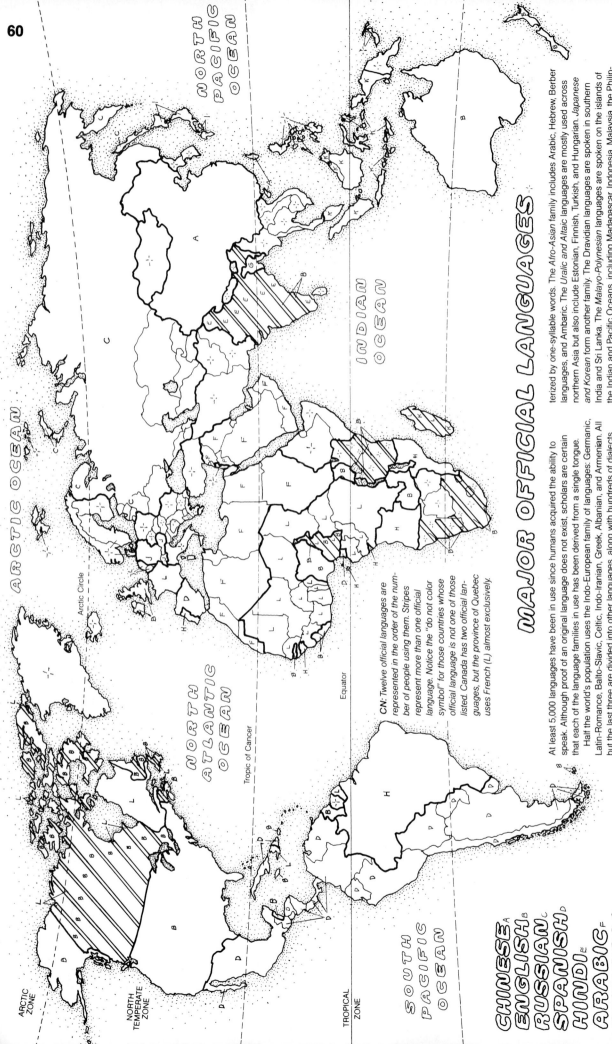

60

ARCTIC OCEAN

NORTH PACIFIC OCEAN

Arctic Circle

NORTH ATLANTIC OCEAN

Tropic of Cancer

Equator

INDIAN OCEAN

SOUTH PACIFIC OCEAN

ARCTIC ZONE

NORTH TEMPERATE ZONE

TROPICAL ZONE

CN: Twelve official languages are represented in the order of the number of people using them. Stripes represent more than one official language. Notice the "do not color symbol" for those countries whose official language is not one of those listed. Canada has two official languages, but the province of Quebec uses French (L) almost exclusively.

CHINESE A
ENGLISH B
RUSSIAN C
SPANISH D
HINDI E
ARABIC F
BENGALI G
PORTUGUESE H
JAPANESE I
GERMAN J
MALAYO-POLYNESIAN K
FRENCH L

MAJOR OFFICIAL LANGUAGES

At least 5,000 languages have been in use since humans acquired the ability to speak. Although proof of an original language does not exist, scholars are certain that each of the language families in use has been derived from a single tongue.

Half the world's population uses the Indo-European family of languages: Germanic, Latin-Romance, Balto-Slavic, Celtic, Indo-Iranian, Greek, Albanian, and Armenian. All but the last three are divided into other languages, along with hundreds of dialects (variations based on regional or social differences). The point at which a dialect becomes a separate language is often unclear. The Germanic languages include English, German, Dutch, and the Scandinavian languages. The Latin-Romance languages, an outgrowth of Latin (imposed by the Roman Empire), include Spanish, French, Italian, Portuguese, and Romanian. The Balto-Slavic family includes Russian, Ukrainian, Polish, Czech, Serbo-Croatian, Slovenian, Bulgarian, Lithuanian, and Latvian. The Celtic group includes Irish Gaelic, Scottish Gaelic, Welsh, and Breton. The Indo-Iranian division includes languages spoken across southern Asia: Farsi (Iran), Pashto (Afghanistan), Urdu (Pakistan), Hindi (India), and Bengali (Bangladesh).

The Sino-Tibetan languages (Chinese, Thai, Burmese, and Tibetan) are all charac-terized by one-syllable words. The Afro-Asian family includes Arabic, Hebrew, Berber languages, and Amharic. The Uralic and Altaic languages are mostly used across northern Asia but also include Estonian, Finnish, Turkish, and Hungarian. Japanese and Korean form another family. The Dravidian languages are spoken in southern India and Sri Lanka. The Malayo-Polynesian languages are spoken on the islands of the Indian and Pacific Oceans, including Madagascar, Indonesia, Malaysia, the Philip-pines, Hawaii, and New Zealand. The Mon-Khmer family is spoken in southeast Asia. The most widely used of 2,000 languages of black Africa is the Bantu group, which includes Swahili, the lingua franca of eastern Africa. Over 1,000 American Indian lan-guages are spoken in isolated parts of the Americas. The map reflects five centuries of European colonialism. The Indo-European languages are the official languages (and the most widely used) in North and South America. European-imposed lan-guages are spoken by only a minority of black Africans but continue to be used as official languages. These languages are the closest thing to a common tongue among the many ethnic groups, and the ruling and business classes have been educated in European languages.

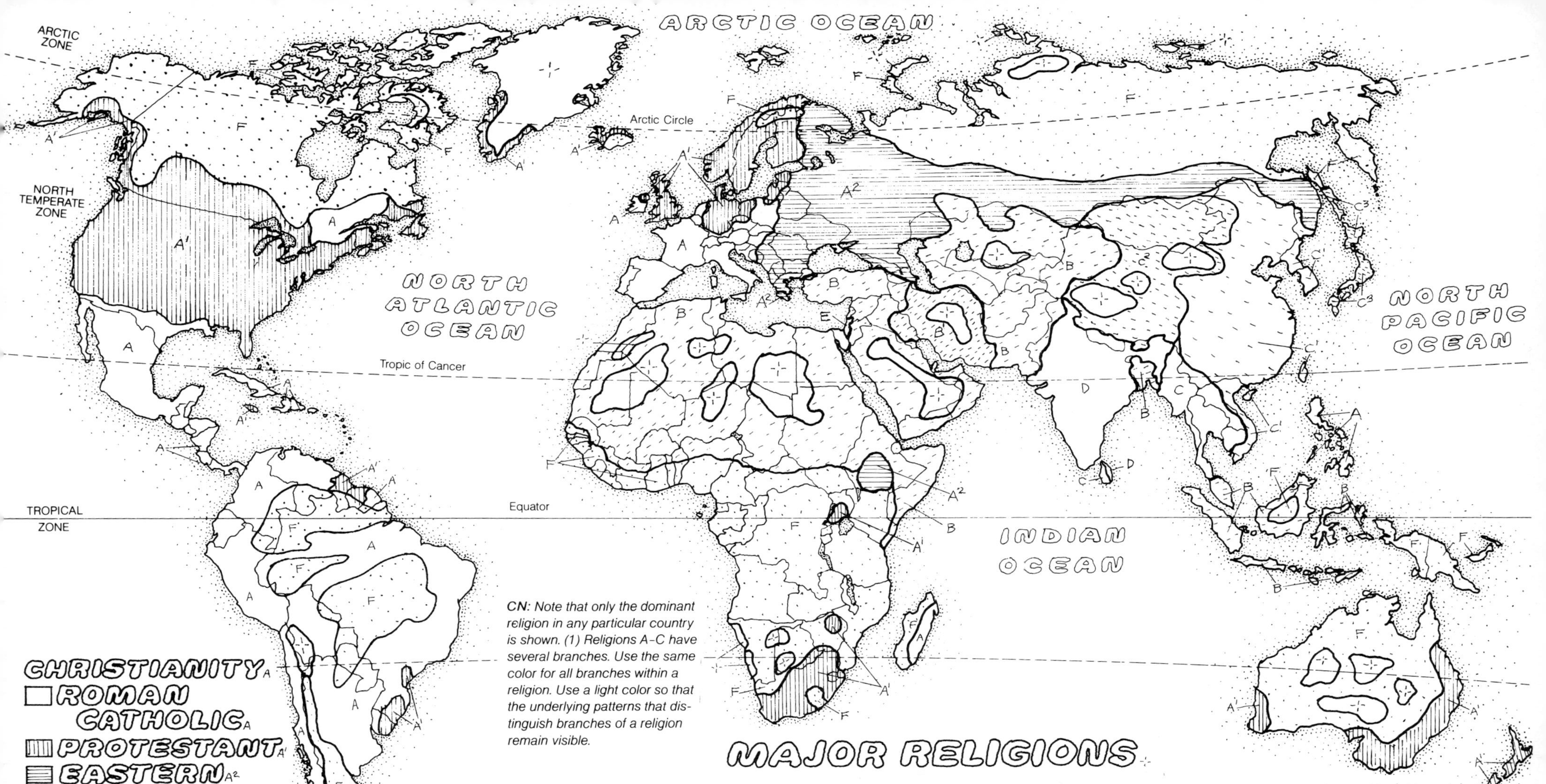

MAJOR RELIGIONS

Judaism is the oldest of the monotheistic religions. The tribal leader Abraham was believed to have received land from God about 3,400 years ago. Jews, Christians, and Muslims believe that Moses was a prophet who received God's law. The Torah of the Jews (sacred writings of the word of God) is the Old Testament of the Christian Bible. The 14 million followers of Judaism are mainly ethnic Jews. Founded 2,000 years ago by Jesus Christ, a Jew whom Christians regard as Savior and Messiah, *Christianity* is the largest group, with about 2 billion adherents. In 1054, Christianity split into the *Roman Catholic Church*, led by the Pope of Rome, and the *Eastern (Orthodox)* group of churches, which followed early Christian tenets. Five hundred years later, the *Protestant* Reformation split the Roman Catholic Church; Protestant denominations spread across the Germanic-speaking nations of northern Europe, and later to the United States and most of Canada. Protestants objected to the practices of the Papacy and the priesthood, and sought revelation solely from the Bible. There are twice as many Catholics as there are Protestants. *Islam*, the third major religion born in the Middle East, was founded about 1,400 years ago by Muhammed, who regarded himself as the last great prophet in a line including Moses and Christ. Five times each day, over 1 billion Muslims kneel in prayer toward Mecca, the holy city in Saudi Arabia. Their God is Allah and their scripture is the Koran. Most Muslims are *Sunnis*;

members of the fundamentalist *Shiite* minority are concentrated in Iran and Iraq. Islam is growing rapidly and has been more successful than Christianity in converting black Animists in Africa. *Animism* is the worship of spirits believed to inhabit all natural forms and forces. *Hinduism*, the dominant religion of India and Nepal, involves the worship of thousands of deities. It has no clergy and no formal scripture other than religious writings called the Vedas. Hinduism has no single founder and has had many changes in its 5,000-year existence. Over half a billion believers seek purification and an end to the cycle of rebirths required by one's "Karma" (deeds in a previous life). Reincarnation takes many forms; hence, Hindus revere all manner of life. *Buddhism*, an offshoot of Hinduism, was founded about 1,500 years ago by Gautama Buddha. The Buddha ("the enlightened one") advocated moderation as a way to overcome desire and ambition and thus reach a state of contentment ("nirvana"), ending the cycle of reincarnations. Meditation is more important to nearly a half billion Buddhists than participation in ritualistic ceremonies. Traditional Buddhism is followed only in Southeast Asia. Elsewhere, it has been modified by regional folk religions—in China by the *Confucian* respect for elders, ancestry, and authority, and the *Taoist* philosophy of passivity and simplicity; in Tibet and Mongolia by *Lamaistic* monasticism and demonology; and in Japan by the spiritualism and rituals of *Shintoism* and the contemplative nature of Zen.

CN: Note that only the dominant religion in any particular country is shown. (1) Religions A–C have several branches. Use the same color for all branches within a religion. Use a light color so that the underlying patterns that distinguish branches of a religion remain visible.

CHRISTIANITY A
- **ROMAN CATHOLIC** A
- **PROTESTANT** A¹
- **EASTERN** A²

ISLAM B
- **SUNNI** B
- **SHIITE** B¹

BUDDHISM C

CONFUCIANIST-TAOIST C¹

LAMAIST C²

SHINTO C³

HINDUISM D

JUDAISM E

ANIMISM F

ARCTIC OCEAN

NORTH PACIFIC OCEAN

NORTH ATLANTIC OCEAN

INDIAN OCEAN

Arctic Circle

Tropic of Cancer

Equator

ARCTIC ZONE

NORTH TEMPERATE ZONE

TROPICAL ZONE

61

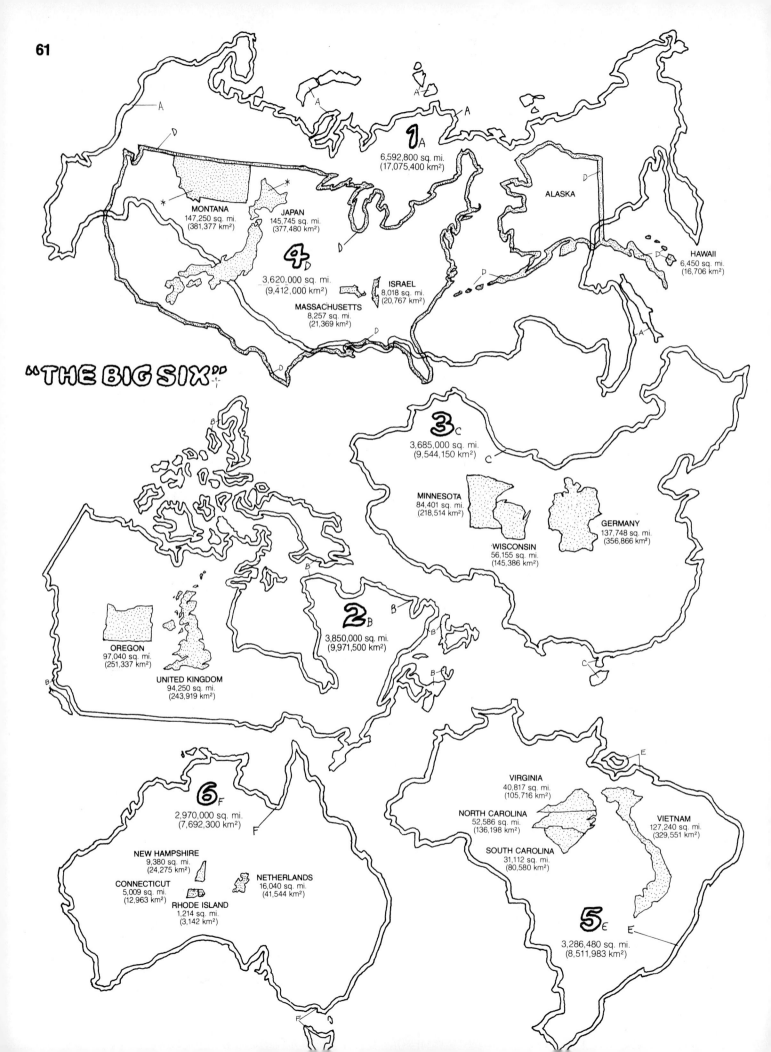

"THE BIG SIX"

1A
6,592,800 sq. mi.
(17,075,400 km²)

ALASKA

MONTANA
147,250 sq. mi.
(381,377 km²)

JAPAN
145,745 sq. mi.
(377,480 km²)

HAWAII
6,450 sq. mi.
(16,706 km²)

4D
3,620,000 sq. mi.
(9,412,000 km²)

ISRAEL
8,018 sq. mi.
(20,767 km²)

MASSACHUSETTS
8,257 sq. mi.
(21,369 km²)

3C
3,685,000 sq. mi.
(9,544,150 km²)

MINNESOTA
84,401 sq. mi.
(218,514 km²)

GERMANY
137,748 sq. mi.
(356,866 km²)

WISCONSIN
56,155 sq. mi.
(145,386 km²)

OREGON
97,040 sq. mi.
(251,337 km²)

UNITED KINGDOM
94,250 sq. mi.
(243,919 km²)

2B
3,850,000 sq. mi.
(9,971,500 km²)

VIRGINIA
40,817 sq. mi.
(105,716 km²)

NORTH CAROLINA
52,586 sq. mi.
(136,198 km²)

VIETNAM
127,240 sq. mi.
(329,551 km²)

SOUTH CAROLINA
31,112 sq. mi.
(80,580 km²)

6F
2,970,000 sq. mi.
(7,692,300 km²)

NEW HAMPSHIRE
9,380 sq. mi.
(24,275 km²)

CONNECTICUT
5,009 sq. mi.
(12,963 km²)

NETHERLANDS
16,040 sq. mi.
(41,544 km²)

RHODE ISLAND
1,214 sq. mi.
(3,142 km²)

5E
3,286,480 sq. mi.
(8,511,983 km²)

COMPARATIVE SIZES

CN: The world's largest 25 states are presented in order of their size. (1) Color each name, outline, and the number of their position. (2) On the opposite page, color gray the smaller nations and the states of the United States.

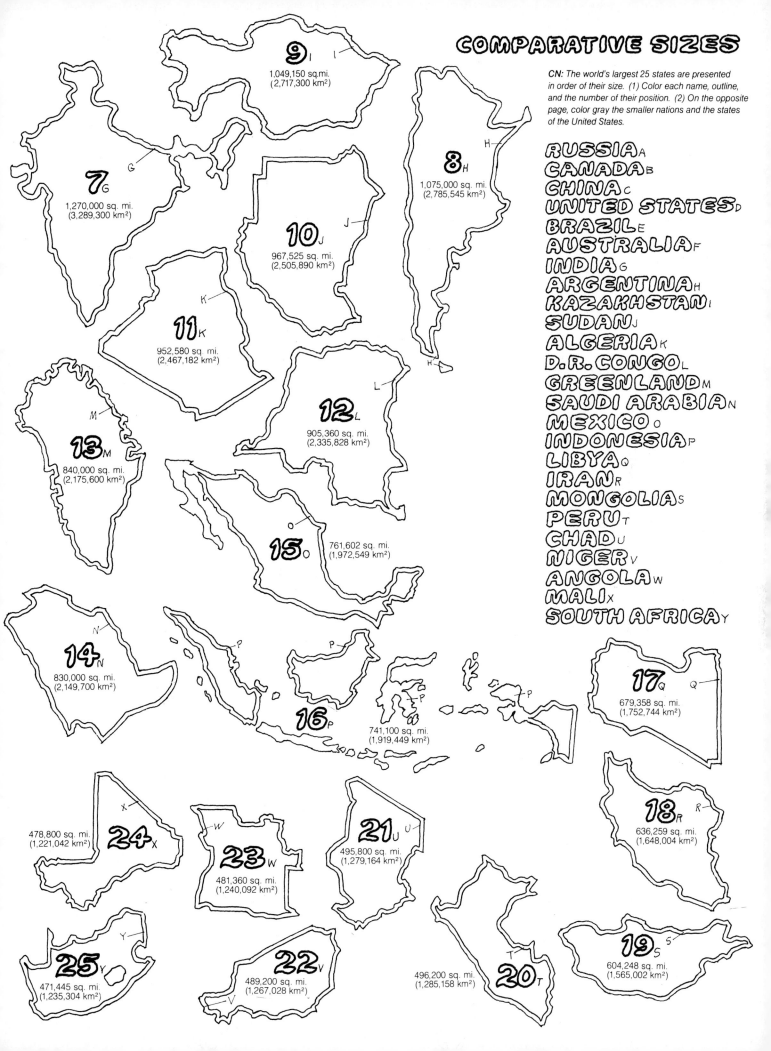

9 I
1,049,150 sq.mi.
(2,717,300 km²)

7 G
1,270,000 sq. mi.
(3,289,300 km²)

8 H
1,075,000 sq. mi.
(2,785,545 km²)

10 J
967,525 sq. mi.
(2,505,890 km²)

11 K
952,580 sq. mi.
(2,467,182 km²)

12 L
905,360 sq. mi.
(2,335,828 km²)

13 M
840,000 sq. mi.
(2,175,600 km²)

15 O
761,602 sq. mi.
(1,972,549 km²)

14 N
830,000 sq. mi.
(2,149,700 km²)

16 P
741,100 sq. mi.
(1,919,449 km²)

17 Q
679,358 sq. mi.
(1,752,744 km²)

18 R
636,259 sq. mi.
(1,648,004 km²)

24 X
478,800 sq. mi.
(1,221,042 km²)

23 W
481,360 sq. mi.
(1,240,092 km²)

21 U
495,800 sq. mi.
(1,279,164 km²)

25 Y
471,445 sq. mi.
(1,235,304 km²)

22 V
489,200 sq. mi.
(1,267,028 km²)

20 T
496,200 sq. mi.
(1,285,158 km²)

19 S
604,248 sq. mi.
(1,565,002 km²)

RUSSIA A
CANADA B
CHINA C
UNITED STATES D
BRAZIL E
AUSTRALIA F
INDIA G
ARGENTINA H
KAZAKHSTAN I
SUDAN J
ALGERIA K
D. R. CONGO L
GREENLAND M
SAUDI ARABIA N
MEXICO O
INDONESIA P
LIBYA Q
IRAN R
MONGOLIA S
PERU T
CHAD U
NIGER V
ANGOLA W
MALI X
SOUTH AFRICA Y

INDEX / DICTIONARY / QUIZ

HOW TO USE THE INDEX

A plate number listed in this index refers to two pages: the map page, on the left, and the page of text opposite it. You will find number in the upper left hand corner of the map page.

Where more than one plate number is shown, the italicized number indicates the principal reference: the plate on which the subject is to be colored, given prominent display, mentioned in the text, or all of the aforementioned.

HOW TO USE THE DICTIONARY

Each index entry is followed by a very brief explanation identifying the subject, describing its location, and often including a significant fact. The name of the applicable continent is either mentioned in the definition or added in parentheses.

For further information on the subject/entry, consult the relevant plate in the book. Even though all of the entries appear somewhere in the book (most often on a map), they may not always be covered in the text.

HOW TO USE THE QUIZ

There are two ways in which you can turn this index/dictionary into a challenging quiz.

1. By placing a sheet of paper over the column of subject entries (the left column), you can test your geographical knowledge by trying to identify the entry/name that corresponds to the definition in the right hand column. Try to visualize the location of the subject when thinking of the answer—the descriptions almost always give you some idea of the environs. Even after finding out the answer, continue to try to visualize the region involved. You might wish to turn to the page mentioned in order to clarify your mental picture. Visualization plays a vital role in learning where things are. As an additional aid in coming up with the answer, always consider the obvious alphabetical clue of where you are in the index.

2. The second method is far more challenging; it involves the covering of the definition column. This gives you a chance to create your own definitions. Begin by reading over a number of definitions in order to get an idea of what is expected. Although this second approach is much more difficult, it will provide greater learning dividends.

Don't be discouraged if you draw a blank on many, or even most, of the questions. Because this quiz also serves as an index, it contains far more detail than you could possibly be expected to know. On the other hand, every place mentioned does exist somewhere on the planet, and sooner or later you may learn about it from the media or the internet or, even better, have the pleasure of visiting it yourself.

ABBREVIATIONS

N—north
NW—northwest   NE—northeast
W—west   WC—west central   C—central   EC—east central   E—east
SW—southwest   SE—southeast
S—south

A

Abidjan, 40	The largest city in Côte d'Ivoire (formerly the Ivory Coast); an Atlantic port in west Africa
Abu Dhabi, 28, 31	The capital of the United Arab Emirates; a port on the southern Persian Gulf . . . (SW Asia)
Accra, 37, 40	The capital of Ghana; a Gulf of Guinea port on west Africa's "Gold Coast"
Aconcagua, 15, 17	Tallest peak in the Western Hemisphere (22,835 ft., 6,960 m); in the Argentinian Andes . . . (S Am)
Addis Ababa, 37, 42	The capital of Ethiopia; situated on the 8,000 ft. (2,440 m) Ethiopian Plateau . . . (E Africa)
Adelaide, 35	The capital of South Australia; a state on Australia's south coast . . . (Oceania)
Aden, 31	The main port of Yemen; on the southwest coast of the Arabian Peninsula . . . (SW Asia)
Adirondack Mts., 7	A branch of the Appalachian range in upper New York State . . . (N Am)
Adriatic Sea, 18, 23, 24	A northcentral sea of the Mediterranean, between Italy and the Balkan Peninsula . . . (S Europe)
Aegean Sea, 19, 23	An island-filled sea between Greece and Turkey . . . (Europe & Asia)
Afghanistan, 28, 32	A mountainous country between Iran and Pakistan . . . (C Asia)
Africa, 1, 2, 37, 38	The second-largest continent, with the third-largest population
African race, 59	Another name for the Negroid race of Subsaharan Africa

B

E

F

Faeroe Islands, 20 . . . . A group of Danish islands, in the northeast Atlantic between Iceland and Scotland

Fairbanks, 9 . . . . Alaska's largest interior city; in the sparsely populated, eastcentral region . . . (N Am)

Falkland Islands, 17 . . . . British islands off southern Argentina, which claims ownership despite losing a war over them . . . (S Am)

Far East, 30 . . . . Traditional western name for a region including China, Mongolia, the two Koreas, and Japan

Fiji, 36 . . . . Southwest Pacific islands, north of New Zealand, in Melanesia . . . (Oceania)

Finland, 18, 20 . . . . A nation lying between Sweden and Russia, mistakenly called Scandinavian . . . (N Europe)

Fjord, see Glossary . . . . A narrow, winding ocean inlet that penetrates a coastal mountain range

Florence, 23 . . . . A city in northcentral Italy known for palaces and museums filled with Renaissance art . . . (S Europe)

Florida, 6, 8 . . . . A fast-growing southeastern tourist state, on a low, flat peninsula in the subtropical Atlantic . . . (N Am)

Florida Keys, 8, 13 . . . . A chain of coral islands arcing west from the tip of Florida into the Gulf of Mexico . . . (N Am)

Forestry, 58 . . . . An industry that manages and replenishes the timber wealth of the temperate zones

Formosa Strait, 33 . . . . A strait that separates the island of Taiwan from the Chinese mainland . . . (E Asia)

Fort-de-France, 13 . . . . The capital of Martinique; an island port in the eastern West Indies . . . (N Am)

France, 18, 21 . . . . The largest country in western Europe, with an uncommon mixture of provincialism and sophistication

Frankfort, 6, 8 . . . . The capital of Kentucky; north centrally located, just south of the Ohio river border . . . (N Am)

Frankfurt, 22 . . . . A southwest German city on the Main River; a commercial center with a long history . . . (C Europe)

Franz Josef Land, 26 . . . . Russian islands in the Arctic Ocean; the most northerly lands in the Eastern Hemisphere . . . (N Europe)

Fraser River, 4, 5 . . . . British Columbia river that empties into the Pacific at Vancouver; origin in Alberta's Rockies . . . (N Am)

Fredericton, 5 . . . . Capital of New Brunswick, an eastern Canadian province on Maine's border . . . (N Am)

Freetown, 37, 40 . . . . Capital of Sierra Leone; west African Atlantic port created as a haven for slaves freed by British Fleet

French Guiana, 14, 16 . . . . Last foreign colony on the South American continent; on the north coast, between Suriname and Brazil

French language, 60 . . . . One of the Latin-Romance languages; was the language of diplomacy

French Polynesia, 36 . . . . Encompasses 130 French islands in the South Pacific; the most famous of these is Tahiti . . . (Oceania)

Ft. Lauderdale, 8 . . . . A southeast Florida coastal city known for its huge yacht harbor and network of canals . . . (N Am)

Ft. Worth, 9 . . . . The western half (with Dallas) of the twin centers of commerce in east Texas . . . (N Am)

Fuzhou, 30 . . . . An important Chinese seaport on the Formosa Strait, opposite Taiwan . . . (E Asia)

G

Gabarone, 37, 43 . . . . The capital of Botswana; east of the Kalahari Desert, on the southern border with South Africa

Gabon, 37, 41 . . . . Oil-rich country on central Africa's Atlantic coast, between Cameroon and the Republic of the Congo

Galapagos Islands, 14 . . . . Ecuadorian Pacific Islands, 600 mi. (960 km) from mainland; visited by Charles Darwin . . . (S Am)

Gambia (The), 37, 39 . . . . Continental Africa's smallest nation; on the Atlantic coast, almost surrounded by Senegal . . . (W Africa)

Gambia River, 38, 39 . . . . A west African river that flows down the middle of The Gambia on its way to the Atlantic . . . (W Africa)

Ganges River, 29, 32 . . . . Hinduism's holiest river; crosses northern India on its way east to the Bay of Bengal . . . (S Asia)

Gaza Strip, 30, 39 . . . . A strip on the Mediterranean between Egypt and Israel; self-governed by Palestinians . . . (SW Asia)

Gdansk, 22 . . . . Poland's main port on the Baltic Sea; was the Free City of Danzig before WW II . . . (C Europe)

Geneva, 22 . . . . A city in the southwest thumb of Switzerland, surrounded by France . . . (C Europe)

Genoa, 23 . . . . Italy's principal seaport, on the northwest coast; the birthplace of Columbus . . . (S Europe)

Georgetown, 14, 16 . . . . Capital of Guyana; a dike-protected, sub-sea level Atlantic port on the north coast . . . (S Am)

Georgia (Eur.), 18, 27 . . . . A west Caucasus nation on the Black Sea; a former republic of the Soviet Union . . . (E Europe)

Georgia (U.S.), 6, 8 . . . . The largest state east of the Mississippi; on the Atlantic, between South Carolina and Florida . . . (N Am)

German language, 60 . . . . The tenth most popular language; speakers are mostly in central Europe

Germany, 18, 22 . . . . Europe's industrial giant; started the two World Wars of the Twentieth Century . . . (C Europe)

Ghana, 37, 40 . . . . A west African coastal nation, between Côte d'Ivoire and Benin; was called the Gold Coast

Ghats, 29 . . . . Mountains on the east and west coasts of India; converge at the tip of the Deccan Peninsula . . . (S Asia)

Gibraltar, 21 . . . . Britain's rock on the Spanish side of a strait linking the Atlantic and the Mediterranean . . . (SW Europe)

Glacier, see Glossary . . . . A river of ice that moves very slowly down a mountain slope

Glasgow, 21 . . . . Scotland's largest city; a port on the west coast, facing Northern Ireland . . . (W Europe)

Gobi Desert, 33 . . . . A high desert in southern Mongolia, subject to extremes of temperature . . . (C Asia)

Godthab, 3, 5 . . . . The capital of Greenland; a port on the island's southwest coast, on the Davis Strait . . . (N Am)

Golan Heights, 30 . . . . Mountains on Syria's southwest border, overlooking Israel; currently under Israel's control . . . (SW Asia)

Gold Coast, 40 . . . . A name for the coast of Ghana; recalls the time when British colonialists were mining gold . . . (W Africa)

Goteborg, 20 . . . . Sweden's second-largest city; the nation's main port, on a strait opposite Denmark . . . (N Europe)

Gotland, 20 . . . . The largest island in the Baltic Sea; played a significant role in Sweden's medieval past . . . (N Europe)

Gran Chaco, 17 . . . . The western half of Paraguay; a harsh region that can be either a desert or a swamp . . . (S Am)

Grand Canyon, 1 . . . . A spectacular gorge in northern Arizona; eroded by six million years of Colorado River activity . . . (N Am)

Grasslands (temperate), 58 . . . . Grassy, treeless regions called "prairies" in the U.S. midwest and "pampas" in central Argentina

Grasslands (tropical), 58 . . . . Grassy regions called "savannas" in Africa, "llanos" in Venezuela, and "campos" in Brazil

Great Barrier Reef, 35 . . . . World's longest coral reef (1,250 mi.; 2,000 km); lies off the northeast coast of Australia . . . (Oceania)

Great Bear Lake, 4, 5 . . . . A lake in the Canadian Northwest Territories; North America's fourth-largest lake

H

ℳ

ℕ

INDEX / DICTIONARY / QUIZ

T

W

X

Y

Z